16. $\displaystyle\int \cosh ax\,dx = \frac{1}{a} \sinh ax + c.$

17. $\displaystyle\int \frac{dx}{a^2 + x^2} = \frac{1}{a} \tan^{-1} \frac{x}{a} + c.$

18. $\displaystyle\int \frac{dx}{\sqrt{a^2 - x^2}} = \sin^{-1} \frac{x}{a} + c.$

19. $\displaystyle\int \frac{dx}{\sqrt{x^2 \pm a^2}} = \ln |x + \sqrt{x^2 \pm a^2}| + c.$

20. $\displaystyle\int (ax + b)^n dx = \frac{(ax + b)^{n+1}}{a(n + 1)} + c, \quad n \neq -1.$

21. $\displaystyle\int \frac{dx}{ax + b} = \frac{1}{a} \ln |ax + b| + c.$

22. $\displaystyle\int \frac{dx}{x^2 - a^2} = \frac{1}{2a} \ln \left| \frac{x - a}{x + a} \right| + c.$

23. $\displaystyle\int \frac{dx}{x(ax + b)} = \frac{1}{b} \ln \left| \frac{x}{ax + b} \right| + c.$

24. $\displaystyle\int \frac{x\,dx}{ax + b} = \frac{x}{a} - \frac{b}{a^2} \ln |ax + b| + c.$

25. $\displaystyle\int \frac{x\,dx}{(ax + b)^2} = \frac{1}{a^2} \left( \ln |ax + b| + \frac{b}{ax + b} \right) + c.$

26. $\displaystyle\int x(ax + b)^n dx = \frac{(ax + b)^{n+1}}{a^2} \left( \frac{ax + b}{n + 2} - \frac{b}{n + 1} \right) + c, \quad n \neq -1, -2.$

27. $\displaystyle\int \frac{x\,dx}{\sqrt{ax + b}} = \frac{2(ax - ab)}{3a^2} \sqrt{ax + b} + c.$

28. $\displaystyle\int \frac{dx}{x\sqrt{ax + b}} = \begin{cases} \dfrac{1}{\sqrt{b}} \ln \left| \dfrac{\sqrt{ax + b} - \sqrt{b}}{\sqrt{ax + b} + \sqrt{b}} \right| + c, & b > 0 \\[3mm] \dfrac{2}{\sqrt{-b}} \tan^{-1} \sqrt{\dfrac{ax + b}{-b}} + c, & b < 0. \end{cases}$

29. $\displaystyle\int \frac{dx}{x^2\sqrt{ax + b}} = -\frac{\sqrt{ax + b}}{bx} - \frac{a}{2b} \int \frac{dx}{x\sqrt{ax + b}} + c, \quad b \neq 0.$

30. $\displaystyle\int \frac{dx}{ax^2 + bx + c} = \begin{cases} \dfrac{2}{\sqrt{4ac - b^2}} \tan^{-1} \dfrac{2ax \mp b}{\sqrt{4ac - b^2}}, & b^2 - 4ac < 0. \\[3mm] \dfrac{1}{\sqrt{b^2 - 4ac}} \ln \left| \dfrac{2ax + b - \sqrt{b^2 - 4ac}}{2ax + b + \sqrt{b^2 - 4ac}} \right|, & b^2 - 4ac > 0. \end{cases}$

# Ordinary Differential Equations with Modern Applications

SECOND EDITION

**N. Finizio and G. Ladas**
University of Rhode Island

**Wadsworth Publishing Company**
**Belmont, California**

A division of Wadsworth, Inc.

*ABOUT THE COVER:* "Hyperbolic Cross Currents" (Miles Color Art A24) is the work of Professor E. P. Miles, Jr., and associates of Florida State University using an InteColor 8051 computer. Photograph by John Owen. The graph displays in dotted character form the hyperbolic family $xy = k$ and its associated family of orthogonal trajectories $x^2 - y^2 = c$. The rectangular shape of the characters causes some distortion of the right angle at which members of the two families cross.

*Mathematics Editor:* RICHARD JONES
*Signing Representative:* RICHARD GIGGEY

ISBN 0-534-00898-4

Printed in the United States of America

4 5 6 7 8 9 10—86 85 84

**Library of Congress Cataloging in Publication Data**

Finizio, N.
  Ordinary differential equations with modern applications.

  Includes index.
  1. Differential equations.  I. Ladas, G. E.
II. Title.
QA372.F56  1981        515.3'52        81-1853
ISBN 0-534-00898-4                      AACR2

*To our wives,*
> Joan Finizio and Theodora Ladas

*and our children,*
> Victoria, Norman, Greg, Gina, and Daryl Finizio
> *and*
> Homer and Andreas Ladas

# List of Applications

Each entry in this list is coded according to whether the application appears in detail (A), little or no detail (B), or is an exercise or example associated with an application (C).

# Preface to the Second Edition

In this new edition, in order to give the student further practice in using the methods demonstrated in each chapter and applying them to problem-solving, we have added review exercises at the end of every chapter. These exercises will help the student review material presented in the chapter. Some of the review exercises have been taken directly from physics and engineering textbooks, in order to further emphasize that differential equations are very much present in applications and that the student is quite apt to encounter them in areas other than mathematics.

Other new additions, which we hope will be helpful to both students and instructors, are: the variation of parameters method for nonhomogeneous systems in Chapter 3; the solution of systems by the method of Laplace transforms in Chapter 4; the unit step function and the unit impulse function in Chapter 4; the addition of flowcharts and computer and calculator exercises in Chapter 7; the solution of polynomial equations in Appendix C; and the proof of the existence and uniqueness theorem in Appendix D.

Some of the applications are presented in considerable detail and some with little or no detail, as indicated in the list of applications. Every application is intended to illustrate the relevance of differential equations outside of their intrinsic value as mathematical topics. We suggest that the instructor present only a few of the applications, while the rest will indicate to the student the relevance of differential equations in real-life situations.

The additions and other modifications and rearrangements of material in this edition were made after careful consideration of recommendations made by users of the first edition. We are grateful to these people for their advice.

Special thanks are due to Richard Jones, the Mathematics Editor of Wadsworth Publishing Company, for his continued support, advice, and active interest in this project.

We would also like to thank the following reviewers for their helpful comments: Martin Billik, California State University, San Jose; Theodore Cullen, California State Polytechnic University, Pomona; Gary L. Eerkes, Gonzaga University; Paul A. Haeder, University of Nebraska at Omaha; N. Jhunjhunwala, California State University, Los Angeles; David A. Smith, Duke University; Bruce Swenson, Foothill College; William Voxmen, University of Idaho; D. D. Weddington, California State University, San Jose; and H. C. Wiser, Washington State University.

**N. Finizio**
**G. Ladas**

# Preface to the First Edition

This book is intended for an introductory course in ordinary differential equations. Its prerequisite is elementary calculus.

Perusal of the table of contents and the list of applications shows that the book contains the theory, techniques, and applications covered in the traditional introductory courses in ordinary differential equations. A major feature of this text is the quantity and variety of applications of current interest in physical, biological, and social sciences. We have furnished a wealth of applications from such diverse fields as astronomy, bioengineering, biology, botany, chemistry, ecology, economics, electric circuits, finance, geometry, mechanics, medicine, meteorology, pharmacology, physics, psychology, seismology, sociology, and statistics.

Our experience gained in teaching ordinary differential equations on the elementary, intermediate, and graduate levels at the University of Rhode Island convinced us of the need for a book at the elementary level which emphasizes to the students the relevance of the various equations that they are exposed to in the course. That is to say that the various types of differential equations encountered are not merely the product of some mathematician's imagination but rather that the equations occur in the course of scientific investigations of real-world phenomena.

The goal of this book, then, is to make elementary ordinary differential equations more useful, more meaningful, and more exciting to the student. To accomplish this, we strive to demonstrate that ordinary differential equations are very much "alive" in present-day applications. This approach has indeed had a satisfying effect in the courses we have taught recently.

During the preparation and class testing of this text we continuously kept in mind both the student and the teacher. We have tried to make the presentation direct, yet informal. Definitions and theorems are stated precisely and rigorously, but theory and rigor have been minimized in favor of comprehension of technique. The general approach is to use a large number of routine examples to illustrate the new concepts, definitions, methods of solution, and theorems. Thus, it is intended that the material will be easily accessible to the student. The presence of modern applications in addition to the traditional applications of geometry, physics, and chemistry should be refreshing to the teacher.

Numerous routine exercises in each section will help to test and strengthen the student's understanding of the new methods under discussion. There are over 1100 exercises in the text with answers to odd-numbered exercises provided. Some thought-provoking exercises from *The American Mathematical*

*Monthly, Mathematics Magazine*, and The William Lowell Putnan Mathematics Competition are inserted in many sections, with references to the source. These should challenge the students and help to train them in searching the literature.

Every type of differential equation studied and every method presented is illustrated by real-life applications which are incorporated in the same section (or chapter) with the specific equation or method. Thus, the students will see immediately the importance of each type of differential equation that they learn how to solve. We feel that these "modern" applications, even if the student only glances at some of them, will help to stimulate interest and enthusiasm toward the subject of differential equations specifically and mathematics in general.

Many of the applications are integrated into the main development of ideas, thus blending theory, technique, and application. Frequently, the mathematical model underlying the application is developed in great detail. It would be impossible in a text of this nature to have such development for every application cited. Therefore, some of the models are only sketched, and in some applications the model alone is presented. In practically all cases, references are given for the source of the model. Additionally, a large number of applications appear in the exercises; these applications are also suitably referenced. Consequently, applications are widespread throughout the book, and although they vary in depth and difficulty, they should be diverse and interesting enough to whet the appetite of every reader.

We are grateful to Katherine MacDougall, who so skillfully typed the manuscript of this text, and to Barbara Zeiders of Service to Publishers, Inc., who did a superb job of copyediting the manuscript.

A special word of gratitude goes to Professors Gerald Bradley, John Haddock, Thomas Hallam, Ken Kalmanson, Gordon McLeod, and David Wend, who painstakingly reviewed this book and offered numerous valuable suggestions for its improvement.

Thanks are also due to Lynnell Stern, Professor Clement McCalla, and to our students Carl Bender, Thomas Buonanno, Michael Fascitelli, and especially Neal Jamnik and Brian McCartin, who proofread parts of the material and doublechecked the solutions to some of the exercises.

We are especially appreciative of the constant encouragement and friendly assistance that we received from Richard Jones, the Mathematics Editor of Wadsworth Publishing Company, and former editor Donald Dellen.

**N. Finizio**
**G. Ladas**

# Contents

## 3  LINEAR SYSTEMS

## 4  THE LAPLACE TRANSFORM

## 5  SERIES SOLUTIONS OF SECOND-ORDER LINEAR EQUATIONS

# 6 BOUNDARY VALUE PROBLEMS

# 7 NUMERICAL SOLUTIONS OF DIFFERENTIAL EQUATIONS

# 8 NONLINEAR DIFFERENTIAL EQUATIONS AND SYSTEMS

# CHAPTER 1

# Elementary Methods— First-Order Differential Equations

## 1.1 INTRODUCTION AND DEFINITIONS

*Differential equations* are equations that involve derivatives of some unknown function(s). Although such equations should probably be called "derivative equations," the term "differential equations" (*aequatio differentialis*) initiated by Leibniz in 1676 is universally used. For example,

$$y' + xy = 3 \tag{1}$$

$$y'' + 5y' + 6y = \cos x \tag{2}$$

$$y'' = (1 + y'^2)(x^2 + y^2) \tag{3}$$

$$\frac{\partial^2 u}{\partial t^2} - \frac{\partial^2 u}{\partial x^2} = 0 \tag{4}$$

are differential equations. In Eqs. (1)–(3) the unknown function is represented by $y$ and is assumed to be a function of the single independent variable $x$, that is, $y = y(x)$. The argument $x$ in $y(x)$ (and its derivatives) is usually suppressed for notational simplicity. The terms $y'$ and $y''$ in Eqs. (1)–(3) are the first and second derivatives, respectively, of the function $y(x)$ with respect to $x$. In Eq. (4) the unknown function $u$ is assumed to be a function of the two independent variables $t$ and $x$, that is, $u = u(t, x)$, $\partial^2 u/\partial t^2$ and $\partial^2 u/\partial x^2$ are the second partial derivatives of the function $u(t, x)$ with respect to $t$ and $x$, respectively. Equation (4) involves partial derivatives and is a *partial differential equation*. Equations (1)–(3) involve ordinary derivatives and are *ordinary differential equations*.

In this book we are primarily interested in studying ordinary differential equations.

### DEFINITION 1

*An ordinary differential equation of order n is an equation that is, or can be put, in the form*

$$y^{(n)} = F(x, y, y', \ldots, y^{(n-1)}), \tag{5}$$

*where $y, y', \ldots, y^{(n)}$ are all evaluated at $x$.*

The independent variable $x$ belongs to some interval $I$ ($I$ may be finite or infinite), the function $F$ is given, and the function $y = y(x)$ is unknown. For the most part the functions $F$ and $y$ will be real valued. Thus, Eq. (1) is an ordinary differential equation of order 1 and Eqs. (2) and (3) are ordinary differential equations of order 2.

### DEFINITION 2

*A solution of the ordinary differential equation (5) is a function $y(x)$ defined over a subinterval $J \subset I$ which satisfies Eq. (5) identically over the interval $J$.*

Clearly, any solution $y(x)$ of Eq. (5) should have the following properties:
1. $y$ should have derivatives at least up to order $n$ in the interval $J$.
2. For every $x$ in $J$ the point $(x, y(x), y'(x), \ldots, y^{(n-1)}(x))$ should lie in the domain of definition of the function $F$, that is, $F$ should be defined at this point.
3. $y^{(n)}(x) = F(x, y(x), y'(x), \ldots, y^{(n-1)}(x))$ for every $x$ in $J$.

As an illustration we note that the function $y(x) = e^x$ is a solution of the second-order ordinary differential equation $y'' - y = 0$. In fact.

$$y''(x) - y(x) = (e^x)'' - e^x = e^x - e^x = 0.$$

Clearly, $e^x$ is a solution of $y'' - y = 0$ valid for all $x$ in the interval $(-\infty, +\infty)$. As another example, the function $y(x) = \cos x$ is a solution of $y'' + y = 0$ over the interval $(-\infty, +\infty)$. Indeed,

$$y''(x) + y(x) = (\cos x)'' + \cos x = -\cos x + \cos x = 0.$$

In each of the illustrations the solution is valid on the whole real line $(-\infty, +\infty)$. On the other hand, $y = \sqrt{x}$ is a solution of the first-order ordinary differential equation $y' = 1/2y$ valid only in the interval $(0, +\infty)$ and $y = \sqrt{x(1-x)}$ is a solution of the ordinary differential equation $y' = (1 - 2x)/2y$ valid only in the interval $(0, 1)$.

As we have seen, $y = e^x$ is a solution of the ordinary differential equation $y'' - y = 0$. We further observe that $y = e^{-x}$ is also a solution and moreover $y = c_1 e^x + c_2 e^{-x}$ is a solution of this equation for arbitrary values of the constants $c_1$ and $c_2$. It will be shown in Chapter 2 that $y = c_1 e^x + c_2 e^{-x}$ is the "general solution" of the ordinary differential equation $y'' - y = 0$. By the general solution we mean a solution with the property that any solution of $y'' - y = 0$ can be obtained from the function $c_1 e^x + c_2 e^{-x}$ for some special values of the constants $c_1$ and $c_2$. Also, in Chapter 2 we will show that the general solution of the ordinary differential equation $y'' + y = 0$ is given by $y(x) = c_1 \cos x + c_2 \sin x$ for arbitrary values of the constants $c_1$ and $c_2$.

In this chapter we present elementary methods for finding the solutions of some first-order ordinary differential equations, that is, equations of the form

$$y' = F(x, y), \tag{6}$$

together with some interesting applications.

The differential of a function $y = y(x)$ is by definition given by $dy = y'dx$.

With this in mind, the differential equation (6) sometimes will be written in the differential form $dy = F(x,y)dx$ or in an algebraically equivalent form. For example, the differential equation

$$y' = \frac{3x^2}{x^3 + 1}(y + 1)$$

can be written in the form

$$dy = \left[\frac{3x^2}{x^3 + 1}(y + 1)\right]dx \quad \text{or} \quad y' - \frac{3x^2}{x^3 + 1}y = \frac{3x^2}{x^3 + 1}.$$

There are several types of first-order ordinary differential equations whose solutions can be found explicitly or implicitly by integrations. Of all tractable types of first-order ordinary differential equations, two deserve special attention: differential equations with *variables separable,* that is, equations that can be put into the form

$$y' = \frac{P(x)}{Q(y)} \quad \text{or} \quad P(x)dx = Q(y)dy,$$

and *linear equations,* that is, equations that can be put into the form

$$y' + a(x)y = b(x).$$

Both appear frequently in applications, and many other types of differential equations are reducible to one or the other of these types by means of a simple transformation.

## APPLICATIONS 1.1.1

Differential equations appear frequently in mathematical models that attempt to describe real-life situations. Many natural laws and hypotheses can be translated via mathematical language into equations involving derivatives. For example, derivatives appear in physics as velocities and accelerations, in geometry as slopes, in biology as rates of growth of populations, in psychology as rates of learning, in chemistry as reaction rates, in economics as rates of change of the cost of living, and in finance as rates of growth of investments.

It is the case with many mathematical models that in order to obtain a differential equation that describes a real-life problem, we usually assume that the actual situation is governed by very simple laws—which is to say that we often make idealistic assumptions. Once the model is constructed in the form of a differential equation, the next step is to solve the differential equation and utilize the solution to make predictions concerning the behavior of the real problem. In case these predictions are not in reasonable agreement with reality, the scientist must reconsider the assumptions that led to the model and attempt to construct a model closer to reality.

First-order ordinary differential equations are very useful in applications. Let the function $y = y(x)$ represent an unknown quantity that we want to study. We know from calculus that the first derivative $y' = dy/dx$ represents the rate of change of $y$ per unit change in $x$. If this rate of change is known (say, by experience or by a physical law) to be equal to a function $F(x, y)$, then the quantity $y$ satisfies the first-order ordinary differential equation $y' = F(x, y)$. We next give some specific illustrations.

**Biology**      ■ It has long been observed that some large colonies of bacteria tend to grow at a rate proportional to the number of bacteria present. For such a colony, let $N = N(t)$ be the number of bacteria present at any time $t$. Then, if $k$ is the constant of proportionality, the function $N = N(t)$ satisfies the first-order ordinary differential equation[1]

$$\dot{N} = kN. \tag{7}$$

This equation is called the *Malthusian law* of population growth. T. R. Malthus observed in 1798 that the population of Europe seemed to be doubling at regular time intervals, and so he concluded that the rate of population increase is proportional to the population present. In Eq. (7) $\dot{N}$ stands for $dN/dt$. (As is customary, derivatives with respect to $x$ will be denoted by primes and derivatives with respect to $t$ by dots.) In this instance it is the time that is the independent variable. Equation (7) is a separable differential equation and its solution $N(t) = N(0)e^{kt}$ is computed in Example 3 of Section 1.3. Here $N(0)$ is the number of bacteria present initially, that is, at time $t = 0$. The solution $N(t)$ can be represented graphically as in Figure 1.1.

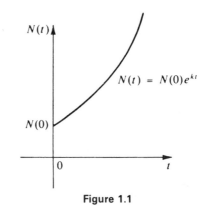

**Figure 1.1**

It should be emphasized that Eq. (7) is a mathematical model describing a colony of bacteria that grows according to a very simple, perhaps oversimplified, law: It grows at a rate proportional to the number of bacteria present at any time $t$. On the other hand, assuming this very simple law of growth leads us to a very simple differential equation. The solution, $N(t) = N(0)e^{kt}$, of Eq. (7) provides us with an approximation to the actual size of this colony of bacteria.

---

[1]Since the function $N(t)$ takes on only integral values, it is not continuous and so not differentiable. However, if the number of bacteria is very large, we can assume that it can be approximated by a differentiable function $N(t)$, since the changes in the size of the population occur over short time intervals.

Clearly, a more realistic mathematical model for the growth of this colony of bacteria is obtained if we take into account such realistic factors as overcrowding, limitations of food, and the like. Of course, the differential equation will then become more complex. It goes without saying that a mathematical model that is impossible to handle mathematically is useless, and consequently some simplifications and modifications of real-life laws are often necessary in order to derive a mathematically tractable model.

■ It is well known in pharmacology[2] that penicillin and many other drugs administered to patients disappear from their bodies according to the following simple rule: If $y(t)$ is the amount of the drug in a human body at time $t$, then the rate of change $\dot{y}(t)$ of the drug is proportional to the amount present. That is, $y(t)$ satisfies the separable differential equation

**Pharmacology**
Drug Dosages

$$\dot{y} = -ky, \tag{8}$$

where $k > 0$ is the constant of proportionality. The negative sign in (8) is due to the fact that $y(t)$ decreases as $t$ increases, and hence the derivative of $y(t)$ with respect to $t$ is negative. For each drug the constant $k$ is known experimentally.

The solution of the differential equation (8) is (see Example 3 of Section 1.3)

$$y(t) = y_0 e^{-kt}, \tag{9}$$

where $y_0 = y(0)$ is the initial amount (initial dose) of the drug. As we see from Eq. (9) (see also Figure 1.2), the amount of the drug in the patient's body tends

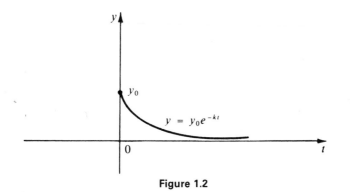

**Figure 1.2**

to zero as $t \to \infty$. However, in many cases it is necessary to maintain (approximately) a constant concentration (and therefore approximately a constant amount) of the drug in the patient's body for a long time. To achieve this it is necessary to give the patient an initial booster dose $y_0$ of the drug and then at equal intervals of time, say every $\tau$ hours, give the patient a dose $D$ of the drug.

[2]This model, as well as other mathematical models in medicine, is discussed by J. S. Rustagi in *Int. J. Math. Educ. Sci. Technol.* 2 (1971): 193–203.

Equation (9) indicates the amount of the drug in the patient's body at any time $t$; hence, it is simple to determine the amount of the dose $D$. In fact, at time $\tau$, and before we administer the dose $D$, the amount of the drug present in the body is

$$y(\tau) = y_0 e^{-k\tau}.$$

If we want to maintain the initial amount $y_0$ of the drug in the body at the times $\tau, 2\tau, 3\tau, \ldots$, the dose $D$ should satisfy the equation

$$y_0 e^{-k\tau} + D = y_0.$$

Hence, the desired dose is given by the equation

$$D = y_0(1 - e^{-k\tau}). \tag{10}$$

**Operations Research**     ■ Southwick and Zionts[3] developed an optimal control-theory approach to the education-investment decision which led them to the first-order linear (also separable) differential equation

$$\dot{x} = 1 - kx, \tag{11}$$

where $x$ denotes the education of an individual at time $t$ and the constant $k$ is the rate at which education is being made obsolete or forgotten.

**Psychology**     ■ In learning theory the separable first-order differential equation

$$\dot{p}(t) = a(t)G(p(t)) \tag{12}$$

is a basic model of the instructor/learner interaction.[4] Here $G$ is known as the characteristic learning function and depends on the characteristics of the learner and of the material to be learned, $p(t)$ is the state of the learner at time $t$, and $a(t)$ is a measure of the intensity of instruction [the larger the value of $a(t)$ the greater the learning rate of the learner, but also, the greater the cost of the instruction].

**Mechanics**     ■ Newton's second law of motion, which states that "the time rate of change of momentum of a body is proportional to the resultant force acting on the body and is in the direction of this resultant force," implies immediately that the motion of any body is described by an ordinary differential equation. Recall that the momentum of a body is the product $mv$ of its mass and its velocity $v$. If $F$ is the resultant force acting on the body, then

$$\frac{d}{dt}(mv) = kF, \tag{13}$$

[3]L. Southwick and S. Zionts, *Operations Res.* 22 (1974): 1156–1174.
[4]V. G. Chant, *J. Math. Psychol.* 11 (1974): 132–158.

where $k$ is a constant of proportionality. Equation (13) is an ordinary differential equation in $v$ whose particular form depends on $m$ and $F$. The mass $m$ can be constant or a function of $t$. Also, $F$ can be constant, a function of $t$, or even a function of $t$ and $v$.

■ Kirchhoff's voltage law states that, "the algebraic sum of all voltage drops **Electric Circuits** around an electric circuit is zero." This law applied to the $RL$-series circuit in Figure 1.3 gives rise to the first-order linear differential equation (see also Section 1.4)

$$L\dot{I} + RI = V(t), \tag{14}$$

where $I = I(t)$ is the current in the circuit at time $t$.

**Figure 1.3**

■ Consider the one-parameter family of curves given by the equation **Orthogonal Trajectories**

$$F(x,y) = c. \tag{15}$$

Computing the differential of Eq. (15), we obtain

$$F_x dx + F_y dy = 0,$$

where $F_x$ and $F_y$ are the partial derivatives of $F$ with respect to $x$ and $y$, respectively. Thus,

$$\frac{dy}{dx} = -\frac{F_x}{F_y} \tag{16}$$

gives the slope of each curve of the family (15). We want to compute another family of curves such that each member of the new family cuts each member of the family (15) at right angles; that is, we want to compute the *orthogonal trajectories* of the family (15). In view of Eq. (16), the slope of the orthogonal trajectories of the family (15) is given by [the negative reciprocal of (16)]

$$\frac{dy}{dx} = \frac{F_y}{F_x}. \tag{17}$$

The general solution of Eq. (17) gives the orthogonal trajectories of the family (15).

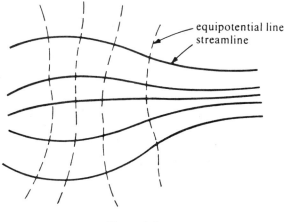

equipotential line
streamline

**Figure 1.4**

There are many physical interpretations and uses of orthogonal trajectories:

1. In *electrostatic fields* the *lines of force* are orthogonal to the *lines of constant potential*.

2. In *two-dimensional flows of fluids* the lines of motion of the flow—called *streamlines*—are orthogonal to the *equipotential* lines of the flow (see Figure 1.4).

3. In *meterology* the orthogonal trajectories of the *isobars* (curves connecting all points that report the same barometric pressure) give the direction of the wind from high- to low-pressure areas.

We have given only a few of the many applications of first-order ordinary differential equations. Many others are developed in detail in subsequent sections. Additionally, we present a number of applications in the exercises. The emphasis there is twofold: first, to expose the reader to the diversity of models incorporating ordinary differential equations; and second, to sharpen the reader's skill at solving differential equations while simultaneously emphasizing the fact that many differential equations are not products of the instructor's imagination but rather are extracted from real-life models.

## EXERCISES

In Exercises 1 through 8, answer true or false.

**1.** $y = e^x + 3e^{-x}$ is a solution of the differential equation $y'' - y = 0$.

**2.** $y = 5 \sin x + 2 \cos x$ is a solution of the differential equation $y'' + y = 0$.

**3.** $y = \sin 2x$ is a solution of the differential equation $y'' - 4y = 0$.

**4.** $y = \cos 2x$ is a solution of the differential equation $y'' + 4y = 0$.

**5.** $y = \left( \dfrac{1}{2x} \right) e^{x^2}$ is a solution of the differential equation $xy' + y = xe^{x^2}$.

**6.** $y = e^x$ is a solution of the differential equation $y' + y = 0$.

**7.** $y = e^{-x}$ is a solution of the differential equation $y' - y = 0$.

**8.** $y = 2 \ln x + 4$ is a solution of the differential equation $x^2 y'' - xy' + y = 2 \ln x$.

In Exercises 9 through 14, a differential equation and a function are listed. Show that the function is a solution of the differential equation. If more than one function is listed, show that both functions are solutions of the differential equation.

**9.** $y'' + y' = 2$; $2x$, $2x - 3$

**10.** $y'' - (\tan x)y' - \dfrac{\tan x}{x} y = \dfrac{1}{x^2} y^3$; $x \sec x$

**11.** $y''' - 5y'' + 6y' = 0$; $e^{3x}$

**12.** $y''' - 5y'' + 6y' = 0$; $e^{2x}$, $1$

**13.** $m\ddot{s} = \frac{1}{2} g t^2$; $\dfrac{g}{24 m} t^4 \left( \ddot{s} = \dfrac{d^2 s}{dt^2} \right)$

**14.** $y'' - 2y' + y = \dfrac{1}{x}(y - y')$; $e^x \ln x$

**15.** Show that the functions $y_1(x) = e^{-x}$ and $y_2(x) = xe^{-x}$ are solutions of the ordinary differential equation $y'' + 2y' + y = 0$.

**16.** Show that the function $y(x) = c_1 e^{-x} + c_2 xe^{-x} + 1$ is a solution of the ordinary differential equation $y'' + 2y' + y = 1$ for any values of the constants $c_1$ and $c_2$.

**17.** Show that the functions $y_1(x) \equiv 0$ and $y_2(x) = x^2/4$, $x \geq 0$, are solutions of the ordinary differential equation

$$y' = y^{1/2}.$$

**18.** Show that the function

$$y(x) = \begin{cases} 0 & \text{for } x \leq c \\ \dfrac{(x - c)^2}{4} & \text{for } x > c, \end{cases}$$

for any real number $c$, is a solution of the differential equation

$$y' = y^{1/2}.$$

(*Warning:* Don't forget to show that the solution is differentiable everywhere and in particular at $x = c$.)

**19.** Show that the function

$$x(t) = \frac{1}{k} + \frac{kx_0 - 1}{k} e^{-k(t-t_0)},$$

where $x_0$ and $t_0$ are constants, is a solution of Eq. (11). Show also that this solution passes through the point $(t_0, x_0)$; that is, $x(t_0) = x_0$.

**20.** Show that $N(t) = ce^{kt}$ for any constant $c$ is a solution of Eq. (7). What is the meaning of the constant $c$?

**21.** Find the differential equation of the orthogonal trajectories of the family of straight lines $y = cx$. Using your geometric intuition, guess the solution of the resulting differential equation.

In Exercises 22 through 28, answer true or false.

**22.** $y = xe^x$ is a solution of the differential equation $y' - 2y = e^x(1 - x)$.

**23.** $y = x^2$ is a solution of the differential equation $y''' = 0$.

**24.** $y = x + 1$ is a solution of the differential equation $yy' - y^2 = x^2$.

**25.** $y = \sin^2 x$ is a solution of the differential equation $y'' + y = \cos^2 x$.

**26.** $y = -1 + e^{-x}$ is a solution of the differential equation $y'' = y'(y' + y)$.

**27.** $y = -\sqrt{\dfrac{4 - x^3}{3x}}$ is a solution of the differential equation $y' = \dfrac{x^2 + y^2}{2xy}$.

**28.** $y = -\sqrt{\dfrac{4 - x^3}{3x}}$ is a solution of the differential equation $y' = -\dfrac{x^2 + y^2}{2xy}$.

**29.** Verify that each member of the one-parameter family of curves

$$x^2 + y^2 = 2cx$$

cuts every member of the one-parameter family of curves

$$x^2 + y^2 = 2ky$$

at right angles and vice versa. (*Hint:* Show that slopes are negative reciprocals.)

**30.** Prove that if the family of solutions of the differential equation

$$y' + p(x)y = q(x), \qquad p(x)q(x) \neq 0$$

is cut by the line $x = k$, the tangents to each member of the family at the points of intersection are concurrent.[5] [*Hint:* The equation of the tangent line to the solution at the point $(k, c)$ is $y - c = [q(k) - cp(k)](x - k)$, and this equation passes through the point $[k + 1/p(k), q(k)/p(k)]$.]

---

[5]From the William Lowell Putnam Mathematical Competition, 1954. See *Amer. Math. Monthly* 61 (1954): 545.

# 1.2 EXISTENCE AND UNIQUENESS

We saw in the last section that first-order differential equations occur in many diverse mathematical models. Naturally, the models are useful if the resulting differential equation can be solved explicitly or at least if we can predict some of the properties of its solutions. Before we attempt to discover any ingenious technique to solve a differential equation, it would be very useful to know whether the differential equation has any solutions at all. That is, do there *exist* solutions to the differential equation? For example, the differential equation $(y')^2 + y^2 + 1 = 0$ has no real solution since the left-hand side is always positive.

The general form of a first-order ordinary differential equation is

$$y' = F(x,y). \tag{1}$$

For the sake of motivation, assume that Eq. (1) represents the motion of a particle whose velocity at time $x$ is given by $F(x,y)$, where $y$ is the position (the state) of the particle at time $x$. Since we imagine that we are observing the motion of the particle, it is plausible that if we know its position $y_0$ at time $x_0$, that is, if

$$y(x_0) = y_0 \tag{2}$$

we should be able to find its position at a later time $x$. Starting at time $x_0$ the particle will move and will move in a unique way. That is, the differential equation (1) *has* a solution that satisfies the condition (2), and moreover it has *only one* solution, that is, the motion that we are observing. The condition (2) is called an *initial condition* and Eq. (1) together with the initial condition (2) is called an *initial value problem (IVP)*. The term "initial" has been adopted from physics. In Eq. (2) $y_0$ is the initial position of the particle at the initial time $x_0$.

On the basis of the motivation above we may expect that an IVP which is a reasonable mathematical model of a real-life situation should have a solution (existence), and in fact it should have only one solution (uniqueness). We shall now state (see Appendix D for a proof)[6] a basic existence and uniqueness theorem for the IVP (1)–(2) which is independent of any physical considerations and which covers a wide class of first-order differential equations.

**THEOREM 1**

*Consider the IVP*

$$y' = F(x,y), \qquad y(x_0) = y_0.$$

---

[6]This theorem and many others that appear in this text are truly important for an over-all appreciation of differential equations. Often the proofs of these theorems require mathematical sophistication which is customarily beyond the scope of an elementary treatment. The reader can rest assured that these theorems are accurately presented and that their proofs have been satisfactorily scrutinized. Such proofs can be found in almost any advanced text on differential equations.

*Assume that the functions F and ∂F/∂y are continuous in some rectangle*

$$\mathcal{R} = \left\{(x,y): \quad \begin{array}{c} |x - x_0| \le A \\ |y - y_0| \le B \end{array}\right\}, \quad A > 0, B > 0$$

*about the point $(x_0, y_0)$. Then there is a positive number $h \le A$ such that the IVP has one and only one solution in the interval $|x - x_0| \le h$.*

Thus, in the notation of Definition 1 of Section 1.1

$$I = \{x: |x - x_0| \le A\} \quad \text{and} \quad J = \{x: |x - x_0| \le h\}.$$

Let us illustrate this theorem by a few examples.

**EXAMPLE 1**  Show that the IVP

$$y' = x^2 + y^2 \tag{3}$$

$$y(0) = 0 \tag{4}$$

has a unique solution in some interval of the form $-h \le x \le h$.

**Proof**  Here $F(x,y) = x^2 + y^2$ and $\partial F/\partial y = 2y$ are continuous in any rectangle $\mathcal{R}$ about $(0, 0)$. By Theorem 1 there exists a positive number $h$ such that the IVP (3)–(4) has a unique solution in the interval $|x - 0| \le h$; that is, $-h \le x \le h$.

**EXAMPLE 2**  Assume that the coefficients $a(x)$ and $b(x)$ of the first-order linear differential equation

$$y' + a(x)y = b(x) \tag{5}$$

are continuous in some open interval $I$. Show that Eq. (5) has a unique solution through any point $(x_0, y_0)$ where $x_0 \in I$.

**Solution**  Choose a number $A$ such that the interval $|x - x_0| \le A$ is contained in the interval $I$. Then $F(x, y) = b(x) - a(x)y$ and $F_y(x, y) = -a(x)$ are continuous in

$$\mathcal{R} = \left\{(x, y): \quad \begin{array}{c} |x - x_0| \le A \\ -\infty < y < +\infty \end{array}\right\}.$$

By Theorem 1, Eq. (5) has a unique solution satisfying the initial condition $y(x_0) = y_0$ [in other words, through the point $(x_0, y_0)$], for any real number $y_0$.

**REMARK 1**  For linear equations such as Eq. (5), one can prove that its solutions exist throughout any interval about $x_0$ in which the coefficients $a(x)$ and $b(x)$ are continuous. For example, the IVP

$$y' + \frac{1}{x - 2} y = \ln x$$

$$y(1) = 3$$

has a unique solution in the interval $(0, 2)$, the IVP

$$y' + \frac{1}{x - 2} y = \ln x$$

$$y(3) = -1$$

has a unique solution in the interval $(2, \infty)$, and the IVP

$$y' + \frac{1}{x - 2} y = \ln | x |$$

$$y(-1) = 5$$

has a unique solution in the inverval $(-\infty, 0)$.

**EXAMPLE 3**   The functions $y_1(x) \equiv 0$ and $y_2(x) = x^2/4$ are two different solutions of the IVP

$$y' = y^{1/2} \tag{6}$$

$$y(0) = 0. \tag{7}$$

Is this in violation of Theorem 1?

**Solution**   Here $F(x, y) = y^{1/2}$ and $F_y(x, y) = \frac{1}{2} y^{-1/2}$. Since $F_y$ is not continuous at the point $(0, 0)$, one of the hypotheses of Theorem 1 is violated, and consequently there is no guarantee that the IVP (6)–(7) has a unique solution. In fact, in addition to $y_1$ and $y_2$, the IVP (6)–(7) has infinitely many solutions through the point $(0, 0)$. For each $c > 0$, these solutions are given by

$$y(x) = \begin{cases} 0, & x \le c \\ \dfrac{(x - c)^2}{4}, & c < x \end{cases}$$

and their graphs are given in Figure 1.5.

**REMARK 2**   As we have seen in Example 3 for the IVP (6)–(7), we have existence of solutions but not uniqueness. Uniqueness of solutions of initial value problems is very important. When we know that an IVP has a unique solution we can apply any method (including guessing) to find its solution. For example, we know that the IVP

$$y' + xy = 0 \tag{8}$$

$$y(0) = 0 \tag{9}$$

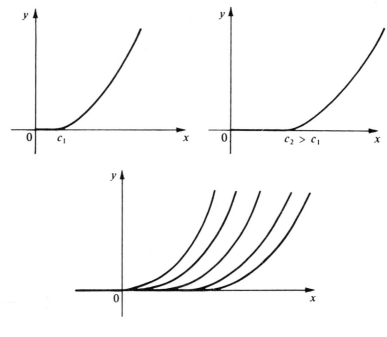

**Figure 1.5**

has a unique solution. Just by inspection we see that $y = 0$ satisfies the IVP. Thus, $y(x) = 0$ is the only solution of the IVP (8)–(9) and no further work is required in solving this IVP.

Theorem 1 gives conditions under which a first-order differential equation possesses a unique solution. Knowing that a solution exists, we desire to develop methods for determining this solution either exactly or approximately. The following method, called the method of *isoclines,* is useful in finding an approximate solution.

Consider again the first-order differential equation

$$y' = F(x, y). \tag{10}$$

For a better understanding of the differential equation (10) and its solutions, it is useful to interpret $y'$ as the slope of the tangent line drawn to the solution at the point $(x, y)$. Then at each point $(x_0, y_0)$ in the domain $D$ of $F$, the differential equation (10) assigns a slope $y'$ equal to $F(x_0, y_0)$. Hence, if we know that (10) has a solution through a given point $(x_0, y_0)$, then (without knowing the solution explicitly) we know immediately the equation of the tangent line to this solution at the point $(x_0, y_0)$. In particular, its slope is $F(x_0, y_0)$. A short segment along this tangent about $(x_0, y_0)$ will give us, graphically, an approximation to the true solution of (10) near the point $(x_0, y_0)$. Thus for graphical approximations of the solutions of (10) it is useful to draw a short line segment at each point $(a, b) \in D$ with slope $F(a, b)$. Such line segments are called *linear elements.* The totality of all linear elements is called the *direction field* of the differential equation (10). The construction of the direction field can

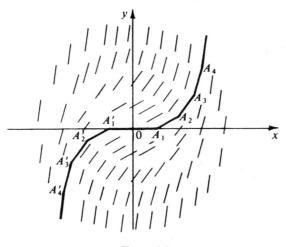

Figure 1.6

be carried out more efficiently by setting $F(x, y) = c$ and realizing that for any constant $c$ all points on the (level) curve $F(x, y) = c$ have the same slope $c$. This is the *method of isoclines* (equal slopes).

For example, the isoclines of the differential equation

$$y' = x^2 + y^2 \tag{11}$$

are the circles $x^2 + y^2 = c$, centered at the origin $(0, 0)$ with radius $\sqrt{c}$ (see Figure 1.6). At each point on the circle $x^2 + y^2 = c$ the differential equation defines the slope $c$. Therefore, the linear elements of the direction field for all these points are parallel, and each one of them has slope $c$. In Figure 1.6 we have drawn a few isoclines (for $c = 1, 4, 9, 16$) and linear elements which give a good idea of the direction field of the differential equation (11).

Now suppose that we want to find a graphical approximation of the solution of the differential equation (10) which goes through the point $(x_0, y_0)$. First we draw a few isoclines $C_1, C_2, C_3, \ldots$ corresponding to the values $c = c_1, c_2, c_3, \ldots$ [$C_i$ is the curve $F(x, y) = c_i$; see Figure 1.7]. Through the point $(x_0, y_0)$ we

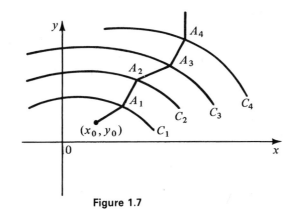

Figure 1.7

draw the linear element with slope $F(x_0, y_0)$ and extend it until it meets the isocline $C_1$ at a point $A_1$. Through $A_1$ we draw the linear element with slope $c_1$ until it meets the isocline $C_2$ at a point $A_2$. Proceeding in this fashion we construct a polygonal curve through $(x_0, y_0)$ which is an approximation of the true solution of the differential equation through that point.

In Figure 1.6 we have drawn an approximate solution of the differential equation (11) through the point $(0, 0)$. Note that, from Example 1, the differential equation (11) has a unique solution through $(0, 0)$.

### EXERCISES

In Exercises 1 through 6, check whether the hypotheses of Theorem 1 are satisfied.

**1.** $y' + xy = 3$

$y(0) = 0$

**2.** $xy' + y = 3$

$y(0) = 1$

**3.** $y' = \dfrac{x - y}{x + y}$

$y(0) = -1$

**4.** $y' = y^{1/2}$

$y(1) = 1$

**5.** $y' = y^{2/3}$

$y(0) = 0$

**6.** $y' = \dfrac{x - y}{x + y}$

$y(1) = -1$

**7.** Review Remark 1 of this section and apply it to the differential equation

$$xy' + \frac{1}{2x + 3}y = \ln|x - 2|$$

for its solutions through the points $(-3, 0)$, $(-1, 5)$, $(1, -7)$, and $(3, 0)$.

**8. Astronomy** In an article[7] concerning the accumulation processes in the primitive solar nebula, the following first-order differential equation was obtained:

$$\frac{dx}{dt} = \frac{ax^{5/6}}{(b - Bt)^{3/2}},$$

where $a$, $b$, and $B$ are constants. Find all points $(t_0, x_0)$ through which this differential equation has unique solutions.

**9.** Show that the only solution of the IVP

$$xy' - y = 1$$
$$y(2) = 3$$

is $y(x) = 2x - 1$.

In Exercises 10 through 13, answer true or false.

---

[7]A. G. W. Cameron, *Icarus* 18 (1973): 407–450.

**10.** The unique solution of the IVP

$$y' - xy = 1 - x^2$$
$$y(0) = 0$$

is $y(x) = -x$.

**11.** The only solution of the IVP

$$y' - xy = 1 - x^2$$
$$y(0) = 0$$

is $y(x) = x$.

**12.** The unique solution of the IVP

$$xy' + y^2 = 1$$
$$y(-2) = 1$$

is $y(x) = 1$.

**13.** The differential equation $y' = y^{3/4}$ has a unique solution through the point $(0, 0)$ but not through the point $(2, 0)$.

Use the method of isoclines to find graphical approximations to the solutions of the following IVPs.

**14.**  $y' = y - x$          **15.** $y' = xy$          **16.**  $y' = x^2 + y^2$

  $y(0) = 0$                $y(1) = 2$                  $y(0) = 1$

---

# 1.3  VARIABLES SEPARABLE

A separable differential equation is characterized by the fact that the two variables of the equation together with their respective differentials can be placed on opposite sides of the equation. In such equations the equality sign "separates" one variable from the other. Algebraic manipulations enable us to write separable differential equations in the form $y' = P(x)/Q(y)$ or, more explicitly,

$$Q(y)dy = P(x)dx. \tag{1}$$

To obtain the general solution of a separable differential equation we first separate the two variables as in Eq. (1) and then integrate both sides, to obtain

$$\int P(x)dx = \int Q(y)dy + c, \tag{2}$$

where $c$ is an arbitrary constant of integration. After performing the integrations in Eq. (2), it is desirable to solve the resulting expression for the dependent variable $y$, thereby obtaining $y$ explicitly in terms of $x$. If this is not possible or convenient, then Eq. (2) gives implicitly the general solution of Eq. (1).

The following are examples of differential equations that are separable:

$$x \, dx - (5y^4 + 3) \, dy = 0 \tag{3}$$

$$2x(y^2 + y) \, dx + (x^2 - 1) \, y \, dy = 0 \tag{4}$$

$$e^{x-y}y' = \sin x. \tag{5}$$

In fact, they can be brought, respectively, into the forms (3'), (4'), and (5'):

$$x \, dx = (5y^4 + 3)dy \tag{3'}$$

$$\frac{2x}{x^2 - 1} \, dx = -\frac{1}{y + 1} \, dy, \qquad \text{for } x \neq \pm 1, y \neq 0, y \neq -1 \tag{4'}$$

$$e^{-y} \, dy = e^{-x} \sin x \, dx. \tag{5'}$$

**EXAMPLE 1**  Find the general solution of Eq. (3).

**Solution**  As we have seen, Eq. (3) is separable because it can be written in the form (3'), where the variables $x$ and $y$ are separated. Integrating both sides of Eq. (3'), we obtain

$$\int x \, dx = \int (5y^4 + 3) \, dy + c.$$

Thus,

$$\tfrac{1}{2}x^2 = y^5 + 3y + c$$

is an implicit representation of the general solution of Eq. (3).

**EXAMPLE 2**  Solve the differential equation (4).

**Solution**  The variables of Eq. (4) can be separated as in Eq. (4'). Integrating both sides of Eq. (4') yields

$$\ln | x^2 - 1 | = -\ln | y + 1 | + c.$$

Thus,[8]

$$\ln | x^2 - 1 | + \ln | y + 1 | = c \Rightarrow \ln | (x^2 - 1)(y + 1) | = c.$$

Taking exponentials of both sides and using the fact that $e^{\ln p} = p,$ we have

$$| (x^2 - 1)(y + 1) | = e^c \Rightarrow (x^2 - 1)(y + 1) = \pm e^c.$$

Since $c$ is an arbitrary constant, clearly $\pm e^c$ is again an arbitrary constant, different than zero, and for economy in notation we can still denote it by $c$ with $c \neq 0$. In what follows we shall use this convention frequently.

---

[8]The symbol $\Rightarrow$ stands for "imply" (or "implies"), "it follows," "then," and will be used when appropriate.

$$(x^2 - 1)(y + 1) = c \Rightarrow y = -1 + \frac{c}{x^2 - 1}, \qquad c \neq 0.$$

Now, do not forget that Eq. (4′) is obtained from Eq. (4) by dividing through by the expression $(x^2 - 1)(y^2 + y)$. Thus, we should ensure that this expression is not zero. Hence, we require that $x \neq \pm 1$, $y \neq -1$, and $y \neq 0$. We must finally examine what happens when $x = \pm 1$ and when $y = 0$ or $y = -1$. Going back to the original Eq. (4) we see that the four lines $x = \pm 1$, $y = 0$, and $y = -1$ also satisfy the differential equation (4). If we relax the restriction $c \neq 0$, the curve $y = -1$ will be contained in the formula $y = -1 + [c/(x^2 - 1)]$ for $c = 0$. However, the curves $x = \pm 1$ and $y = 0$ are not contained in the same formula, no matter what the value of $c$ is. Sometimes such curves are called *singular solutions* and the one-parameter family of solutions

$$y = -1 + \frac{c}{x^2 - 1},$$

where $c$ is an arbitrary constant (parameter), is called the *general solution*.

The following initial value problems are frequently encountered in applications of exponential growth and decay, and readers are advised to familiarize themselves with the solutions.

**EXAMPLE 3**   Prove that the solution of the IVP

$$\dot{y} = ky, \qquad k \text{ is a constant}$$

$$y(0) = y_0$$

is $y(t) = y_0 e^{kt}$. Similarly, show that the solution of the IVP

$$\dot{y} = -ky, \qquad k \text{ is a constant}$$

$$y(0) = y_0$$

is $y(t) = y_0 e^{-kt}$.

**Proof**   Clearly, the solution to the second IVP follows from the solution to the first IVP by replacing the constant $k$ by $-k$. Thus, it suffices to solve the first IVP. Since $\dot{y} = dy/dt$, separating the variables $y$ and $t$ we obtain, for $y \neq 0$,

$$\frac{dy}{y} = k \, dt.$$

Integrating both sides, we have

$$\ln|y| = kt + c \Rightarrow |y| = e^{kt+c} = e^c e^{kt} \Rightarrow y = \pm e^c e^{kt} \Rightarrow y = ce^{kt}, \qquad c \neq 0.$$

Since $y = 0$ is also a solution of $\dot{y} = ky$, it follows that the general solution of $\dot{y} = ky$ is $y = ce^{kt}$, where $c$ is an arbitrary constant. Using the initial condition, we find that $y_0 = c$, that is, $y(t) = y_0 e^{kt}$, and the proof is complete. The solutions of the above IVPs for $k > 0$ and $k < 0$ are represented in Figure 1.8.

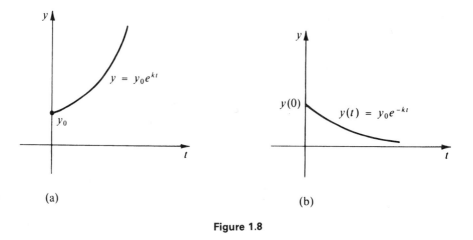

(a)                                                        (b)

**Figure 1.8**

### APPLICATIONS 1.3.1

One frequently encounters separable differential equations in applications. We shall mention a few instances here. Review also the applications in Section 1.1.1 that involved separable differential equations.

**Psychology**  ■ In 1930, Thurstone[9] obtained a differential equation as a mathematical model describing the state of a learner, or the *learning curve* of a learner, while learning a specific task or body of knowledge. If $y(t)$ denotes the state of a learner at time $t$, then Thurstone's equation is the following separable differential equation:

$$\frac{dy}{y^{3/2}(1 - y)^{3/2}} = \frac{2k}{\sqrt{m}}\,dt. \tag{6}$$

Here $k$ and $m$ are positive constants that depend on the individual learner and the complexity of the task, respectively. For the solution of the differential equation (6), see Exercise 16.

**Biology**  ■ Assume that a colony of bacteria increases at a rate proportional to the number present. If the number of bacteria doubles in 5 hours, how long will it take for the bacteria to triple?

**Solution**   Let $N(t)$ be the number of bacteria present at time $t$. Then the assumption that this colony of bacteria increases at a rate proportional to the number present can be mathematically written as

$$\frac{dN}{dt} = kN, \tag{7}$$

[9]L. L. Thurstone, *J. Gen. Psychol.* 3 (1930): 469–493.

where $k$ is the constant of proportionality. From Example 3 we have

$$N(t) = N(0)e^{kt}, \tag{8}$$

where $N(0)$ is the initial number of bacteria in this colony. Since the number of bacteria doubles in 5 hours, we have

$$N(5) = 2N(0)$$

$$\Rightarrow \quad N(0)e^{5k} = 2N(0) \Rightarrow e^{5k} = 2$$

$$\Rightarrow \quad k = \tfrac{1}{5}\ln 2.$$

The time $t$ that is required for this colony to triple must satisfy the equation

$$N(t) = 3N(0)$$

$$\Rightarrow \quad N(0)e^{kt} = 3N(0) \Rightarrow e^{kt} = 3$$

$$\Rightarrow \quad t = \frac{1}{k}\ln 3 = 5\frac{\ln 3}{\ln 2} = 5\frac{1.09}{0.69} = 7.89 \text{ hours.}$$

■ The sum of $5000 is invested at the rate of 8% per year compounded con- **Finance** tinuously. What will the amount be after 25 years?

**Solution**  Let $y(t)$ be the amount of money (capital plus interest) at time $t$. Then the rate of change of the money at time $t$ is given by

$$\frac{dy}{dt} = \frac{8}{100}y. \tag{9}$$

Equation (9) is clearly a separable differential equation. Its solution (by Example 3) is

$$y(t) = y(0)e^{(8/100)t}.$$

Since $y(0) = 5000$ (the initial amount invested), we find that

$$y(25) = 5000e^{(8/100)\cdot 25} = 5000e^2 = \$36,945.28.$$

■ In Section 1.1.1 we observed that according to Newton's second law of motion **Mechanics** a moving body of mass $m$ and velocity $v$ is governed by the differential equation

$$\frac{d}{dt}(mv) = kF, \tag{10}$$

where $F$ is the resultant force acting on the body and $k$ is a constant of proportionality. If it happens that $F$ is a function of the velocity $v$ and does not depend explicitly on time, and if $m$ is a constant, then Eq. (10) is a separable differential equation.

In the cgs (centimeter-gram-second) system and in the English system, whose units are given in Table 1.1, the constant $k$ is equal to 1.

|  | Cgs system | English system |
|---|---|---|
| Distance | centimeter | foot |
| Mass | gram | slug |
| Time | second | second |
| Force | dyne | pound |
| Velocity | cm/sec | ft/sec |
| Acceleration | cm/sec² | ft/sec² |

**Table 1.1** Table of Units

**EXAMPLE 4** From a great height above the earth, a body of mass $m = 2$ kilograms (kg) is thrown downward with initial velocity $v_0 = 10^5$ centimeters per second (cm/sec). In addition to its weight, air resistance is acting upon this body, which is numerically (in dynes) equal to twice its speed at any time. Find the velocity of the body after $t = 10^3$ sec. (Take $g = 900$ cm/sec².)

**Solution** Let $v(t)$ be the velocity of the body at time $t$. The resultant force acting upon this falling body is

$$F = (\text{weight}) - (\text{air resistance}) = (mg - 2v) \text{ dynes.}$$

Thus, by Newton's second law of motion, Eq (10),

$$\frac{d}{dt}(mv) = mg - 2v$$

$$\Rightarrow \quad m\frac{dv}{dt} = mg - 2v$$

$$\Rightarrow \quad \frac{m}{mg - 2v}\,dv = dt. \tag{11}$$

Also,

$$v(0) = 10^5. \tag{12}$$

The solution of the IVP (11)–(12) is $v(t) = \frac{1}{2}[mg - (mg - 2v_0)e^{-2t/m}]$, and so

$$v(10^3) = \left(9 - \frac{8}{e}\right)10^5 \text{ cm/sec} = \left(9 - \frac{8}{e}\right) \text{km/sec} = 6.05 \text{ km/sec.}$$

**Medicine** ■ Here we shall derive a separable differential equation that serves as a mathematical model of the dye-dilution procedure for measuring cardiac output. From the practical point of view this is a reasonable model, in spite of the many simplifications that we introduce.

In the dye-dilution method an amount of dye of mass $D_0$ milligrams (mg) is injected into a vein so near the heart that we shall assume that the heart contains $D_0$ milligrams of dye at time $t = 0$. The dye is mixed with the blood passing through the heart, and at each stroke of the heart the diluted mixture flows out and is replaced by blood from the veins. To derive our model we shall assume that the mixture of blood and dye inside the heart is uniform and flows out at a constant rate of $r$ liters per minute. We also assume that the heart is a container of constant volume $V$ liters.

With these data and assumptions, we can now formulate an initial value problem for the amount $D(t)$ of dye in the heart at any time $t$. This is a special case of mixture problems that we will study in the next section. Since $dD(t)/dt$ represents the rate of change of the dye in the heart at time $t$, we have the equation

$$\frac{dD(t)}{dt} = \text{rate in} - \text{rate out},$$

where "rate in" is the rate at which dye runs into the heart at time $t$, and "rate out" is the rate at which dye runs out of the heart at time $t$. Here

$$\text{rate in} = 0$$

as no dye runs into the heart, and

$$\text{rate out} = \frac{D(t)}{V} r,$$

where $D(t)/V$ is the concentration of dye in the heart (in other words, the amount of dye per liter of diluted mixture). Thus, the differential equation describing this mixture problem is

$$\frac{dD(t)}{dt} = -\frac{r}{V} D(t). \tag{13}$$

We also have the initial condition

$$D(0) = D_0, \tag{14}$$

where $D_0$ is the (initial) amount of dye injected at time $t = 0$. The solution of the IVP (13)–(14) is

$$D(t) = D_0 e^{-(r/V)t}.$$

The quantity (concentration) $D(t)/V$ is useful in measuring the cardiac output.

■ It is known, from experiments, that radioactive substances decay at a rate proportional to the amount of the substance that is present. Thus, if $N(t)$ is the number of atoms of the substance present at any time $t$, then

$$\dot{N}(t) = -kN(t), \tag{15}$$

**Radioactive Dating**
Archaeology,
Paleontology,
Geology, Arts

where the positive constant $k$ is called the *decay* constant of the substance. [The negative sign is due to the fact that $N(t) > 0$ and decreasing, so it has negative "derivative" $\dot{N}(t)$.] If

$$N(0) = N_0 \tag{16}$$

is the (initial) number of atoms of the substance at time $t = 0$, the solution of the IVP (15)–(16) is given by

$$N(t) = N_0 e^{-kt}. \tag{17}$$

If we know the values of $k$, $N(t)$, and $N_0$, then from Eq. (17) we can find the time $t$ that it took the radioactive substance to decay from its initial value $N_0$ to its present value $N(t)$. In fact, from Eq. (17) we obtain $e^{-kt} = N(t)/N_0$ and (taking logarithms of both sides)

$$t = -\frac{1}{k} \ln \frac{N(t)}{N_0} = \frac{1}{k} \ln \frac{N_0}{N(t)}. \tag{18}$$

Equation (17) is the basic idea behind radioactive dating when we try to determine the age of materials of archaeology, rocks, fossils, old paintings, and so on. The value of $N(t)$ is easily computed from the present amount of the substance. To find the decay constant $k$, we utilize the *half-life* of the radioactive substance, that is, the time required for half of a given sample of a radioactive substance to decay. Half-life is easily measured in the laboratory, and tables are available that give the half-life of many radioactive substances. Using Eq. (17), we see that if $T$ is the half-life of a radioactive substance, then

$$\frac{1}{2} = e^{-kT} \Rightarrow k = \frac{\ln 2}{T};$$

that is, the decay constant of a radioactive substance is obtained by dividing the natural logarithm of 2 by the half-life of the substance. Although the value $N_0$ is in general unknown, we can utilize some additional information known about some radioactive substances to compute their age. For example, it is known that each decaying uranium 238 atom gives rise to a single lead 206 atom. Thus, the number $P$ of lead 206 atoms found in a uranium mineral containing $N(t)$ atoms of uranium 238 is given by

$$P = N_0 - N(t) = N(t)(e^{kt} - 1),$$

and solving for $t$ we find the age of the mineral to be

$$t = \frac{1}{k} \ln \left[ 1 + \frac{P}{N(t)} \right].$$

Those with some interest in radioactive dating of old paintings should read the work of Coreman,[10] where it is proved beyond any scientific doubt that the paintings sold by Van Meegeren are faked Vermeers and De Hooghs. (Vermeer and De Hooghs are famous seventeenth-century Dutch painters.)

[10]P. Coreman, *Van Meegeren's Faked Vermeers and De Hooghs* (Amsterdam: Meulenhoff, 1949). See also the articles in *Science* 155 (1967): 1238–1241 and *Science* 160 (1968): 413–415.

■ Certain diseases[11] that affect humans can be considered to impart immunity **Medicine** for life. That is, an individual who has had the disease and survived is protected from ever having that disease again. This is the case (rather it is very close to being the case, since there are exceptions to all rules in these matters) for such common diseases as measles, mumps, and chicken pox, as well as for smallpox, which is now believed to be completely eradicated.

In trying to assess the effect of such a disease, we can consider one cohort group, that is, all the individuals born in one specific year, and define $N(t)$ to be the number who have survived to age $t$, and $S(t)$ the number who have not had the disease and are still susceptible to it at age $t$.

If $p$ is the probability of a susceptible's getting the disease ($0 < p < 1$), and $1/m$ the proportion of those who die due to the disease, then the following relation can be derived:

$$dS(t)/dt = -pS(t) + (S(t)/N(t))dN(t)/dt + pS^2(t)/mN(t). \qquad (19)$$

This equation was first derived by Daniel Bernoulli (in 1760) in his work on the effects of smallpox. The equation is actually less formidable than it looks. If we multiply both sides by $N/S^2$ and regroup terms, we get

$$(1/S)dN/dt - (N/S^2)dS/dt = pN/S - p/m. \qquad (20)$$

Recall that $(d/dt)(N/S) = (1/S)dN/dt - (N/S^2)dS/dt$, and substitute in the left-hand side of this last equation to get

$$d(N/S)/dt = pN/S - p/m.$$

Since $N/S$ occurs as a variable, define $y = N/S$ and rewrite the equation as

$$dy/dt = py - p/m. \qquad (21)$$

Equation (21) is a differential equation with separable variables, so we can write this relation for its solution:

$$\int_0^a \frac{dy/dt}{py - p/m} \, dt = a.$$

Or

$$\int_{y(0)}^{y(a)} \frac{dy}{py - p/m} = a.$$

So,

$$\frac{1}{p} \log \left| \frac{y(a) - 1/m}{y(0) - 1/m} \right| = a.$$

From this we get

$$\frac{y(a) - 1/m}{y(0) - 1/m} = e^{ap}.$$

[11]The authors wish to thank Stavros Busenberg for bringing this model to their attention.

Solving for $my(a)$, we have

$$my(a) = 1 + (my(0) - 1)e^{ap}. \tag{22}$$

Since, at birth, every member of the cohort group is susceptible, $S(0) = N(0)$, yielding $y(0) = 1$. Equation (22) now gives $my(a) = 1 + (m - 1)e^{ap}$. Finally, recalling that $y(a) = N(a)/S(a)$, we get

$$S(a) = mN(a)/[1 + (m - 1)e^{ap}]. \tag{23}$$

This expression gives the number of susceptibles in terms of the number of survivors to age $a$ of the cohort group, and the two basic constants $m$ and $p$. The values of these parameters in the right-hand side of the equation can be determined from census records and statistics on the disease.

**EXAMPLE 5**   Using the population data that were available to him, Bernoulli estimated $p = 1/8$ and $m = 8$ for the case of smallpox in Paris of the 1760s. Using these constants in Eq. (23), the number of susceptibles at age $a$ among a cohort group with $N(a)$ survivors of that age would be

$$S(a) = 8N(a)/(1 + 7e^{a/8}).$$

So, by age 24 ($a = 24$), $S(a)/N(a)$ would be about 0.056. In other words, only one in about eighteen of those twenty-four years old would not have had small-pox.

**REMARK 1**   Formula (23) was derived on the basis of what one believes is a reasonable, but certainly not foolproof, relation (19). To see that (19) is reasonable, let us examine what its terms mean. The left-hand side, $dS(t)/dt$, is the rate at which the number of susceptibles is changing (decreasing). The terms on the right can be collected into two groups. The first, $-pS(t)$, is the rate at which we expect susceptibles at age $t$ to be infected with the disease. This term includes those who will die from the disease. Now we need an expression for the rate of removal of susceptibles due to death from all *other* causes. This is given by the second group of terms $[dN(t)/dt + pS(t)/m](S(t)/N(t))$, where $dN(t)/dt$ is the rate of change of the whole cohort group due to death for all reasons, $pS(t)/m$ is the rate of death due to the particular disease, and $S(t)/N(t)$ is the proportion of susceptibles among the cohort group. Thus, $[dN(t)/dt + pS(t)/m](S(t)/N(t))$ is indeed the rate of decrease of susceptibles due to death for all reasons other than the disease. So. Eq. (19) is an equality between two different ways of expressing the rate of change of susceptibles.

All these assumptions leading to Eq. (19) are reasonable on the intuitive level and lead to the mathematical formulation, or model, of the situation we are examining. Once we have Eq. (19), we can use differential equations to get relation (23).

There are special cases where we can have reasonable expectations for the ratio $S(a)/N(a)$, and we should make sure that (23) fulfills these expectations. For example, if the mortality rate of the disease is very high ($m \to 1$), we could expect just about all survivors $N(a)$ to also be susceptible (that is, only people who have not had the disease survive). Here we would expect that $S(a)/N(a)$

→ 1 as $m \to 1$. In the opposite circumstance of very low mortality ($m \to \infty$), we can easily verify that $S(a)/N(a) = e^{-ap}$. Is this what you would expect in this case?

## EXERCISES

In Exercises 1 through 14, some of the differential equations are separable and some are not. Solve those that are separable.

**1.** $(1 + x)y\, dx + x\, dy = 0$

**2.** $(x^2 + y^2)dx - 2xy\, dy = 0.$

**3.** $y' = y^{1/2}$

**4.** $xy' + y = 3$

**5.** $y' + xy = 3$

**6.** $\dfrac{dp}{dt} = Ap - Bp^2;\ A, B \neq 0$

**7.** $xy' - \dfrac{y}{\ln x} = xy^2$

**8.** $yy' = y + x^2$

**9.** $y' = x - xy - y + 1$

**10.** $3xy\, dx + (x^2 + 4)dy = 0$

**11.** $\cos x \sin y\, dy - (\cos y \cos x + \cos x)dx = 0$

**12.** $yy' = y^2x^3 + y^2x$

**13.** $y^4dx + (x^2 - 3y)dy = 0$

**14.** $(1 + y^2) \cos x\, dx = (1 + \sin^2 x)2y\, dy$

**15.** Compute and graph a curve passing through the point $(0, 1)$ and whose slope at any point $(x, y)$ on the curve is equal to the ordinate $y$.

**16.** Solve Thurstone's differential equation (6). (*Hint:* Set $y = \sin^2\theta$.)

**17. Biology** Volterra[12] obtained differential equations of the forms

$$\frac{dx}{dt} = x(a_1 + a_2y)$$

$$\frac{dy}{dt} = y(b_1 + b_2x)$$

as a mathematical model describing the competition between two species coexisting in a given environment. From this system one obtains (by the chain rule, in other words, $dy/dx = dy/dt \cdot dt/dx$) the separable differential equation

$$\frac{dy}{dx} = \frac{y(b_1 + b_2x)}{x(a_1 + a_2y)}.$$

Solve this differential equation.

[12]Vito Volterra, *Lecons sur la théorie mathématique de la lutte pour la vie* (Paris: Gauthier-Villars, 1931).

**18.** Assume that the rate at which a radioactive substance decays is proportional to the amount present. In a certain sample 50% of the substance disappears in a period of 1600 years. (This is true, for example, for radium 226.)

(a) Write the differential equation that describes the decay of the substance.
(b) Compute the decay constant of the substance.
(c) What percentage of the original sample will disappear in 800 years?
(d) In how many years will only one fifth of the original amount remain?

**19.** Solve the differential equation

$$\frac{dx}{dt} = \frac{ax^{5/6}}{(b - Bt)^{3/2}}$$

given in the astronomy application in Exercise 8 of Section 1.2.

**20. Chemistry** Solve the differential equation

$$\frac{dx}{dt} = k(a - x)(b - x), \quad k, a, b > 0$$

which occurs in *chemical reactions*. What will the value of $x$ be after a very long time, that is, as $t \to \infty$?

**21. Botany** Solve the differential equation

$$\frac{dI}{dw} = 0.088(2.4 - I),$$

which seems to fit the data collected in a botanical experiment,[13] and find the value of $I$ as $w \to \infty$.

**22.** Compute the *orthogonal trajectories* of the one-parameter family of curves $y = cx$.

**23.** A colony of bacteria increases at a rate proportional to the number present. If the number of bacteria triples in 4 hours, how long will it take for the bacteria to be 27 times the initial number?

**24. Biology** Assume that a population grows according to Verhulst's *logistic law* of population growth

$$\frac{dN}{dt} = AN - BN^2,$$

where $N = N(t)$ is the population at time $t$, and the constants $A$ and $B$ are the *vital coefficients* of the population. What will the size of this population be at any time $t$? What will the population be after a very long time, that is, as $t \to \infty$?

---

[13]*Australian J. Bot.* 4–5 (1956–57): 159.

**25.** From the surface of the earth a body of mass 2 slugs is thrown upward with initial velocity of $v_0 = -30$ ft/sec (the convention adopted here is that the positive direction is downward). In addition to its weight, air resistance is acting upon this body which is numerically (in pounds) equal to three times its speed at any time. For how long will the body be moving upward?

**26.** A tank contains 50 gallons of brine in which 5 pounds of salt is dissolved. Pure water runs into the tank at the rate of 3 gallons per minute. The mixture, kept uniform by stirring, runs out of the tank at the same rate as the inflow. How much salt is in the tank after 15 minutes?

**27.** The sum of $15,000 is invested at the rate of 6% per year compounded continuously. In how many years will the money double?

**Statistics** The solutions of the separable differential equation

$$y' = \frac{A - x}{B + Cx + Dx^2} y$$

give most of the important distributions of statistics for appropriate choices of the constants $A$, $B$, $C$, and $D$. Solve the differential equation in the following cases.

**28.** $C = D = 0$, $B > 0$, and $A$ arbitrary          (normal distribution).

**29.** $A = B = D = 0$ and $C > 0$          (exponential distribution).

**30.** $B = D = 0$, $C > 0$, and $A > -C$          (gamma distribution).

**31.** $B = 0$, $C = -D$, $(A - 1)/C < 1$, and
$A/C > -1$          (beta distribution).

**32.** In the medicine application dealing with cardiac output assume that $V = 500$ milliliters ($\frac{1}{2}$ liter), $r = 120$ liters per minute, and $D_0 = 2$ milligrams. Find the amount of dye in the heart after: 1 second, 2 seconds, $2\frac{1}{2}$ seconds, and 10 seconds.

**33.** How long will it take for $2000 to grow to $8000 if this sum is invested at 5.5% annual interest compounded continuously?

**34.** A sum of $1000 is invested at $i$% annual interest compounded continuously. Find $i$ if it is desired that the money be doubled at the end of 10 years.

**35.** Repeat Exercise 34 if it is desired that the money be tripled at the end of 10 years.

**36.** A person borrows $5000 at 18% annual interest compounded continuously. How much will this person owe the lender at the end of 1 year?

**37.** How long does it take to become a millionaire if $1000 is invested at 8% annual interest compounded continuously? What if the initial investment is $10,000?

# 1.4 FIRST-ORDER LINEAR DIFFERENTIAL EQUATIONS

A first-order linear differential equation is an equation of the form

$$a_1(x)y' + a_0(x)y = f(x).$$

We always assume that the coefficients $a_1(x)$, $a_0(x)$, and the function $f(x)$ are continuous functions in some interval $I$ and that the leading coefficient $a_1(x) \neq 0$ for all $x$ in $I$. If we divide both sides of the differential equation by $a_1(x)$ and set $a(x) = a_0(x)/a_1(x)$ and $b(x) = f(x)/a_1(x)$, we obtain the equivalent differential equation

$$y' + a(x)y = b(x), \tag{1}$$

where $a(x)$ and $b(x)$ are continuous functions of $x$ in the interval $I$. The general solution of Eq. (1) can be found explicitly by observing that the change of variables

$$w = ye^{\int a(x)dx} \tag{2}$$

transforms Eq. (1) into a separable differential equation. In fact (recall that $(d/dx)[\int a(x)dx] = a(x)$),

$$w' = y' e^{\int a(x)dx} + ya(x)e^{\int a(x)dx}$$
$$= [y' + a(x)y]e^{\int a(x)dx}$$
$$= b(x)e^{\int a(x)dx}. \tag{3}$$

This is a separable differential equation with general solution

$$w(x) = c + \int b(x)e^{\int a(x)dx}dx$$
$$\Rightarrow \qquad ye^{\int a(x)dx} = c + \int b(x)e^{\int a(x)dx} dx$$
$$\Rightarrow \qquad y = e^{-\int a(x)dx}[c + \int b(x)e^{\int a(x)dx} dx].$$

We summarize our results in the form of a theorem.

**THEOREM 1**

*If $a(x)$ and $b(x)$ are continuous functions in some interval I, then the general solution of the differential equation*

$$y' + a(x)y = b(x)$$

*is given by the formula*

$$y(x) = e^{-\int a(x)dx}[c + \int b(x)e^{\int a(x)dx} dx]. \tag{4}$$

**REMARK 1** If you read the method above carefully once more, you will realize that the general solution of a first-order linear differential equation is found in three steps as follows.

**Step 1** Multiplying both sides of Eq. (1) by $e^{\int a(x)dx}$, we obtain

$$[ye^{\int a(x)dx}]' = b(x)e^{\int a(x)dx}.$$

[This is a restatement of Eq. (3).] The term $e^{\int a(x)dx}$ is called an *integrating factor*.

**Step 2** Integrating both sides, we find

$$ye^{\int a(x)dx} = c + \int b(x)e^{\int a(x)dx} \, dx.$$

**Step 3** Dividing both sides by $e^{\int a(x)dx}$ yields

$$y(x) = e^{-\int a(x)dx}[c + \int b(x)e^{\int a(x)dx} \, dx],$$

which is the general solution of Eq. (1).

**REMARK 2** Before you apply the method described above, make sure that the coefficient of $y'$ is 1. In general, you have to divide by the coefficient of $y'$ before you apply the method. That is the case in Exercises 1, 2 and 5, and Example 2.

**REMARK 3** It will be helpful in computations if you recall that for any positive function $p(x)$ the following identities hold:

$$e^{\ln p(x)} = p(x).$$

$$e^{-\ln p(x)} = \frac{1}{p(x)}.$$

**REMARK 4** From Remark 1 of Section 1.2 the solutions of the differential equation (1) exist throughout the interval $I$, where the functions $a(x)$ and $b(x)$ are continuous. This, together with Example 2 of Section 1.2, implies that the IVP

$$y' + a(x)y = b(x)$$

$$y(x_0) = y_0,$$

where $x_0 \in I$, has a unique solution which exists throughout the interval $I$. The reader can verify by direct substitution (see Exercise 18) that the unique solution of the IVP is given by the formula

$$y(x) = y_0 e^{-\int_{x_0}^{x} a(s)ds} + \int_{x_0}^{x} b(s)e^{-\int_{s}^{x} a(t)dt} \, ds. \tag{5}$$

**EXAMPLE 1** Find the general solution of the differential equation

$$\frac{dy}{dx} + (\tan x)\, y = \sin x$$

in the interval $(0, \pi/2)$.

**Solution** This is a first-order linear differential equation with

$$a(x) = \tan x \quad \text{and} \quad b(x) = \sin x,$$

both continuous in the interval $(0, \pi/2)$. Multiplying both sides by

$$e^{\int \tan x\, dx} = e^{-\ln \cos x} = \frac{1}{\cos x},$$

we obtain

$$\left( y\, \frac{1}{\cos x} \right)' = \frac{\sin x}{\cos x}.$$

Integrating both sides, we find

$$y\, \frac{1}{\cos x} = c - \ln \cos x.$$

Dividing both sides by $1/\cos x$ (in other words, multiplying both sides by $\cos x$) yields the solution

$$y(x) = (\cos x)(c - \ln \cos x), \quad 0 < x < \frac{\pi}{2}.$$

**EXAMPLE 2** Solve the IVP

$$xy' - 2y = -x^2$$
$$y(1) = 0.$$

**Solution** The differential equation can be written in the form $y' - (2/x)y = -x$ with $x \neq 0$. Here $a(x) = -2/x$ and $b(x) = -x$. Therefore, the interval $I$, where $a(x)$ and $b(x)$ are continuous, is either $(-\infty, 0)$ or $(0, \infty)$. Since $x_0 = 1$ lies in the interval $(0, \infty)$, it follows from Remark 4 that the IVP has a unique solution that exists in the entire interval $(0, \infty)$. Although we could use formula (5) to immediately obtain the solution of the IVP, it is perhaps more instructive to proceed directly by using the steps described in the text. Multiplying both sides by the integrating factor,

$$e^{-\int (2/x)\, dx} = e^{-2\ln x} = e^{-\ln x^2} = \frac{1}{x^2},$$

we have

$$\left( y\, \frac{1}{x^2} \right)' = -\frac{1}{x}.$$

Integrating both sides, we obtain

$$y\frac{1}{x^2} = c - \ln x.$$

So, the general solution is

$$y(x) = x^2(c - \ln x).$$

Using the initial condition $y(1) = 0$, we find

$$0 = 1^2(c - \ln 1)$$

$$\Rightarrow \quad c = 0,$$

and therefore the solution of the IVP is

$$y(x) = -x^2 \ln x.$$

The graph of the solution is given in Figure 1.9.

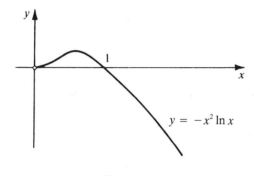

**Figure 1.9**

## APPLICATIONS 1.4.1

In the literature there are many occurrences of the applications of linear differential equations, some of which we illustrate below.

■ Consider the $RL$-series circuit of Figure 1.10, which contains a resistance $R$,    **Circuit Theory**
an inductance $L$, and a generator that supplies a voltage $V(t)$ when the switch

**Figure 1.10**

is closed. The current $I = I(t)$ in the circuit satisfies the linear first-order differential equation

$$L\frac{dI}{dt} + RI = V(t). \tag{6}$$

This differential equation is obtained from Kirchhoff's voltage law (see the electric circuits application in Section 1.1.1) and the following facts:

1. Across the resistor the voltage drop is $RI$.
2. Across the inductor the voltage drop is $L(dI/dt)$.
3. $V(t)$ is the only voltage increase in this circuit.

The reader can verify that the general solution of Eq. (6) is given by (see Exercise 19)

$$I(t) = e^{-(R/L)t}\left[c + \frac{1}{L}\int V(t)e^{(R/L)t}\,dt\right]. \tag{7}$$

Table 1.2 lists the units and conventional symbols of circuit elements that will be used in this book in connection with circuit-theory applications.

| Quantity | Unit | Symbol |
|---|---|---|
| Resistance | ohm | $R$ |
| Inductance | henry | $L$ |
| Capacitance | farad | $C$ |
| Voltage | volt | $V$ |
| Current | ampere | $I$ |
| Charge | coulomb | $Q$ |
| Time | seconds | sec |

*Symbols of circuit elements*

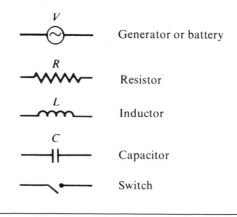

Generator or battery

Resistor

Inductor

Capacitor

Switch

**Table 1.2**  Table of Units

■ With the geometric interpretation of the concept of derivative in mind, it **Geometry**
follows that the solutions of the differential equation

$$y' = F(x, y)$$

are curves $y = y(x)$ whose slope at any point $(x, y)$ is equal to $F(x, y)$.

**EXAMPLE 3** Find a curve in the $xy$ plane, passing through the point $(1, 2)$, whose slope at the point $(x, y)$ is given by $2 - y/x$.

**Solution** The differential equation that describes the curve is

$$y' = 2 - \frac{y}{x}. \tag{8}$$

Since the curve passes through the point $(1, 2)$, we should have

$$y(1) = 2. \tag{9}$$

Let us solve this IVP (8)–(9). The differential equation can be written in the form

$$y' + \frac{1}{x}y = 2$$

and so is a linear first-order differential equation. Multiplying both sides by $e^{\int(1/x)dx} = e^{\ln x} = x$, we obtain

$$(xy)' = 2x$$

$$\Rightarrow \quad xy = c + x^2.$$

From $y(1) = 2$ it follows that $c = 1$, and so $y = x + (1/x)$ is the curve with the desired properties (Figure 1.11).

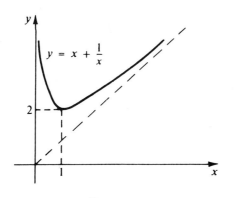

$$y = x + \frac{1}{x}$$

**Figure 1.11**

**Economics**      ■ Assume that the rate of change of the price $P$ of a commodity is proportional to the difference $D - S$ of the demand $D$ and supply $S$ in the market at any time $t$. Thus,

$$\frac{dP}{dt} = \alpha(D - S), \tag{10}$$

where the constant $\alpha$ is called the "adjustment" constant. In the simple case where $D$ and $S$ are given as linear functions of the price, that is,

$$D = a - bP, \quad S = -c + dP \qquad (a, b, c, d \text{ positive constants}),$$

Eq. (10) is a linear first-order differential equation. Let us denote by $\overline{P}$ the "equilibrium" price, that is, the price at which

$$a - b\overline{P} = D = S = -c + d\overline{P}.$$

It is left to the reader (see Exercise 26) to verify that the solutions of Eq. (10) is given by

$$P(t) = \left[P(0) - \overline{P}\right] e^{-\alpha(b+d)t} + \overline{P}. \tag{11}$$

Observe from Eq. (11) that as $t \to \infty$, the price of the commodity $P(t) \to \overline{P}$.

**Mixtures**     ■ First-order linear differential equations arise as mathematical models in rate problems involving mixtures. Let us illustrate this by means of a typical example.

**EXAMPLE 4**   A large tank contains 81 gallons of brine in which 20 pounds of salt is dissolved. Brine containing 3 pounds of dissolved salt per gallon runs into the tank at the rate of 5 gallons per minute. The mixture, kept uniform by stirring, runs out of the tank at the rate of 2 gallons per minute. How much salt is in the tank at the end of 37 minutes?

**Solution**   Let $y(t)$ be the amount of salt in the tank at any time $t$. Then $y'(t)$ is the rate of change of the salt in the tank at time $t$. Clearly,

$$y'(t) = \text{rate in} - \text{rate out},$$

where "rate in" is the rate at which salt runs into the tank at time $t$, and "rate out" is the rate at which salt runs out of the tank at time $t$. Here

$$\text{rate in} = (3 \text{ lb/gal})(5 \text{ gal/min}) = 15 \text{ lb/min},$$

that is, 15 pounds of salt per minute flows into the tank. To compute the "rate out," we should first find the *concentration* of salt at time $t$, that is, the amount of salt per gallon of brine at time $t$. Since

$$\text{concentration} = \frac{\text{pounds of salt in the tank at time } t}{\text{gallons of brine in the tank at time } t}$$

$$= \frac{y(t)}{81 + (5 - 2)t},$$

we have

$$\text{rate out} = \left[\frac{y(t)}{81 + 3t} \text{ lb/gal}\right](2 \text{ gal/min}) \doteq \frac{2y(t)}{81 + 3t} \text{lb/min.}$$

Thus, the differential equation describing this mixture problem is

$$y' = 15 - \frac{2y}{81 + 3t}. \tag{12}$$

We also have the initial condition

$$y(0) = 20, \tag{13}$$

which expresses the fact that initially, that is, at $t = 0$, there is 20 pounds of salt in the tank. We must now solve the IVP (12)–(13). Equation (12) is linear and can be written in the form

$$y' + \frac{2}{81 + 3t}y = 15.$$

Multiplying both sides by

$$\exp\left(\int \frac{2}{81 + 3t} dt\right) = \exp\left(\frac{2}{3}\int \frac{dt}{27 + t}\right) = \exp\left[\frac{2}{3}\ln(27 + t)\right] = (27 + t)^{2/3},$$

we obtain

$$[y(t)(27 + t)^{2/3}]' = 15(27 + t)^{2/3}.$$

Integrating both sides with respect to $t$, we obtain

$$y(t)(27 + t)^{2/3} = 15(27 + t)^{5/3} \cdot \frac{3}{5} + c,$$

and so

$$y(t) = 9(27 + t) + c(27 + t)^{-2/3}.$$

Using the initial condition (13), we find

$$20 = 9(27) + c(27)^{-2/3},$$

and so $c = -2007$. Thus,

$$y(t) = 9(27 + t) - 2007(27 + t)^{-2/3},$$

and the amount of salt in the tank at the end of 37 minutes is

$$y(37) = 9(27 + 37) - 2007(27 + 37)^{-2/3} = 450.6 \text{ lb.}$$

## EXERCISES

In Exercises 1 through 14, some of the differential equations are linear and some are not. Determine those that are linear and solve them.

**1.** $xy' + y = 3$

**2.** $xy' + y = 3x$

**3.** $(x^2 + y^2)dx - 2xy\, dy = 0$

**4.** $(1 + x)y\, dx + x\, dy = 0$

**5.** $xy' - y = 2x^2$

**6.** $y^2dy + y \tan x \, dx = \sin^3 x \, dx$

**7.** $y' - \dfrac{3}{x - 1}y = (x - 1)^4$

**8.** $xy' + 6y = 3x + 1$

**9.** $y' + \dfrac{1}{\sin x}y - y^2 = 0$

**10.** $e^x \, dx + x^3 \, dy + 4x^2y \, dx = 0$

**11.** $xy' + y = x^5$

**12.** $y' - \dfrac{x}{1 + x^2} = -\dfrac{x}{1 + x^2}y$

**13.** $yy' - 7y = 6x$

**14.** $y\dfrac{dx}{dy} + x = y$

**15.** Find a curve in the $xy$ plane that passes through the point $(1, 1 + e)$ and whose slope at any point $(x, y)$ is given by

$$2 - \frac{y}{x} + \frac{e^x}{x}.$$

**16.** Compute the orthogonal trajectories of the family of concentric circles

$$x^2 + y^2 = R^2.$$

**17.** Find the solution of the IVP

$$y' - y = b(x)$$
$$y(0) = 1,$$

where

$$b(x) = \begin{cases} 0, & x < 0 \\ x, & x \geq 0. \end{cases}$$

**18.** Verify formula (5) and apply it to the IVPs of Examples 2 and 3.

**19.** Verify Eq. (7) of this section.

**20.** In an $RL$-series circuit assume that the voltage $V(t)$ is a constant $V_0$. Show that the current at any time $t$ is given by

$$I(t) = \frac{V_0}{R} + \left(I_0 - \frac{V_0}{R}\right)e^{-(R/L)t},$$

where $I_0$ is the current at time $t = 0$.

**21.** In an $RL$-series circuit, $L = 4$ henries, $R = 5$ ohms, $V = 8$ volts, and $I(0) = 0$ amperes. Find the current at the end of 0.1 seconds. What will the current be after a very long time?

**22.** Assume that the voltage $V(t)$ in Eq. (6) is given by $V_0 \sin \omega t$, where $V_0$ and $\omega$ are given constants. Find the solution of the differential equation (6) subject to the initial condition $I(0) = I_0$.

**23.** In an *RL*-series circuit [see Eq. (6)] it is given that $L = 3$ henries; $R = 6$ ohms, $V(t) = 3 \sin t$, and $I(0) = 10$ amperes. Compute the value of the current at any time $t$.

**24.** A large tank contains 40 gallons of brine in which 10 pounds of salt is dissolved. Brine containing 2 pounds of dissolved salt per gallon runs into the tank at the rate of 4 gallons per minute. The mixture, kept uniform by stirring, runs out of the tank at the rate of 3 gallons per minute.
(a) How much salt is in the tank at any time $t$?
(b) Find the amount of salt in the tank at the end of 1 hour.

**25.** A tank initially contains 10 gallons of pure water. Starting at time $t = 0$ brine containing 3 pounds of salt per gallon flows into the tank at the rate of 2 gal/min. The mixture is kept uniform by stirring and the well-stirred mixture flows out of the tank at the same rate as the inflow. How much salt is in the tank after 5 minutes? How much salt is in the tank after a very long time?

**26.** Verify Eq. (11) of this section.

**27.** The *Bernoulli equation* is the differential equation

$$y' + a(x)y = b(x)y^n, \qquad n \neq 0, 1.$$

Show that the transformation $w = y^{1-n}$ reduces the Bernoulli differential equation to the linear differential equation

$$w' + (1 - n)a(x)w = (1 - n)b(x).$$

In Exercises 28 through 31, solve the Bernoulli differential equation (see Exercise 27).

**28.** $y' - \dfrac{1}{x}y = -\dfrac{1}{2y}$

**29.** $y' + \dfrac{1}{x}y = -2xy^2$

**30.** $y' - 2xy = 4xy^{1/2}$

**31.** $xy' - \dfrac{y}{2 \ln x} = y^2$

In Exercises 32 through 37 solve the initial value problem.

**32.** $y' - xy = (1 - x^2)e^{(1/2)x^2}$
$\quad y(0) = 0$

**33.** $xy' + y = 2x$
$\quad y(2) = 2$

**34.** $xy' - \dfrac{y}{\ln x} = 0$
$\quad y(e) = -1$

**35.** $(1 + x^2)y' + 2xy = -2x$
$\quad y(0) = -1$

**36.** $(1 - x)y' + xy = x(x - 1)^2$
$\quad y(5) = 24$

**37.** $(x - 1)y' - 3y = (x - 1)^5$
$\quad y(-1) = 16$

**38. Biophysics** In a study related to the biophysical limitations associated with deep diving,[14] Bradner and Mackay, obtained the following first-order linear ordinary differential equation

$$y' - Ay = B + be^{-ax},$$

where $a$, $b$, $A$, and $B$ are constants. Show that the general solution of this differential equation is given by

$$y(x) = -\frac{B}{A} - \frac{b}{a + A} e^{-ax} + ce^{Ax},$$

where $c$ is an arbitrary constant.

**39.** Show that the general solution of the differential equation

$$y' - 2xy = x^2$$

is of the form

$$y = Y_0(x) + Ae^{x^2},$$

where $Y_0(x)$ is a solution of the differential equation that obeys the inequalities

$$-\frac{x}{2} - \frac{1}{4x} \leq Y_0(x) \leq -\frac{x}{2}, \qquad x \geq 2.$$

[*Hint:* Find the general solution of the differential equation and observe that the improper integral $\int_0^x t^2 e^{-t^2}\, dt$ converges.[15]

**40. Physics** Newton's law of cooling states that the time rate of change of the temperature $T = T(t)$ of a body at time $t$ is proportional to the difference $T - T_m$ in temperature between the body and its surrounding medium. That is,

$$\frac{dT}{dt} = -k(T - T_m),$$

where $k > 0$ is a constant of proportionality. This model gives a good approximation of the true physical situation provided that the temperature difference is small, generally less than about 36°F. Show that the general solution of this differential equation is

$$T(t) = T_m + ce^{-kt}.$$

**41.** A body of temperature 70°F is placed (at time $t = 0$) outdoors, where the temperature is 40°F. After 3 minutes the temperature of the body has fallen to 60°F.
(a) How long will it take the body to cool to 50°F?
(b) What is the temperature of the body after 5 minutes?

[14]H. Bradner and R. S. Mackay, *Bull. Math. Biophys.* 25 (1963): 251–72.
[15]*Amer. Math. Monthly* 63 (1956): 414.

**42.** A *Riccati equation* is a differential equation of the form

$$y' = f(x) + g(x)y + h(x)y^2,$$

where the functions $f$, $g$, and $h$ are continuous in some interval $I$. Show that if $y_1(x)$ is a particular solution of the Riccati differential equation, then the transformation

$$y(x) = y_1(x) + \frac{1}{w(x)}$$

reduces it to the linear first-order differential equation

$$w' + [g(x) + 2y_1(x)h(x)]w = -h(x).$$

**Mass Behavior**  In a research article on the theory of the propagation of a single act in a large population (in other words, an act that is performed at most once in the lifetime of an individual, such as suicide), Rapoport[16] derived the following Riccati differential equation:

$$\dot{y} = (1 - y)[x(t) + by]. \tag{14}$$

Here $y = y(t)$ is the fraction of the population who performed the act at time $t$, $x(t)$ is the external influence or stimulus and the product $by$ is the imitation component. Solve this Riccati differential equation in each of the following cases. [*Hint:* $y = 1$ is a particular solution.[17]]

**43.** $x(t) = \dfrac{1}{t} - \dfrac{1}{10}, b = \dfrac{1}{10}.$

**44.** $x(t) = -\dfrac{1}{t \ln t} - 0.03, b = 0.03.$

---

# 1.5  EXACT DIFFERENTIAL EQUATIONS

---

A differential equation of the form

$$M(x, y)dx + N(x, y)dy = 0 \tag{1}$$

is called *exact* if there is a function $f(x, y)$ whose total differential is equal to $M(x, y)dx + N(x, y)dy$, that is (suppressing the arguments $x$ and $y$)

$$df = M\,dx + N\,dy. \tag{2}$$

[16]A. Rapoport, *Bull. Math. Biophys.* 14 (1952): 159–69.
[17]For a detailed study of the differential equation (14), see the paper by J. Z. Hearon, *Bull. Math. Biophys.* 17 (1965): 7–13.

(Recall that the total differential of a function $f$ is given by $df = f_x dx + f_y dy$, provided that the partial derivatives of the function $f$ with respect to $x$ and $y$ exist.)

If Eq. (1) is exact, then [because of (2) and (1)] it is equivalent to

$$df = 0.$$

Thus, the function $f(x, y)$ is a constant and the general solution of Eq. (1) is given by

$$f(x, y) = c. \tag{3}$$

For example, the differential equation

$$(2x - y)dx + (-x + 4y)dy = 0 \tag{4}$$

is exact because

$$d(x^2 - xy + 2y^2) = \frac{\partial}{\partial x}(x^2 - xy + 2y^2)dx + \frac{\partial}{\partial y}(x^2 - xy + 2y^2)dy$$
$$= (2x - y)dx + (-x + 4y)dy.$$

Thus, the general solution of Eq. (4) is given (implicitly) by

$$x^2 - xy + 2y^2 = c. \tag{5}$$

The following two questions are now in our minds.

**QUESTION 1**  Is there a systematic way to check if the differential equation (1) is exact?

**QUESTION 2**  If we know that the differential equation (1) is exact, is there a systematic way to solve it?

**Answer 1**  The differential equation (1) is exact if and only if

$$M_y = N_x. \tag{6}$$

That is, if the partial derivative of $M$ with respect to $y$ is equal to the partial derivative of $N$ with respect to $x$, then Eq. (1) is exact; conversely, if Eq. (1) is exact, then (6) holds.

**Answer 2**  Choose any point $(x_0, y_0)$ in the region where the functions $M$, $N$ and their partial derivatives $M_y$, $N_x$ are continuous. Then

$$f(x, y) = \int_{x_0}^{x} M(x, y_0)dx + \int_{y_0}^{y} N(x, y)dy = c \tag{7}$$

gives (implicitly) the general solution of the differential equation (1).

To justify the answers above, it suffices to prove that

$$M_y = N_x \quad \text{implies that} \quad df = M\,dx + N\,dy,$$

where $f$ is the sum of the two integrals in (7) and conversely if Eq. (1) is exact, then (6) holds. In fact, from Eq. (7) we have

$$f_x = M(x, y_0) + \int_{y_0}^{y} N_x(x, y)dy = M(x, y_0) + \int_{y_0}^{y} M_y(x, y)dy$$

$$= M(x,y_0) + M(x,y)\Big|_{y_0}^{y} = M(x,y)$$

and similarly $f_y = N(x,y)$. Therefore, $df = f_x dx + f_y dy = M\,dx + N\,dy$, which proves the "if" part of Answer 1. To prove the converse, observe that, since Eq. (1) is exact, there exists a function $f$ such that $df = M\,dx + N\,dy \Rightarrow f_x = M$ and $f_y = N \Rightarrow f_{xy} = M_y$ and $f_{yx} = N_x$, and since $f_{yx} = f_{xy}$, we have $M_y = N_x$.

**REMARK 1** The point $(x_0, y_0)$ in Eq. (7) may be chosen judiciously for the purpose of simplifying $M(x, y_0)$ in the first integral of (7) and the evaluations of both integrals at the lower points of integrations.

**REMARK 2** The precise form of Eq. (7) should be

$$f(x, y) = \int_{x_0}^{x} M(t, y_0)dt + \int_{y_0}^{y} N(x, s)ds = c, \qquad (7')$$

and the assumption should be made that the segments from $(x_0, y_0)$ to $(x, y_0)$ and from $(x, y_0)$ to $(x, y)$ lie in the region where the functions $M$, $N$ and $M_y$, $N_x$ exist and are continuous.

**REMARK 3** As shown, Eq. (7) is useful to establish the existence of the function $f$ and also to verify the test for exactness, that is Eq. (6). Students sometimes find the presence of the "arbitrary" point $(x_0, y_0)$ and the judicious selection of values for $x_0$ and $y_0$ confusing. Consequently alternative systematic methods for the determination of $f$ are given in Examples 3 and 4.

**EXAMPLE 1** Solve the differential equation

$$(3x^2 + 4xy^2)dx + (2y - 3y^2 + 4x^2y)dy = 0.$$

**Solution** Here

$$M(x, y) = 3x^2 + 4xy^2 \quad \text{and} \quad N(x, y) = 2y - 3y^2 + 4x^2y$$
$$\Rightarrow \quad M_y = 8xy \quad \text{and} \quad N_x = 8xy$$
$$\Rightarrow \quad M_y = N_x.$$

Thus, the differential equation is exact and from Eq. (7), with $(x_0, y_0) = (0, 0)$, its general solution is given implicitly by

$$\int_0^x 3x^2\, dx + \int_0^y (2y - 3y^2 + 4x^2 y)dy = c.$$

$$\Rightarrow \quad x^3 + y^2 - y^3 + 2x^2 y^2 = c.$$

**EXAMPLE 2**   Solve the IVP

$$\left[\cos x \ln (2y - 8) + \frac{1}{x}\right] dx + \frac{\sin x}{y - 4}\, dy = 0 \tag{8}$$

$$y(1) = \frac{9}{2}. \tag{9}$$

**Solution**   Here

$$M(x, y) = \cos x \ln (2y - 8) + \frac{1}{x} \quad \text{and} \quad N(x, y) = \frac{\sin x}{y - 4}$$

$$\Rightarrow \quad M_y = \cos x \frac{2}{2y - 8} = \frac{\cos x}{y - 4} \quad \text{and} \quad N_x = \frac{\cos x}{y - 4}$$

$$\Rightarrow \quad M_y = N_x.$$

Thus, the differential equation is exact and from Eq. (7) its solutions are given by

$$f(x, y) = \int_{x_0}^x \left[\cos x \ln (2y_0 - 8) + \frac{1}{x}\right] dx + \int_{y_0}^y \frac{\sin x}{y - 4}\, dy = c,$$

where $x_0$ is any number $\neq 0$ and $y_0 > 4$. Let us make the judicious choice $x_0 = 1$ and $y_0 = \frac{9}{2}$. Then

$$f(x, y) = \ln x + \sin x \ln (y - 4) + \sin x \ln 2 = c.$$

Using the initial condition (9), we find that $c = 0$. Thus, the solution of the IVP is implicitly given by

$$\ln x + \sin x \ln 2(y - 4) = 0.$$

Solving with respect to $y$, we find that

$$y = 4 + \tfrac{1}{2} e^{-\ln x/\sin x}.$$

**REMARK 4**   In the differential equation above, the largest interval of existence of the solution through $(1, \frac{9}{2})$ is $(0, \pi)$ (by the existence and uniqueness theorem, Theorem 1 of Section 1.2), and so we assume that $x > 0$.

**REMARK 5**   Sometimes it is easier to solve an exact differential equation by regrouping its terms into two groups of terms—one being the group of terms of the form $p(x)dx$ and $q(y)dy$ and the other the remaining terms—and recognizing that each group is (in fact) a total differential of a function.

**EXAMPLE 3** Solve the differential equation

$$(3x^2 + 4xy^2)dx + (2y - 3y^2 + 4x^2y)dy = 0.$$

**Solution** From Example 1 we know that this equation is exact. Regrouping terms in accordance with Remark 5, we have

$$[3x^2\,dx + (2y - 3y^2)dy] + (4xy^2\,dx + 4x^2y\,dy) = 0$$
$$\Rightarrow \qquad d(x^3 + y^2 - y^3) + d(2x^2y^2) = 0$$
$$\Rightarrow \qquad d(x^3 + y^2 - y^3 + 2x^2y^2) = 0$$
$$\Rightarrow \qquad x^3 + y^2 - y^3 + 2x^2y^2 = c.$$

**EXAMPLE 4** Solve the differential equation

$$(3x^2 + 4xy^2)dx + (2y - 3y^2 + 4x^2y)dy = 0.$$

**Solution** From Example 1 we know that this equation is exact, and therefore there exists a function $f$ such that $f_x = M$ and $f_y = N$. Consider

$$f_x = M = 3x^2 + 4xy^2.$$

Integrating with respect to $x$ we obtain

$$f = x^3 + 2x^2y^2 + h(y).$$

The "constant" of integration is a function of $y$ due to the partial integration with respect to $x$. To determine $h(y)$ we utilize the fact that $f_y = N$. Taking the partial derivative with respect to $y$, this last expression yields

$$f_y = 4x^2y + h'(y).$$

Setting this equal to $N$ ($N = 2y - 3y^2 + 4x^2y$), we have

$$h'(y) = 2y - 3y^2 \Rightarrow h(y) = y^2 - y^3.$$

Thus,

$$f(x, y) = x^3 + 2x^2y^2 + y^2 - y^3,$$

and the general solution is given (implicitly) by

$$x^3 + 2x^2y^2 + y^2 - y^3 = c.$$

Note that the solution for $h(y)$ omitted a constant of integration which normally would appear. This omission is permissible since otherwise it would be combined with the constant $c$ when we set $f = c$.

As an alternative construction we could have integrated the expression $f_y = N$ with respect to $y$. In this case the "constant" of integration would be a function of $x$, say $g(x)$, and $g(x)$ would be determined by setting $f_x = M$.

**APPLICATION 1.5.1**

**Geometry** ■ Compute the orthogonal trajectories of the one-parameter family of curves

$$x^2 + 2xy - y^2 + 4x - 4y = c. \tag{10}$$

**Solution** First, we should compute the slope of the curves (10). Differentiating with respect to $x$, we obtain

$$2x + 2y + 2xy' - 2yy' + 4 - 4y' = 0,$$

and so

$$y' = -\frac{x + y + 2}{x - y - 2}.$$

Thus, the slope of the orthogonal trajectories of (10) is

$$y' = \frac{x - y - 2}{x + y + 2}. \tag{11}$$

Equation (11) is the differential equation of the orthogonal trajectories of the curves (10). This equation is not separable or linear, but after writing it in the form

$$(-x + y + 2)dx + (x + y + 2)dy = 0, \tag{12}$$

we recognize it as an exact differential equation. Regrouping the terms in Eq. (12) in the form

$$(-x\,dx + y\,dy + 2dx + 2dy) + (y\,dx + x\,dy) = 0, \tag{13}$$

we recognize that Eq. (13) is equivalent to (the total differential)

$$\frac{1}{2}d(-x^2 + y^2 + 4x + 4y + 2xy) = 0$$
$$\Rightarrow \quad -\frac{1}{2}d(x^2 - y^2 - 4x - 4y - 2xy) = 0.$$

Thus, the general solution of the differential equation (13) is

$$x^2 - 2xy - y^2 - 4x - 4y = \text{const.} = k. \tag{14}$$

The curves given by Eq. (14) are the orthogonal trajectories of (10). We recall from analytic geometry that an equation of the form $Ax^2 + Bxy + Cy^2 + Dx + Ey + F = 0$ represents a hyperbola, parabola, or ellipse depending on whether $B^2 - 4AC$ is positive, zero, or negative.[18] For the family (10), $B^2 - 4AC = (2)^2 - 4(1)(-1) = 8$, and for the family (14), $B^2 - 4AC = (-2)^2 - 4(1)(-1) = 8$. Thus, both of these families consist of hyperbolas. The standard forms for hyperbolas are $\alpha x^2 - \beta y^2 = \alpha\beta$, $\beta y^2 - \alpha x^2 = \alpha\beta$, and $xy = \gamma$ with $\alpha > 0$, $\beta > 0$, and $\gamma$ arbitrary. Equations (10) and (14) can be written in these forms by the transformations of variables known as rotation of axes and translation of axes (in that order). In general, the rotation angle $\theta$ is

---

[18]G. B. Thomas, Jr., *Calculus and Analytic Geometry*, Alternate Edition (Reading, Mass.: Addison–Wesley, 1972), p. 508.

chosen so as to eliminate the cross-product term and hence must satisfy the condition

$$\cot 2\theta = \frac{A - C}{B}.$$

For the family (10), we have

$$\cot 2\theta = \frac{1 - (-1)}{2} = 1$$

$$\Rightarrow \quad 2\theta = 45°$$

$$\theta = 22\tfrac{1}{2}°.$$

If the variables corresponding to the rotated $x$ and $y$ axes are denoted by $u$ and $v$, respectively, we have

$$x = u \cos \theta - v \sin \theta \tag{15}$$

$$y = u \sin \theta + v \cos \theta. \tag{16}$$

Now

$$\sin \theta = \sqrt{\frac{1 - \cos 2\theta}{2}} = \sqrt{\frac{1 - 1/\sqrt{2}}{2}} = \sqrt{\frac{\sqrt{2} - 1}{2\sqrt{2}}}$$

$$\cos \theta = \sqrt{\frac{1 + \cos 2\theta}{2}} = \sqrt{\frac{1 + 1/\sqrt{2}}{2}} = \sqrt{\frac{\sqrt{2} + 1}{2\sqrt{2}}}.$$

Substituting (15) and (16) into Eqs. (10) and (14), we obtain, respectively,

$$u^2 - v^2 - 2u + 2v = \frac{c}{\sqrt{2}} \tag{17}$$

$$uv - u - v = \frac{k}{\sqrt{2}}. \tag{18}$$

By completing the square Eq. (17) can be written in the form

$$(u - 1)^2 - (v - 1)^2 = \frac{c}{\sqrt{2}}. \tag{19}$$

By factoring, Eq. (18) can be written in the form

$$(u - 1)(v - 1) = \frac{k}{\sqrt{2}} + 1. \tag{20}$$

The translation of axes $s = u - 1$, $t = v - 1$ enables us to write (19) and (20) in the respective standard forms

$$s^2 - t^2 = \frac{c}{\sqrt{2}} \tag{21}$$

$$st = \frac{k}{\sqrt{2}} + 1. \tag{22}$$

These families are illustrated in Figure 1.12. $F_1$ refers to the family (21) and $F_2$ refers to the family (22).

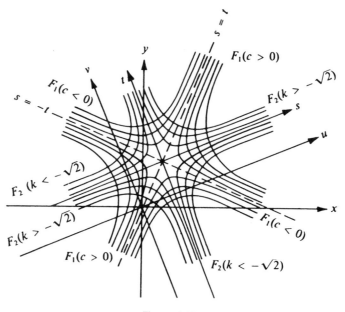

**Figure 1.12**

## EXERCISES

In Exercises 1 through 12, some of the differential equations are exact and some are not. Determine those that are exact and solve them.

**1.** $(x - y)dx + (-x + y + 2)dy = 0$     **2.** $(x + y)dx + (x - y)dy = 0$

**3.** $y' = \dfrac{y - x + 1}{-x + y + 3}$     **4.** $(x^2 + y^2)dx - 2xy\,dy = 0$

**5.** $(x^2 + y^2)dx + 2xy\,dy = 0$     **6.** $y' = y^{1/2}$

**7.** $ye^{xy}\,dx + (xe^{xy} + 1)\,dy = 0$     **8.** $\cos y\,dx + \sin x\,dy = 0$

**9.** $(y + \cos x)dx + (x + \sin y)dy = 0$

**10.** $(3x^2y + y^2)dx = (-x^3 - 2xy)dy$

**11.** $(e^x \cos y - x^2)dx + (e^y \sin x + y^2)dy = 0$

**12.** $(2x - y \sin xy)dx + (6y^2 - x \sin xy)dy = 0$

**13.** Compute the orthogonal trajectories of the one-parameter family of curves $x^3 - 3xy^2 + x + 1 = c$.

**14.** If $u(x, y)$ is a harmonic function, that is, $u_{xx} + u_{yy} = 0$, show that the orthogonal trajectories of the one-parameter family of curves $u(x, y) = c$ satisfy an exact differential equation.

**15.** Consider the one-parameter family of curves

$$Ax^2 + Bxy + Cy^2 + Dx + Ey = c, \tag{23}$$

where $c$ is the parameter and $A$, $B$, $C$, $D$, and $E$ are arbitrary (but fixed) coefficients. Show that the differential equation for the orthogonal trajectories of this family is

$$(Bx + 2Cy + E)dx + (-2Ax - By - D)dy = 0. \tag{24}$$

**16.** (a) Under what conditions will the differential equation (24) be exact?
(b) Using the condition obtained in part (a), find the equation of orthogonal trajectories and show that they and the family (23) are hyperbolas.

Solve the following initial value problems:

**17.** $(x - y)\, dx + (-x + y + 2)dy = 0$
$\qquad y(1) = 1$

**18.** $(x + y)\, dx + (x - y)\, dy = 0$
$\qquad y(0) = 2$

**19.** $(x^2 + y^2)dx + 2xy\, dy = 0$
$\qquad y(1) = -1$

**20.** $\quad y' = \dfrac{y - x + 1}{-x + y + 3}$
$\qquad y(1) = 2$

**21. Integrating Factors** If the differential equation

$$M(x, y)\, dx + N(x, y)\, dy = 0 \tag{25}$$

is not exact, that is, $M_y \neq N_x$, we can sometimes find a (nonzero) function $\mu$ that depends on $x$ or $y$ or both $x$ and $y$ such that the differential equation

$$\mu M\, dx + \mu N\, dy = 0 \tag{26}$$

is exact, that is, $(\mu M)_y = (\mu N)_x$. The function $\mu$ is then called an *integrating factor* of the differential equation (25). Since (26) is exact, we can solve it, and its solutions will also satisfy the differential equation (25).
   Show that a function $\mu = \mu(x, y)$ is an integrating factor of the differential equation (25) if and only if it satisfies the partial differential equation

$$N\mu_x - M\mu_y = (M_y - N_x)\mu. \tag{27}$$

**22.** In general, it is very difficult to solve the partial differential equation (27) without some restrictions on the functions $M$ and $N$ of Eq. (25). In this and the following exercise the restrictions imposed on $M$ and $N$ reduce Eq. (27) into a first-order linear differential equation whose solutions can be found explicitly. If it happens that the expression

$$\frac{1}{N}(M_y - N_x)$$

is a function of $x$ alone, it is always possible to choose $\mu$ as a function of $x$ only. Show that with these assumptions the function

$$\mu(x) = e^{\int (1/N)(M_y - N_x)dx}$$

is an integrating factor of the differential equation $M\, dx + N\, dy = 0$.

**23.** If it happens that the expression

$$\frac{1}{M}(M_y - N_x)$$

is a function of $y$ alone, it is always possible to choose $\mu$ as a function of $y$ only. Show that with these assumptions the function

$$\mu(y) = e^{-\int (1/M)(M_y - N_x)dy}$$

is an integrating factor of the differential equation $M \, dx + N \, dy = 0$.

For each of the following differential equations, find an integrating factor and then use it to solve the differential equation. [*Hint:* Use Exercise 22 or 23]

**24.** $y \, dx - x \, dy = 0$             **25.** $(x^2 - 2y)dx + x \, dy = 0$

**26.** $y \, dx + (2x - y^2)dy = 0$          **27.** $(y - 2x)dx - x \, dy = 0$

**28.** $y \, dx - (x - 2y)dy = 0$          **29.** $(x^4 + y^4)dx - xy^3 dy = 0$

**30.** $(x^2 - y^2 + x)dx + 2xy \, dy = 0$

**31.** Verify that $\mu(x) = e^{\int a(x)dx}$ is an integrating factor of the first-order linear differential equation

$$y' + a(x)y = b(x),$$

and then use it to find its solution. [*Hint:* Write the differential equation in the equivalent form $[a(x)y - b(x)]dx + dy = 0$].

Verify that each of the following functions is an integrating factor of the differential equation $y \, dx - x \, dy = 0$ and then use the function to solve the equation.

**32.** $\mu(y) = \dfrac{1}{y^2}$   for   $y \neq 0$

**33.** $\mu(x, y) = \dfrac{1}{xy}$   for   $x \neq 0$   and   $y \neq 0$

**34.** $\mu(x, y) = \dfrac{1}{x^2 + y^2}$   for   $x \neq 0$   or   $y \neq 0$

**35.** $\mu(x) = \dfrac{1}{x^2}$   for   $x \neq 0$

**36.** $\mu(x, y) = \dfrac{1}{x^2 - y^2}$   for   $x \neq \pm y$

**37.** $\mu(x, y) = \dfrac{1}{(x - y)^2}$   for   $x \neq y$

**38.** $\mu(x, y) = \dfrac{1}{(x + y)^2}$   for   $x \neq -y$

**39.** Verify that the function $\mu(x, y) = 1/(x^2 + y^2)$ is an integrating factor of the differential equation

$$(2x^2 + 2y^2 + x)dx + (x^2 + y^2 + y)dy = 0,$$

and then use it to find its solution.

## 1.6  HOMOGENEOUS EQUATIONS

A homogeneous differential equation is an equation of the form

$$y' = \frac{g(x, y)}{h(x, y)}, \tag{1}$$

where the functions $g$ and $h$ are homogeneous of the same degree. Recall that a function $g(x, y)$ is called *homogeneous of degree n* if $g(\alpha x, \alpha y) = \alpha^n g(x, y)$. For example, the functions $x/y$, $\sin x/y$, $x - y$, $x + y + \sqrt{x^2 + y^2}$, $x^2 + \sqrt{x^4 + y^4}$, $x^3 + y^3 + xy^2$ are homogeneous of degree zero, zero, 1, 1, 2, and 3, respectively. The following are examples of homogeneous differential equations:

$$y' = \frac{x - y}{x + y}, \qquad y' = \frac{x^2 + y^2}{x^2 + 3xy - y^2}, \qquad y' = \sin\frac{x}{y},$$

$$y' = \frac{x + \sqrt{x^2 + y^2}}{x - \sqrt{x^2 + y^2}}.$$

To solve the homogeneous differential equation (1), set $y = wx$ and observe that this substitution reduces Eq. (1) to a separable differential equation. In fact,

$$(wx)' = \frac{g(x, wx)}{h(x, wx)} \Rightarrow w'x + w = \frac{x^n g(1, w)}{x^n h(1, w)},$$

where $n$ is the degree of homogeneity of $g$ and $h$. Thus,

$$w'x = \frac{g(1, w)}{h(1, w)} - w,$$

which is easily recognized as a separable differential equation.

**EXAMPLE 1**  Solve the differential equation

$$y' = \frac{x^3 + y^3}{xy^2}. \tag{2}$$

**Solution**  Here $g(x, y) = x^3 + y^3$ and $h(x, y) = xy^2$ are homogeneous of degree 3. $[g(\alpha x, \alpha y) = (\alpha x)^3 + (\alpha y)^3 = \alpha^3(x^3 + y^3) = \alpha^3 g(x, y)$, and $h(\alpha x, \alpha y) =$

$(\alpha x)(\alpha y)^2 = \alpha^3 x y^2 = \alpha^3 h(x, y)$.] Thus, Eq. (2) is a homogeneous differential equation. Set $y = wx$. Then, from Eq. (2),

$$(wx)' = \frac{x^3 + w^3 x^3}{x w^2 x^2} \Rightarrow w'x + w = \frac{1 + w^3}{w^2} = \frac{1}{w^2} + w$$

$$\Rightarrow \qquad w'x = \frac{1}{w^2} \Rightarrow w^2 \, dw = \frac{1}{x} dx \quad \text{(separable)}$$

$$\Rightarrow \qquad \frac{1}{3} w^3 = \ln|x| + c \Rightarrow \frac{1}{3} \frac{y^3}{x^3} = \ln|x| + c.$$

Thus,

$$y = x(3 \ln|x| + c)^{1/3}.$$

**EXAMPLE 2**   Solve the IVP

$$(x^2 + y^2) \, dx + 2xy \, dy = 0 \qquad (3)$$

$$y(1) = -1. \qquad (4)$$

**Solution**   The differential equation (3) can be written in the form

$$y' = -\frac{x^2 + y^2}{2xy}, \qquad (5)$$

which is clearly homogeneous. [The reader can verify that the hypotheses of the existence theorem, Theorem 1 of Section 1.2, are satisfied for the IVP (3)–(4). Also, we are looking for a solution through the point $(1, -1)$, and so the division by $2xy$ in Eq. (5) is permissible in a small rectangle about the point $(1, -1)$.] Setting $y = wx$ in Eq. (5), we obtain

$$(wx)' = -\frac{x^2 + w^2 x^2}{2xwx} \Rightarrow w'x + w = -\frac{1 + w^2}{2w} \Rightarrow w'x = -\frac{1 + 3w^2}{2w}$$

$$\Rightarrow \qquad \frac{2w}{1 + 3w^2} dw + \frac{dx}{x} = 0 \Rightarrow \frac{1}{3} \ln(1 + 3w^2) + \ln|x| = c$$

$$\Rightarrow \qquad \ln[(1 + 3w^2)|x|^3] = c \Rightarrow \left(1 + 3\frac{y^2}{x^2}\right) x^3 = c.$$

Using the initial condition (4), we find that $c = 4$

$$\Rightarrow \qquad \left(1 + 3\frac{y^2}{x^2}\right) x^3 = 4 \Rightarrow y^2 = \frac{4 - x^3}{3x} \Rightarrow y = \pm \sqrt{\frac{4 - x^3}{3x}}.$$

Clearly, one of the two signs should be rejected. In this example the right sign is the negative one, because $y(1) = -1$. So the solution is

$$y = -\sqrt{\frac{4 - x^3}{3x}}.$$

**APPLICATION 1.6.1**

■ Compute the orthogonal trajectories of the one-parameter family of circles    **Geometry**
$x^2 + y^2 = 2cx$.

**Solution**  By taking derivatives of both sides, we find that $2x + 2yy' = 2c$. Since $2c = (x^2 + y^2)/x$, it follows that

$$2x + 2yy' = \frac{x^2 + y^2}{x} \Rightarrow y' = \frac{y^2 - x^2}{2xy} .$$

Thus, the slope of each curve in this family at the point $(x, y)$ is

$$y' = \frac{y^2 - x^2}{2xy} .$$

The differential equation of the orthogonal trajectories is, therefore,

$$y' = \frac{2xy}{x^2 - y^2} . \tag{6}$$

Equation (6) is a homogeneous differential equation. (It is not separable, it is not linear, it is not exact.) Setting $y = wx$ in (6), we obtain

$$(wx)' = \frac{2x^2 w}{x^2 - w^2 x^2} \Rightarrow w'x + w = \frac{2w}{1 - w^2} \Rightarrow w'x = \frac{w + w^3}{1 - w^2}$$

$$\Rightarrow \quad \frac{(1 - w^2)dw}{w(1 + w^2)} = \frac{dx}{x} \Rightarrow \left( \frac{1}{w} - \frac{2w}{1 + w^2} \right) dw = \frac{dx}{x} .$$

Integrating both sides and replacing $w$ by $y/x$, we find the one-parameter family of circles $x^2 + y^2 = 2ky$. Clearly, every circle in the family $x^2 + y^2 = 2cx$ has center at the point $(c, 0)$ of the $x$ axis and radius $| c |$, while every circle in the family of orthogonal trajectories $x^2 + y^2 = 2ky$ has center at the point $(0, k)$ of the $y$ axis and radius $| k |$ (see Figure 1.13).

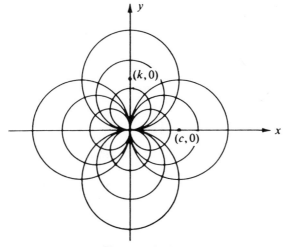

**Figure 1.13**

**EXERCISES**

In Exercises 1 through 12, some differential equations are homogeneous and some are not. Solve those that are homogeneous.

**1.** $(5x - y)dx + 3x\,dy = 0$

**2.** $xy' + y^2 = 3$

**3.** $(x^2 + y^2)dx - 2xy\,dy = 0$

**4.** $(x^2 + y^2 + 1)dx - 2xy\,dy = 0$

**5.** $-x^2y\,dx + (x^3 + y^3)dy = 0$

**6.** $(2x - 3y)dx + (7y^2 + x^2)dy = 0$

**7.** $3y\,dx + (7x - y)\,dy = 0$

**8.** $e^{y/x} + y' - \dfrac{y}{x} = 0$

**9.** $xy\,dx - (x^2 - y^2)dy = 0$

**10.** $(xy + 1)\,dx + y^2\,dy = 0$

**11.** $(x - y)\,dx + (2x + y)\,dy = 0$

**12.** $y' = \dfrac{x}{y} + \dfrac{y}{x}$

**13.** Compute the orthogonal trajectories of the one-parameter family of curves

$$x^2 + 3xy + y^2 = c.$$

**14.** Find a curve in the $xy$ plane passing through the point $(1, 2)$ and whose slope at any point $(x, y)$ is given by $-(x + 2y)/y$.

**15.** Show that any differential equation of the form

$$y' = \frac{ax + by + c}{Ax + By + C}, \qquad \text{with } aB - bA \neq 0,$$

can be reduced to a homogeneous differential equation by means of the transformation

$$x = X + x_0$$

$$y = Y + y_0,$$

where $(x_0, y_0)$ is the solution of the system

$$ax + by + c = 0$$

$$Ax + By + C = 0.$$

Using the method of Exercise 15, solve each of the following differential equations.

**16.** $y' = \dfrac{x - y}{x + y + 2}$

**17.** $y' = \dfrac{2x + y - 4}{x - y + 1}$

**18.** $y' = \dfrac{3x - 2y + 7}{2x + 3y + 9}$

**19.** $y' = \dfrac{5x - y - 2}{x + y + 4}$

**20.** $y' = \dfrac{x - y + 5}{2x - y - 3}$

**21.** $y' = \dfrac{-x + y + 1}{3x - y - 1}$

**22.** $y' = \dfrac{y}{x - y + 1}$

**23.** $y' = \dfrac{2x}{x - y + 1}$

Solve the following initial value problems.

**24.** $y' = -\dfrac{x + 2y}{y}$

$y(1) = 1$

**25.** $(x^2 + y^2)dx - 2xy\,dy = 0$

$y(1) = 1$

**26.** $y' = \sqrt{\dfrac{x + y}{2x}}$

$y(1) = 2$

**27.** $y' = \dfrac{2x + y - 4}{x - y + 1}$

$y(2) = 2$

---

# 1.7 EQUATIONS REDUCIBLE TO FIRST ORDER

---

In this section we study two types of higher-order ordinary differential equations that can be reduced to first-order equations by means of simple transformations.

**A.** Equations of the form

$$y^{(n)} = F(x, y^{(n-1)}), \tag{1}$$

containing only two consecutive derivatives $y^{(n)}$ and $y^{(n-1)}$ can be reduced to first order by means of the transformation

$$w = y^{(n-1)}. \tag{2}$$

In fact, differentiating both sides of Eq. (2) with respect to $x$ we find $w' = y^{(n)}$, and using (1) we obtain

$$w' = F(x, w).$$

**EXAMPLE 1** Compute the general solution of the differential equation

$$y''' - \frac{1}{x}y'' = 0. \tag{3}$$

**Solution** Setting $w = y''$, Eq. (3) becomes

$$w' - \frac{1}{x}w = 0. \tag{4}$$

Equation (4) is separable (and linear) with general solution

$$w(x) = c_1x.$$

Thus, $y'' = c_1x$. Integrating with respect to $x$, we obtain $y' = \frac{1}{2}c_1x^2 + c_2$.

Another integration yields $y = \frac{1}{6} c_1 x^3 + c_2 x + c_3$. As $c_1$ is an arbitrary constant, the general solution of Eq. (3) is

$$y(x) = c_1 x^3 + c_2 x + c_3. \tag{5}$$

**REMARK 1**  The general solution of an ordinary differential equation of order $n$ contains $n$ arbitrary constants. For example, Eq. (3) is of order 3, and as we have seen, its general solution (5) contains the three arbitrary constants $c_1$, $c_2$, and $c_3$.

**B.** Second-order differential equations of the form

$$y'' = F(y, y') \tag{6}$$

(they should not contain $x$) can be reduced to first order by means of the transformation

$$w = y'. \tag{7}$$

In fact, from (7) we obtain (using the chain rule)

$$y'' = \frac{dw}{dx} = \frac{dw}{dy} \cdot \frac{dy}{dx} = \frac{dw}{dy} w.$$

Thus, Eq. (6) becomes

$$w \frac{dw}{dy} = F(y, w),$$

which is first order with independent variable $y$ and unknown function $w$. Sometimes this latter equation can be solved by one of our earlier methods.

**EXAMPLE 2**  Solve the IVP

$$y'' = y'(y' + y) \tag{8}$$

$$y(0) = 0$$

$$y'(0) = -1.$$

**Solution**  Here the differential equation (8) does not contain $x$ and therefore can be reduced to a first-order differential equation by means of the transformation $w = y'$. In fact, Eq. (8) becomes

$$\frac{dw}{dy} - w = y, \tag{9}$$

which is linear. Multiplying both sides of (9) by $e^{-y}$ we obtain $(d/dy)(we^{-y}) = ye^{-y}$. Integrating with respect to $y$ we find $we^{-y} = -ye^{-y} - e^{-y} + c_1$, and so

$$y'(x) = w = -y - 1 + c_1 e^y.$$

Using the initial conditions, we find that $c_1 = 0$. Thus,

$$y' = -y - 1$$

and $y(x) = -1 + ce^{-x}$. Using the initial condition $y(0) = 0$, we find that $c = 1$, and the solution of Eq. (8) is

$$y(x) = -1 + e^{-x}.$$

## APPLICATION 1.7.1

■ Assume that two compartments $A_1$ and $A_2$, of volumes $V_1$ and $V_2$, respectively, are separated by a barrier. Through the barrier a solute can diffuse from one compartment to the other at a rate proportional to the difference $c_1 - c_2$ in concentration of the two compartments, from the higher concentration to the lower. Find the concentration in each compartment at any time $t$.

**Chemistry**
Diffusion
between
Two
Compartments

**Solution**  Let $y_1(t)$ and $y_2(t)$ be the amount of the solute in the compartments $A_1$ and $A_2$, respectively, at time $t$. Then

$$c_1(t) = \frac{y_1(t)}{V_1} \quad \text{and} \quad c_2(t) = \frac{y_2(t)}{V_2}$$

are the concentrations in compartments $A_1$ and $A_2$, respectively. The differential equations describing the diffusion are

$$\dot{y}_1(t) = k(c_2 - c_1) \quad \text{and} \quad \dot{y}_2 = k(c_1 - c_2),$$

where $k$ is a constant of proportionality ($k > 0$). Dividing by the volumes of the compartments, we get

$$\dot{c}_1 = \frac{k}{V_1}(c_2 - c_1) \quad \text{and} \quad \dot{c}_2 = \frac{k}{V_2}(c_1 - c_2) = -\frac{\dot{y}_1}{V_2} = -\frac{V_1}{V_2}\dot{c}_1. \tag{10}$$

Differentiating both sides of the first equation and using the second equation, we find that

$$\ddot{c}_1 = \frac{k}{V_1}(\dot{c}_2 - \dot{c}_1) = -k\left(\frac{1}{V_1} + \frac{1}{V_2}\right)\dot{c}_1$$

or

$$\ddot{c}_1 + k\left(\frac{1}{V_1} + \frac{1}{V_2}\right)\dot{c}_1 = 0. \tag{11}$$

Equation (11) is a differential equation of form A and can be reduced to a first-order differential equation. We leave the computational details for Exercises 5, 6, and 7.

## EXERCISES

Compute the general solution of the following differential equations.

**1.** $y'' + y' = 3$                        **2.** $y^{(5)} - y^{(4)} = 0$

Solve the following initial value problems.

**3.**  $y'' = \dfrac{1 + y'^2}{2y}$

$y(0) = 1$

$y'(0) = -1$

**4.**  $y'' + y = 0$

$y(0) = 0$

$y'(0) = 1$

In Exercises 5 through 7, let $c_1(0)$ and $c_2(0)$ be the initial concentrations in the two compartments in the chemistry application. Show that:

**5.**  $c_1(t) = c_1(0) - \dfrac{V_2}{V_1 + V_2}[c_2(0) - c_1(0)]\left\{\exp\left[-k\left(\dfrac{1}{V_1} + \dfrac{1}{V_2}\right)t\right] - 1\right\}$

**6.**  $c_2(t) = c_2(0) - \dfrac{V_1}{V_1 + V_2}[c_1(0) - c_2(0)]\left\{\exp\left[-k\left(\dfrac{1}{V_1} + \dfrac{1}{V_2}\right)t\right] - 1\right\}$

**7.**  $c_1(\infty) = c_2(\infty) = \dfrac{V_1 c_1(0) + V_2 c_2(0)}{V_1 + V_2}$. That is, after a long time, the con-

centrations in the two compartments are equal, and so equilibrium would be reached.

Find the general solution of the following differential equations.

**8.**  $xy'' + y' = 3$

**9.**  $y^{(5)} - \dfrac{1}{x}y^{(4)} = 0$

**10.**  $y''' + y'' = 1$

**11.**  $y'' - y = 0$

**12.**  $y'' = 2y' + 2y'y$

**13.**  $y'' - 2y^{-3}y' = 0$

**Astrophysics**   Differential equations of the form

$$y'' = (1 + y'^2)\,f(x, y, y')$$

have been obtained in connection with the study of orbits of satellites.[19] Solve this differential equation in each of the following cases.

**14.**  $f(x, y, y') = \dfrac{1}{y}$

**15.**  $f(x, y, y') = y'$

**16.**  $f(x, y, y') = 1$

## REVIEW EXERCISES

In Exercises 1 through 7, solve the initial value problem.

**1.**  $(1 + x)y\,dx + x\,dy = 0$

$y(1) = 2$

**2.**  $y' = 1 + y^2, \quad \dfrac{-\pi}{2} < x < \dfrac{\pi}{2}$

$y\left(\dfrac{\pi}{4}\right) = 1$

[19]See *Notices AMS* (Jan. 1975): A142.

**3.** $(1 + x^2)y' - 2xy = (x^2 + 1)^2$
$$y(0) = 0$$

**4.** $2xyy' - 2y^2 + x = 0$
$$y(1) = 1$$

**5.** $(2x^2 + y^2)dx + 2xy\,dy = 0$
$$y(1) = -1$$

**6.** $y' = \dfrac{y^2 - x^2 - x}{2xy}$
$$y(1) = 1$$

**7.** $xy'' + y' = 2x$
$$y(1) = 2$$
$$y'(1) = 0$$

In Exercises 8 through 21, solve the differential equation.

**8.** $\dfrac{dx}{dt} = (x - 2)(x - 3)$

**9.** $\dfrac{dx}{dt} = \dfrac{x - 2}{x - 3}$

**10.** $\dfrac{dx}{dt} = x - x^2$

**11.** $\dfrac{dx}{dt} = \dfrac{\cos t}{\cos x}$

**12.** $y' + y = \dfrac{x}{y}$ (Bernoulli)

**13.** $y' = (1 + x + x^2) - (1 + 2x)y + y^2$
(Riccati with particular solution $y_1(x) = x$)

**14.** $\dfrac{dx}{dt} = \dfrac{3x^{5/6}}{(1 - t)^{3/2}}$

**15.** $(3x - 2y)y\,dx = x^2dy$

**16.** $(x^2 + 2y)\,dx - x\,dy = 0$

**17.** $xy' - y = x^2e^x$

**18.** $xy'' - y' = x^2e^x$

**19.** $(x^4 + y^4)\,dx + (4x - y)y^3dy = 0$

**20.** $(x + y + 1)\,dx + (3x + y + 1)\,dy = 0$

**21.** $(x - 2y)\,dx + (y - 2x)\,dy = 0$

**22.** Show that the IVP

$$y' = \dfrac{x^2 + 2y}{x}$$
$$y(1) = 0$$

has a unique solution in some interval of the form $1 - h < x < 1 + h$ with $h > 0$.

**23.** Find the largest interval of existence of the solution of the IVP in Exercise 22.

**24.** Compute the orthogonal trajectories of the one-parameter family of curves $x^2 + y^2 = 2cy$.

**25.** Find the solution of the IVP

$$y' + \frac{1}{x}y = b(x)$$

$$y(1) = 0,$$

where

$$b(x) = \begin{cases} 1, & x \le 1 \\ \dfrac{1}{x}, & x > 1. \end{cases}$$

**26.** Compute the orthogonal trajectories of the one-parameter family of curves $x^2 + b^2y^2 = 1$.

**27. Circuits** In the $RL$-series circuit shown in Figure 1.14, assume that $V(t) = V_0 \sin \omega t$ and $I(0) = 0$. Show that

$$I(t) = \frac{V_0 \omega L}{R^2 + \omega^2 L^2} e^{-Rt/L} + \frac{V_0}{(R^2 + \omega^2 L^2)^{1/2}} \sin (\omega t - \phi),$$

where $\phi$ is an angle defined by

$$\sin \phi = \frac{\omega L}{(R^2 + \omega^2 L^2)^{1/2}}, \quad \cos \phi = \frac{R}{(R^2 + \omega^2 L^2)^{1/2}}.$$

Note that the current $I(t)$ is the sum of two terms; as $t \to \infty$, the first term dies out and the second term dominates. For this reason the first term is called the *transient* and the second term the *steady state*.

**Figure 1.14**

**28.** A colony of bacteria increases at a rate proportional to the number present. If the number of bacteria doubles in 3 hours, how long will it take for the bacteria to triple?

**29.** A colony of bacteria increases at a rate proportional to the number present. In 1 hour their number increases from 2000 to 5000. How long does it take this colony of bacteria to double?

**30.** The sum of $4000 is invested at the rate of 12% compounded continuously. What will the amount be after 6 years?

**31.** It is known that the half-life of radiocarbon is 5568 years. How old is a wooden archaeological specimen which has lost 15% of its original radiocarbon?

**32.** Find a curve in the $xy$ plane, passing through the origin and whose slope at any point $(x, y)$ is given by $y + e^x$.

**33.** A large tank contains 30 gallons of brine in which 10 pounds of salt is dissolved. Brine containing 2 pounds of dissolved salt per gallon runs into the tank at the rate of 3 gallons per minute. The mixture, kept uniform by stirring, runs out of the tank at the rate of 2 gallons per minute.
(a) How much salt is in the tank at any time $t$?
(b) How much salt will be in the tank after 10 minutes?

**34.** In the circuit[20] shown in Figure 1.15, calculate and sketch the current $i$ for $t \geq 0$, knowing that $i(0) = -10$ mA and given $R = 500\ \Omega$, $L = 10$ mH, and $e = 8$ volts. [*Note:* 1mA $= 10^{-3}$ A and 1mH $= 10^{-3}$ H]

**Figure 1.15**

**35.** Radioactive $^{14}C$ is produced[21] in the earth's atmosphere as a result of bombardment with cosmic rays from outer space. The radioactive carbon is taken up by living things, with the result that they contain 10 disintegrations per minute of $^{14}C$ per g of carbon. On their death, this $^{14}C$ decays with a half-life of 5568 years, and no fresh $^{14}C$ is absorbed. Carbon derived from wood taken from an Egyptian tomb proved to have a $^{14}C$ content of 7.62 disintegrations per min. per g. What was the age of the tomb?

**36.** In Figure 1.16, find and sketch[22] the voltage $v_2$ across $C_2$ as a function of time after the switch is closed. Initial voltage across $C_1$ and $C_2$ is zero. [*Note:* 1μf $= 10^{-6}$ f; 1 kilohm $= 10^3\ \Omega$.]

[20]This is Exercise 12a in C. A. Desoer and E. S. Kuh, *Basic Circuit Theory* (New York: McGraw-Hill Book Co., 1969), p. 169. Reprinted by permission of McGraw-Hill Book Company.
[21]This is Exercise 9 in B. G. Harvey, *Introduction to Nuclear Physics and Chemistry*, 2nd ed. (Englewood Cliffs, N.J.: Prentice-Hall, 1969), p. 54. Reprinted by permission of Prentice-Hall, Inc.
[22]This is Exercise 5–33 in J. L. Shearer, A. T. Murphy, and H. H. Richardson, *Introduction to System Dynamics* (Reading, Mass.: Addison-Wesley, 1971), p. 143. © 1971 Addison-Wesley Publishing Company, Inc. Reprinted with permission of Addison-Wesley Publishing Company.

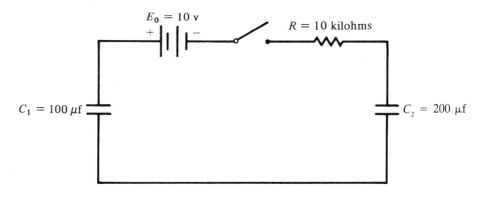

**Figure 1.16**

**37.** The capacitor $C$ in the circuit[23] shown in Figure 1.17 is charged to 10 volts when the switch is thrown. Derive the differential equation for the capacitor voltage. Solve the voltage as a function of time and sketch a plot of it, given that $R = 1$ kilohm and $C = 1$ μf. (*Note:* 1μf $= 10^{-6}$ f; 1 kilohm $= 10^3$ $\Omega$.)

**38.** Show[24] that the general solution of the differential equation

$$C\frac{dv}{dt} + \frac{1}{R}v = A_1 \cos(\omega t + \phi_1)$$

is of the form

$$v(t) = K_1 e^{-(1/RC)t} + A_2 \cos(\omega t + \phi_2),$$

where $K_1$ is an arbitrary constant,

$$A_2 = \frac{A_1}{\sqrt{(1/R)^2 + (\omega C)^2}},$$

and

$$\phi_2 = \phi_1 - \tan^{-1}\omega RC.$$

**39.** A yacht[25] having a displacement (weight) of 15,000 lb is drifting toward a collision with a pier at a speed of 2 ft/sec. Over what minimum distance can a 150 lb crewman hope to stop the yacht by exerting an average force equal to his weight?

---

[23]This is Exercise 5–28 in Shearer et al., *System Dynamics*, p. 141. © 1971 Addison-Wesley Publishing Company, Inc. Reprinted with permission of Addison-Wesley Publishing Company.
[24]This exercise is developed from material found in Desoer and Kuh, *Basic Circuit Theory*, pp. 121–22. Used by permission of McGraw-Hill Book Company.
[25]This is Exercise 2-1.3 in J. Norwood, Jr., *Intermediate Classical Mechanics* (Englewood Cliffs, N.J.: Prentice-Hall, 1979), p. 49. Reprinted by permission of Prentice-Hall, Inc.

**Figure 1.17**

**40.** A particle[26] is moving on a straight line with constant acceleration $f$. If $x$ is the distance from some fixed point on the straight line, show that the equation $d^2x/dt^2 = f$ can be written as

$$v \frac{dv}{dx} = f,$$

where $v$ is the particle's speed. If $v = u$ at $x = 0$, integrate this equation and show that

$$v^2 = u^2 + 2f \cdot x.$$

**41.** A ball[27] is thrown vertically downwards from the top of a tall building. Assuming a model with constant gravity and air resistance proportional to its speed, show that if the building is sufficiently tall, the ball's velocity on hitting the ground is approximately independent of its initial speed.

**42.** A hawk, eagle, and sparrow are in the air.[28] The eagle is 50 feet above the sparrow and the hawk is 100 feet below the sparrow. The sparrow flies straight forward in a horizontal line. Both hawk and eagle fly directly towards the sparrow. The hawk flies twice as fast as the sparrow. The hawk and eagle reach the sparrow at the same time. How far does each fly and at what rate does the eagle fly?

**43.** Solve the differential equations[29]

$$y' = \frac{x^2 - y^2 + 1}{x^2 - y^2 - 1}, \qquad y' = -\frac{y(2x + y - 1)}{x(x + 2y - 1)}.$$

[*Hint:* The first has $(x + y)^{-1}$ as an integrating factor; the second is exact.]

---

[26]This is Problem 2.4 in D. N. Burghes and A. M. Downs, *Modern Introduction to Classical Mechanics and Control* (West Sussex, England: Ellis Horwood, 1975), p. 31. Reprinted by permission of Ellis Horwood Limited, Publishers.

[27]This is Example 5.1 in Burghes and Downs, *Classical Mechanics,* p. 109. Reprinted by permission of Ellis Horwood Limited, Publishers.

[28]From *Amer. Math. Monthly* 40 (1933): 436–37.

[29]From *Amer. Math. Monthly* 50 (1943): 572.

**44.** A cat[30] is running along a straight edge of a garden. A dog, sitting in the garden at a distance $b$ from the edge, sees the cat when it is at its nearest point. The dog immediately chases the cat with twice the cat's speed in such a way that it is always running towards the cat. Find the time that elapses before the cat is caught and show that the cat runs a distance $(2/3)b$ before being caught.

**45.** In an $RL$-series circuit, $L = 2$ henries, $R = 4$ ohms, $V(t) = e^{-t}$ volts, and $I(0) = 5$ amperes. Find the current at any time $t$.

---

[30]This is Exercise 2.17 in Burghes and Downs, *Classical Mechanics*, p. 34. Reprinted by permission of Ellis Horwood Limited, Publishers.

# CHAPTER 2

# Linear Differential Equations

## 2.1 INTRODUCTION AND DEFINITIONS

A linear differential equation of order $n$ is a differential equation of the form

$$a_n(x)\, y^{(n)} + a_{n-1}(x)\, y^{(n-1)} + \cdots + a_1(x)\, y' + a_0(x)\, y = f(x). \tag{1}$$

We always assume that the coefficients $a_n(x), a_{n-1}(x), \ldots, a_0(x)$ and the function $f(x)$ are continuous functions in some interval $I$ and that the leading coefficient $a_n(x) \neq 0$ for all $x \in I$. The interval $I$ is called the *interval of definition of the differential equation.* When the function $f$ is identically zero, we say that Eq. (1) is *homogeneous.* When $f(x)$ is not identically zero, Eq. (1) is called *nonhomogeneous.* If all the coefficients $a_n(x), a_{n-1}(x), \ldots, a_0(x)$ are constants, we speak of Eq. (1) as a linear differential equation with *constant coefficients;* otherwise, it is a linear differential equation with *variable coefficients.* The following are examples of linear differential equations:

$$xy' - 2y = x^3, \quad x \neq 0 \tag{2}$$

$$y'' + 2y' + 3y = \cos x \tag{3}$$

$$y^{(4)} - y = 0. \tag{4}$$

Equation (2) is a nonhomogeneous linear differential equation of order 1 with variable coefficients. Equation (3) is a nonhomogeneous linear differential equation of order 2 with constant coefficients. Equation (4) is a homogeneous linear differential equation of order 4 with constant coefficients. The term *linear* refers to the fact that each expression in the differential equation is of degree *one* or degree *zero* in the variables $y, y', \ldots, y^{(n)}$. The following are nonlinear differential equations:

$$y'' + y^2 = \sin x$$

$$y''' + yy' = x$$

$$y'' + \sin y = 0.$$

The first differential equation is nonlinear because of the term $y^2$, the second because of the term $yy'$, and the third because of the term $\sin y$ [recall that $\sin y = y - (y^3/3!) + (y^5/5!) - \cdots$].

Our main concern in this chapter, aside from applications, is to develop the elements of the theory of solutions of linear differential equations and to discuss methods for obtaining their general solution.

Of course, we already know how to find the general solution of any linear differential equation of order 1 with variable or constant coefficients. This was done in Section 1.4 of Chapter 1. For example, the general solution of Eq. (2) is easily found to be

$$y(x) = x^3 + cx^2, \tag{5}$$

where $c$ is an arbitrary constant.

Furthermore (subject to some algebraic difficulties only), we shall learn in Sections 2.4, 2.5, and 2.11 how to find the general solution of any linear differential equation with constant coefficients, no matter what the order of the differential equation is.

However, matters are not so simple when we try to solve linear differential equations with variable coefficients of order 2 or higher. In fact, there is *no way* to find explicitly the general solution of a second-order (or higher) differential equation with *variable coefficients* unless the differential equation is of a very special form (for example, of the Euler type, Section 2.7) or unless we know (or can guess) one of its solutions (see Section 2.8). A practical way out of this difficulty is to approximate the solutions of differential equations with variable coefficients. For example, the variable coefficients of a differential equation can be approximated by constant coefficients, in "small" intervals, and hopefully the solution of the differential equation with constant coefficients is an "approximation" to the solution of the original differential equation with variable coefficients. Another way, which will give a better approximation, is to look for a power-series solution of the differential equation with variable coefficients (see Section 2.9 and Chapter 5). A direct substitution of the power-series solution into the differential equation will enable us to compute as many coefficients of the power series as we please in our approximation. Finally, the reader should be told that in more advanced topics on differential equations, called "qualitative theory of differential equations," one can discover many properties of the solutions of differential equations with variable coefficients directly from the properties of these coefficients and without the luxury of knowing explicitly the solutions of the differential equation.

For the sake of motivation, let us consider the second-order linear differential equation

$$\ddot{y} + a^2 y = 0, \tag{6}$$

together with the initial conditions

$$y(0) = y_0, \qquad \dot{y}(0) = v_0. \tag{7}$$

As is customary, dots stand for derivatives with respect to time $t$. It will be shown in the applications that the *initial value problem (IVP)* (6)–(7) describes the motion of a vibrating spring [Eq. (6) is Eq. (10) with $a^2 = K/m$]. It is easy to verify that the function $y_1(t) = \cos at$ when substituted into Eq. (6) reduces the differential equation to an identity. In fact, $y_1 = \cos at \Rightarrow \dot{y}_1 = -a \sin at \Rightarrow \ddot{y}_1 = -a^2 \cos at$, and substitution of the results into Eq. (6) yields $-a^2 \cos at + a^2 \cos at = 0$, or $0 = 0$. Hence, $y_1$ is a solution. In a similar fashion we can verify that the function $y_2(t) = \sin at$ is also a solution of the differential equation (6). Furthermore, the sum of the two solutions is also a solution, and any constant multiple of any one of these solutions is again a solution. Thus, for any constants $c_1$ and $c_2$, the linear combination

$$y(t) = c_1 y_1(t) + c_2 y_2(t) \tag{8}$$

of two solutions is also a solution.

The preceding comments serve to distinguish linear homogeneous differential equations from other types of differential equations. For linear homogeneous differential equations we have the principle that if $y_1$ and $y_2$ are solutions and if $c_1$ and $c_2$ are arbitrary constants, then the *linear combination* $c_1 y_1 + c_2 y_2$ is also a solution. This result and its proof are presented as Theorem 1 of Section 2.2.

Note that the functions $y_1 = \cos at$ and $y_2 = \sin at$, although they satisfy the differential equation (6), *do not*, in general, satisfy the initial conditions, since $\dot{y}_1(0) = 0 (\neq v_0)$ and $y_2(0) = 0 (\neq y_0)$. It seems reasonable, however, to suspect that if the constants $c_1$ and $c_2$ are chosen properly, then $y(t)$ as given in (8) will be a solution of the initial value problem (6)–(7). Now $y(t)$ satisfies the differential equation (6) and will satisfy the initial conditions provided that $c_1$ and $c_2$ satisfy the system of algebraic equations

$$y_0 = y(0) = c_1 y_1(0) + c_2 y_2(0)$$
$$v_0 = \dot{y}(0) = c_1 \dot{y}_1(0) + c_2 \dot{y}_2(0).$$

Since $y_1(0) = 1$, $\dot{y}_1(0) = 0$, $y_2(0) = 0$, and $\dot{y}_2(0) = a$, this system takes the form

$$y_0 = c_1$$
$$v_0 = ac_2.$$

Hence, a solution to the initial value problem is

$$y(t) = y_0 \cos at + \frac{v_0}{a} \sin at. \tag{9}$$

In Section 2.3 we establish that Eq. (9) is the *only* solution of the IVP (6)–(7).

It would be natural to ask what motivated us to try $\cos at$ as a solution. Or we might ask whether there are simple solutions other than $\cos at$ and $\sin at$. [The reader can verify that $e^{iat}$ and $e^{-iat}$, where $i^2 = -1$, are solutions of Eq. (6).] How many "different" solutions are there to the differential equation (6)? Is (9) the only solution to the initial value problem, or would some other choice of $y_1$ and $y_2$ produce different values for $c_1$ and $c_2$? [For example, if $y_1 = e^{iat}$, $y_2 = e^{-iat}$, $c_1 = \frac{1}{2}y_0 - (i/2a)v_0$, $c_2 = \frac{1}{2}y_0 + (i/2a)v_0$, then $c_1 y_1 + c_2 y_2$ also solves the initial value problem (6)–(7).]

The answers to these questions will be obtained as we proceed through the remainder of this chapter.

## APPLICATIONS 2.1.1

Linear differential equations occur in many mathematical models of real-life situations. Newton's second law of motion, for example, involves a second derivative (the acceleration), and consequently linear differential equations of second order have been of primary interest in motion problems.

**Mechanics**
The Vibrating
Spring

■ A light coil spring of natural length $L$ is suspended from a point on the ceiling [Figure 2.1(a)].

We attach a mass $m$ to the lower end of the spring and allow the spring to come to an equilibrium position [Figure 2.1(b)]. Assume that at the equilibrium position the length of the spring is $L + l$. Naturally the stretch $l$ is caused by the weight $mg$ of the mass $m$. According to Hooke's law, the force needed to produce a stretch of length $l$ is proportional to $l$. That is, $mg = Kl$, with the constant of proportionality, $K$, known as the *spring constant*. Next we displace the mass vertically (up or down) to a fixed (initial) position, and at that point we give it a vertical (initial) velocity (directed up or down), setting the mass in a vertical motion [Figure 2.1(c)]. It is this motion that we wish to investigate. For convenience, we choose the origin 0 at the equilibrium point, and the positive direction for the vertical ($y$) axis is downward.

Let $y = y(t)$ be the position of the mass at time $t$. Then by Newton's second law of motion, we have

$$m\ddot{y} = F,$$

where the double dots denote second derivative with respect to time (in other words, the acceleration of the mass). There are two forces acting upon the mass: one is the force of gravity, equal to $mg$, and the other is the restoring force of the spring, which by Hooke's law is $K(l + y)$. Thus (since $mg = Kl$),

$$F = mg - K(l + y) = -Ky.$$

The differential equation that describes the motion of the mass $m$ is

$$m\ddot{y} = -Ky,$$

or

$$\ddot{y} + \frac{K}{m}y = 0. \tag{10}$$

Recall that the mass was set to motion by first displacing it to an initial

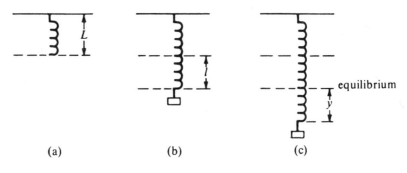

(a)                    (b)                    (c)

**Figure 2.1**

position, say $y_0$, and then giving it an initial velocity $v_0$. Thus, together with the differential equation (10), we have the two initial conditions

$$y(0) = y_0, \qquad \dot{y}(0) = v_0. \tag{11}$$

In Eq. (11) $y_0$ is positive if the initial position is below the equilibrium and negative if it is above. Also, $v_0$ is positive if the initial velocity was directed downward and negative if it was directed upward.

Equation (10) we recognize to be a second-order, homogeneous, linear differential equation with constant coefficients. The differential equation (10), together with the initial conditions (11), constitute an initial value problem.

In Table 2.1 we list the units used in the two most common systems of measurement.

|  | Cgs system | English system |
|---|---|---|
| Distance | centimeter (cm) | foot (ft) |
| Mass | gram (g) | slug |
| Time | seconds (sec) | seconds (sec) |
| Force | dyne (dyn) | pound (lb) |
| Velocity | cm/sec | ft/sec |
| Acceleration | cm/sec$^2$ | ft/sec$^2$ |
| Spring constant | dyn/cm | lb/ft |
| Gravity ($g$) | 980 cm/sec$^2$ | 32 ft/sec$^2$ |

**Table 2.1**   Units of Measurement

**EXAMPLE 1**   A spring is stretched 12 cm by a force equal to 360 dyn. A mass of 300 g is attached to the end of the spring and released 3 cm above the point of equilibrium with initial velocity 5 cm/sec directed downward. Find the position of the mass at any time $t$.

**Solution**   Using Hooke's law we compute the spring constant:

$$K = \tfrac{360}{12} \text{ dyn/cm} = 30 \text{ dyn/cm}.$$

The IVP describing the motion of the mass is [from Eqs. (10) and (11)]

$$\ddot{y} + \tfrac{30}{300}y = 0$$
$$y(0) = -3$$
$$\dot{y}(0) = 5.$$

From Eq. (9) the solution of this IVP is

$$y(t) = -3 \cos \frac{t}{\sqrt{10}} + 5\sqrt{10} \sin \frac{t}{\sqrt{10}}.$$

The reader is urged to read Exercise 40 of Section 2.4, where the vibrating spring is discussed in greater detail.

**Diffusion**  ■ Suppose that a gas diffuses into a liquid in a long and narrow pipe. Suppose that this process takes place for such a long period of time that the concentration $y(x)$ of gas in the pipe depends only on the distance $x$ from some initial point 0 (and is independent of time). Assume that the gas reacts chemically with the liquid and that the amount $H(x)$ of gas that disappears by this reaction is proportional to $y(x)$. That is,

$$H(x) = ky(x). \tag{12}$$

We want to compute the concentration $y(x)$ of gas in the liquid, at any point $x$, given that we know the concentration at the points 0 and $l$, say

$$y(0) = A \quad \text{and} \quad y(l) = 0. \tag{13}$$

To solve this problem we consider a small section $S$ of the pipe between the point $x$ and $x + \Delta x$ (Figure 2.2). Then the amount $G(x)$ of gas diffusing into $S$ at the point $x$ minus the amount $G(x + \Delta x)$ diffusing out of $S$ at the point $x + \Delta x$ must be equal to the amount of gas disappearing in $S$ by the chemical reaction. To compute $G(x)$ and $G(x + \Delta x)$, we use Fick's law of diffusion, which states that the amount of gas that passes through a unit area per unit time is proportional to the rate of change of the concentration. That is,

$$G(x) = -Dy'(x), \tag{14}$$

where $D > 0$ is the *diffusion coefficient* and the negative sign justifies the fact that the gas diffuses from regions of high concentration to regions of low concentration. Similarly,

$$G(x + \Delta x) = -Dy'(x + \Delta x). \tag{15}$$

Equation (12) gives the amount of gas that disappears at the point $x$ because of the chemical reaction of the gas with the liquid. Therefore, the total amount of gas that disappears by the chemical reaction in the section $S$ is given by

$$\int_x^{x+\Delta x} ky(s)ds. \tag{16}$$

Thus,

$$[-Dy'(x)] - [-Dy'(x + \Delta x)] = \int_x^{x+\Delta x} ky(s)ds. \tag{17}$$

Figure 2.2

A good approximation to the integral in (16) is given by $ky(x + \Delta x/2) \cdot \Delta x$. Thus after some rearrangement Eq. (17) becomes

$$\frac{y'(x + \Delta x) - y'(x)}{\Delta x} = \frac{k}{D} y\left(x + \frac{\Delta x}{2}\right), \tag{18}$$

where $y[x + (\Delta x/2)]$ is the value of $y$ at the midpoint between the limits of integration and $|\Delta x|$ the length of the interval of integration. Taking limits on both sides of Eq. (18) as $\Delta x \to 0$, we obtain the second-order linear differential equation

$$y'' - \frac{k}{D} y = 0. \tag{19}$$

The problem consisting of the differential equation (19) and the "boundary" conditions (13) is referred to as a *boundary value problem (BVP)*. Boundary value problems are discussed in Chapter 6. Most of the investigation in this book centers on initial value problems.

## EXERCISES

In Exercises 1 through 19, classify the differential equation as being either linear or nonlinear. Furthermore, classify the linear ones as homogeneous or non-homogeneous, with constant coefficients or with variable coefficients, and state its order. (Note that derivatives are denoted by primes, lowercase roman numerals, or by arabic numbers in parentheses.)

**1.** $y'' + xy = 0$

**2.** $y''' + x^2y = e^x$

**3.** $y'' + yy^{iv} = 5$

**4.** $y'' + \cos y = 0$

**5.** $y^{(5)} - 2y^{(4)} + y = 2x^2 + 3$

**6.** $(1 + x^2)y'' - xy' + y = 0$

**7.** $(\sin x)y' + e^{x^2}y = 1$

**8.** $2y''' + 3y'' - 4y' + xy = 0$

**9.** $x^2y'' - e^xy' - 2 = 0$

**10.** $y' + y^{1/2} = 3x$

**11.** $y'' + xy = x$

**12.** $2y - 3xy'' + 4y' = 0$

**13.** $y''' - 2 = 0$

**14.** $e^x(y')^2 + 3y = 0$

**15.** $y'' + 5y' - 6y = 0$

**16.** $yy' = 3$

**17.** $xy''' + 4xy'' - xy = 1, x > 0$

**18.** $7y' - xy = 0$

**19.** $7y'' + 3xy' - 4y = 0$

**20.** A spring is stretched 2 cm by a force equal to 16 dyn. A mass of 32 g is attached to the end of the spring. The system is then set to motion by pulling the mass 4 cm above the point of equilibrium and releasing it, at time $t = 0$, with an initial velocity of 2 cm/sec directed downward. Derive an IVP that describes the motion of the mass. Use Eq. (9) to find the solution of this IVP.

**21.** A spring is stretched 1.5 in. by a 4-lb weight. The system is then set to motion by pushing the mass (note that $mg$ = weight and that $g = 32$ ft/sec$^2$) 6 in. below the point of equilibrium and releasing it, at time $t = 0$, with an initial velocity of 4 ft/sec directed upward. Use Eq. (9) to find the position of the mass at any time $t$.

The solution of the vibrating spring is given by Eq. (9),

$$y(t) = y_0 \cos \sqrt{\frac{K}{m}} t + \frac{v_0 \sqrt{m}}{\sqrt{K}} \sin \sqrt{\frac{K}{m}} t.$$

The graph of $y(t)$ is periodic of period $2\pi \sqrt{m}/\sqrt{K}$; consequently, the motion of the mass repeats itself after every time interval of length $2\pi \sqrt{m}/\sqrt{K}$. For this reason we say that the motion is *oscillatory* or that *vibrations* occur. If we recall a technique from trigonometry [that an expression of the form $\alpha \cos t + \beta \sin t$ can be written as $A \cos (t - \phi)$, where $A = \sqrt{\alpha^2 + \beta^2}$ and $\phi = \cos^{-1}(\alpha/\sqrt{\alpha^2 + \beta^2}) = \sin^{-1}(\beta/\sqrt{\alpha^2 + \beta^2})$], the solution can be written in the form

$$y(t) = A \cos \left( \sqrt{\frac{K}{m}} t - \phi \right),$$

with $A = \sqrt{y_0^2 + mv_0^2/K}$ and $\phi = \cos^{-1}(y_0/A)$. Hence, we say that the motion is oscillatory of *amplitude A*, period $2\pi\sqrt{m}/\sqrt{K}$, and *phase* $\phi$.

In each of Exercises 22, 23, and 24, evaluate the amplitude and the period of the oscillation and sketch the graph of $y(t)$.

**22.** $y_0 = 5$ in.,    $K = m$,    $v_0 = 0$ in./sec

**23.** $y_0 = 0$ cm,    $K = 5m$,    $v_0 = 10\sqrt{5}$ cm/sec

**24.** $y_0 = 3$ cm,    $K = m$,    $v_0 = 4$ cm/sec

**25.** A function $f$ is said to be bounded[1] as $x \to \infty$ if there exists a positive number $M$ such that $\lim_{x \to \infty} |f(x)| \leq M$. Show that all solutions of the differential equation $y'' + e^x y = 0$ are bounded as $x \to \infty$. *Hint:* Multiply the differential equation by $e^{-x}y'$ and integrate from 0 to $T$, obtaining

$$y^2(T) + 2 \int_0^T e^{-x} y' y'' dx = y^2(0).$$

Note that for some $\rho$ such that $0 \leq \rho \leq T$, we must have

$$2 \int_0^T e^{-x} y' y'' \, dx = 2 \int_0^\rho y' y'' \, dx = [y'(\rho)]^2 - [y'(0)]^2.$$

Hence,

$$y^2(T) + [y'(\rho)]^2 = y^2(0) + [y'(0)]^2.$$

---

[1]Problem 25 and its method of solution are taken from the 1966 William Lowell Putnam Mathematical Competition, *Amer. Math. Monthly* 74, no. 7 (1967).

# 2.2  LINEAR INDEPENDENCE AND WRONSKIANS

In Section 2.1 we posed certain questions concerning the number of solutions to a linear differential equation and to the manner of finding solutions to linear differential equations. In this section and the next we concentrate on answering the question: *How many solutions can a linear differential equation have?* To gain some insight into this question, we consider an example.

**EXAMPLE 1**  Solve the second-order homogeneous linear differential equation

$$y'' = 0. \tag{1}$$

**Solution**  Because of the simplicity of this equation, we can integrate it directly to obtain solutions. Integrating Eq. (1) with respect to $x$, we obtain

$$y' = c_1, \tag{2}$$

where $c_1$ is a constant of integration. Integrating Eq. (2) with respect to $x$, we have

$$y = c_1 x + c_2, \tag{3}$$

where $c_2$ is another constant of integration. Regardless of the values given to the constants $c_1$ and $c_2$, the expression for $y$ given in (3) is a solution of the differential equation (1). Thus, there are an infinity of solutions to the differential equation (1). Notice that $y_1 = x$ and $y_2 = 1$ are solutions of Eq. (1) and Eq. (3) shows that they play a special role in that *every* solution can be expressed as a linear combination of these two solutions. That is, if $y$ is a solution of Eq. (1), then there exist constants $c_1$ and $c_2$ such that $y = c_1 y_1 + c_2 y_2$. Of course, if $y = y_1$, then $c_1 = 1$ and $c_2 = 0$, and if $y = y_2$, then $c_1 = 0$ and $c_2 = 1$. Since it is possible to express every solution in terms of the solutions $y_1$ and $y_2$, we speak of $y_1$ and $y_2$ as constituting a *fundamental set of solutions* for Eq. (1). Equation (3) is referred to as the *general solution* of Eq. (1). For this differential equation the general solution contains two arbitrary constants, $c_1$ and $c_2$. Consequently, if we try to formulate conditions that reduce the infinity of solutions to a single solution, we need two conditions which serve to determine $c_1$ and $c_2$. For example, suppose that we want a solution of Eq. (1) that satisfies the initial conditions $y(1) = 1$, $y'(1) = 2$. From Eq. (3) we have

$$1 = y(1) = c_1(1) + c_2,$$

and, differentiating Eq. (3),

$$2 = y'(1) = c_1.$$

Hence, $c_1 = 2$ and $c_2 = -1$, and the solution we seek is

$$y = 2x - 1.$$

Intuitively, we suspect that to solve an $n$th-order equation requires $n$ integrations, and as a consequence the general solution of such a differential equation involves $n$ constants of integration. Furthermore, we suspect that in this case there should be a set of $n$ special solutions having the property that every solution can be expressed as a linear combination of these special solutions. In what follows these notions will be made more precise.

## DEFINITION 1

*A collection of m functions $f_1, f_2, \ldots, f_m$, each defined and continuous on the interval $a \le x \le b$ is said to be linearly dependent on $a \le x \le b$ if there exist constants $\alpha_1, \alpha_2, \ldots, \alpha_m$, not all of which are zero, such that*

$$\alpha_1 f_1 + \alpha_2 f_2 + \cdots + \alpha_m f_m = 0 \tag{4}$$

*for every x in the interval $a \le x \le b$. Otherwise, the functions are said to be linearly independent on this interval.*

**EXAMPLE 2**   (a) Show that the functions $f_1(x) = 3x + (12/5)$ and $f_2(x) = 5x + 4$ are linearly dependent on the interval $(-\infty, \infty)$.
   (b) Show that the functions $f_1(x) = x$ and $f_2(x) = x^2$ are linearly independent on the interval $-1 \le x \le 1$.

**Solution**   (a) To show that the functions $f_1$ and $f_2$ are linearly dependent we must show that there exist constants $\alpha_1$ and $\alpha_2$, not both zero, such that

$$\alpha_1 \left( 3x + \frac{12}{5} \right) + \alpha_2 (5x + 4) = 0$$

for all $x$ in the interval $-\infty < x < \infty$. This is possible if we choose for example, $\alpha_1 = 5$ and $\alpha_2 = -3$.
   (b) Otherwise $f_1$ and $f_2$ are linearly dependent on the interval $-1 \le x \le 1$. That is, there exist constants $\alpha_1$ and $\alpha_2$, not both zero, such that

$$\alpha_1 x + \alpha_2 x^2 = 0$$

*for all x* in the interval $-1 \le x \le 1$. But for $x = 1$ and $x = -1$ we find

$$\alpha_1 + \alpha_2 = 0 \quad \text{and} \quad -\alpha_1 + \alpha_2 = 0.$$

The only solution of this system is $\alpha_1 = \alpha_2 = 0$. This contradicts our assumption that $f_1$ and $f_2$ are linearly dependent. Hence, $f_1$ and $f_2$ are linearly independent on the interval $-1 \le x \le 1$.
   If the functions $f_1, f_2, \ldots, f_n$ are solutions to a linear homogeneous differential equation, there is a simple test that serves to determine whether they are linearly independent or not.
   Before we state this test, we need the following definition.

## DEFINITION 2

*Let $f_1, f_2, \ldots, f_n$ be n functions which together with their first $n - 1$ derivatives are continuous on the interval $a \le x \le b$. The Wronskian of $f_1, f_2, \ldots, f_n$,* *evaluated at x, is denoted by $W(f_1, f_2, \ldots, f_n; x)$ and is defined to be the determinant[2]*

$$W(f_1, f_2, \ldots, f_n; x) = \begin{vmatrix} f_1 & f_2 & \cdots & f_n \\ f_1' & f_2' & \cdots & f_n' \\ f_1'' & f_2'' & \cdots & f_n'' \\ \cdot & \cdot & \cdot & \cdot \\ \cdot & \cdot & \cdot & \cdot \\ \cdot & \cdot & \cdot & \cdot \\ f_1^{(n-1)} & f_2^{(n-1)} & \cdots & f_n^{(n-1)} \end{vmatrix}$$

*Each of the functions appearing in this determinant is to be evaluated at x.*

**EXAMPLE 3**   Given $f_1(x) = x^2$, $f_2(x) = \cos x$, find $W(f_1, f_2; x)$.

**Solution**   From Definition 2 and the functions given we compute

$$W(x^2, \cos x; x) = \begin{vmatrix} x^2 & \cos x \\ 2x & -\sin x \end{vmatrix} = -x^2 \sin x - 2x \cos x.$$

In the introduction to this chapter we mentioned that linear homogeneous differential equations have the property that linear combinations of solutions are also solutions. This principle, which distinguishes linear from nonlinear differential equations, is embodied in the following theorem.

**THEOREM 1**

*If each of the functions $y_1, y_2, \ldots, y_m$ are solutions to the same linear homogeneous differential equation,*

$$a_n(x)y^{(n)} + a_{n-1}(x)y^{(n-1)} + \cdots + a_1(x)y' + a_0(x)y = 0,$$

*then for every choice of the constants $c_1, c_2, \ldots, c_m$ the linear combination*

$$c_1 y_1 + c_2 y_2 + \cdots + c_m y_m$$

*is also a solution.*

**Proof[3]**   The proof depends on two basic properties of differentiation,

$$(f_1 + f_2 + \cdots + f_k)' = f_1' + f_2' + \cdots + f_k'$$

---

[2]For a brief discussion of determinants, their properties, and their application to linear systems of algebraic equations, see Appendix A.

[3]In Chapter 1 we expressed the philosophy that difficult or sophisticated proofs would be omitted in this text. The proof of this theorem is neither difficult nor sophisticated. Nevertheless, we choose not to give a proof under the most general circumstances but rather demonstrate a proof for a specific version of the theorem, with the implication that the general proof follows along similar lines of reasoning. This format will be followed in many places in the text. That is, the theorem will be stated in its general form and the proof given for a specific case.

and

$$(cf)' = cf',$$

where $c$ is a constant.

We present the details for the case $m = 2$, $n = 3$. We assume then that $y_1$ and $y_2$ are solutions of the differential equation

$$a_3(x)y''' + a_2(x)y'' + a_1(x)y' + a_0(x)y = 0.$$

Substituting $c_1y_1 + c_2y_2$ into this differential equation, we obtain

$$a_3(x)(c_1y_1 + c_2y_2)''' + a_2(x)(c_1y_1 + c_2y_2)'' + a_1(x)(c_1y_1 + c_2y_2)'$$
$$+ a_0(x)(c_1y_1 + c_2y_2) \overset{?}{=} 0$$

$$a_3(x)(c_1y_1''' + c_2y_2''') + a_2(x)(c_1y_1'' + c_2y_2'') + a_1(x)(c_1y_1' + c_2y_2')$$
$$+ a_0(x)(c_1y_1 + c_2y_2) \overset{?}{=} 0$$

$$c_1[a_3(x)y_1''' + a_2(x)y_1'' + a_1(x)y_1' + a_0(x)y_1]$$
$$+ c_2[a_3(x)y_2''' + a_2(x)y_2'' + a_1(x)y_2' + a_0(x)y_2] \overset{?}{=} 0$$
$$c_1(0) + c_2(0) \overset{?}{=} 0$$
$$0 + 0 = 0.$$

Suppose that we are given $n$ functions $y_1, y_2, \ldots, y_n$, each of which is a solution to the same $n$th-order linear homogeneous differential equation,

$$a_n(x)y^{(n)} + a_{n-1}(x)y^{(n-1)} + \cdots + a_1(x)y' + a_0(x)y = 0. \tag{5}$$

We wish to determine whether or not these functions are linearly dependent. These functions will be linearly dependent if we can find $n$ constants $\alpha_1$, $\alpha_2$, $\ldots$, $\alpha_n$ not all zero such that

$$\alpha_1y_1 + \alpha_2y_2 + \cdots + \alpha_ny_n = 0. \tag{6}$$

Since each $y_i$ is a solution of Eq. (5), the derivatives $y_i, y_i', y_i'', \ldots, y_i^{(n-1)}$ all exist. Equation (6) can be differentiated $n - 1$ times to obtain the system of equations

$$\alpha_1y_1 + \alpha_2y_2 + \cdots + \alpha_ny_n = 0$$
$$\alpha_1y_1' + \alpha_2y_2' + \cdots + \alpha_ny_n' = 0$$
$$\alpha_1y_1'' + \alpha_2y_2'' + \cdots + \alpha_ny_n'' = 0 \tag{7}$$
$$\cdots\cdots\cdots\cdots\cdots\cdots\cdots\cdots\cdots\cdots\cdots\cdots$$
$$\alpha_1y_1^{(n-1)} + \alpha_2y_2^{(n-1)} + \cdots + \alpha_ny_n^{(n-1)} = 0.$$

We recognize Eqs. (7) as a system of homogeneous linear algebraic equations for the unknowns $\alpha_1, \alpha_2, \ldots, \alpha_n$. Such a system of equations has solutions[4] other than the trivial solution $\alpha_1 = 0, \alpha_2 = 0, \ldots, \alpha_n = 0$ if and only if the determinant of its coefficients is zero. That is, if the functions $y_1, y_2, \ldots, y_n$ are linearly dependent then $W(y_1, y_2, \ldots, y_n; x) = 0$ for *every* $x$ in the interval of definition of the differential equation. It can be shown that the converse is also

[4]See Appendix A.

true: if the functions are linearly independent, then their Wronskian is nonzero for some $x$. Therefore we have the following criterion for testing linear dependence (and linear independence).

**THEOREM 2**

$n$ solutions $y_1, y_2, \ldots, y_n$ of the nth-order differential equation

$$a_n(x)y^{(n)} + a_{n-1}(x)y^{(n-1)} + \cdots + a_1(x)y' + a_0(x)y = 0$$

defined on an interval $a \le x \le b$ are linearly dependent if and only if $W(y_1, y_2, \ldots, y_n; x) = 0$ for every $x$ in the interval $a \le x \le b$.

For example, the solutions $y_1(x) = \cos x$ and $y_2(x) = \sin x$ of the differential equation $y'' + y = 0$ are linearly independent because their Wronskian

$$\begin{vmatrix} \cos x & \sin x \\ -\sin x & \cos x \end{vmatrix} = \cos^2 x + \sin^2 x = 1$$

is never zero. Also, the solutions $e^x$ and $e^{-x}$ of the differential equation $y'' - y = 0$ are linearly independent because their Wronskian

$$\begin{vmatrix} e^x & e^{-x} \\ e^x & -e^{-x} \end{vmatrix} = -e^0 - e^0 = -2$$

is never zero.

**REMARK 1**   In Example 2 we demonstrated that the functions $x$ and $x^2$ are linearly independent on the interval $-1 \le x \le 1$. However, their Wronskian

$$\begin{vmatrix} x & x^2 \\ 1 & 2x \end{vmatrix} = x^2$$

vanishes at the point $x = 0$ of the interval $-1 \le x \le 1$. Does this contradict Theorem 2? No. Rather, there must not exist a second-order linear differential equation with interval of definition $-1 \le x \le 1$ that has $x$ and $x^2$ as solutions. The reader can verify that $y_1 = x$ and $y_2 = x^2$ are solutions of the second-order linear differential equation

$$x^2y'' - 2xy' + 2y = 0.$$

However, any interval of definition $I$ for this differential equation must exclude $x = 0$, since we have assumed that $a_2(x) \ne 0$ in $I$. Thus, we conclude that Theorem 2 is not contradicted by this example.[5]

An interesting property of the Wronskian is that it satisfies a first-order linear differential equation. Thus, it follows from Section 1.2 that the Wronskian is either identically zero or never vanishes. Equation (8), which is called *Abel's*

[5]For a discussion of the connection between the Wronskian having the value zero and linear dependence of functions, the reader is referred to the article by G. H. Meisters, "Local Linear Dependence and the Vanishing of the Wronskian," *Amer. Math. Monthly* 68, no. 9 (1961): 847–56.

*formula* is also the basis for determining a second linearly independent solution of a second-order linear differential equation when one solution is already known (see Section 2.8). For simplicity we show the details for the second-order linear differential equation, but we state the theorem for the more general case.

### THEOREM 3

*If $y_1, y_2, \ldots, y_n$ are solutions of the differential equation*

$$a_n(x)y^{(n)} + a_{n-1}(x)\,y^{n-1} + \cdots + a_1(x)y' + a_0(x)y = 0,$$

*where each $a_i(x)$ is defined and continuous on $a \le x \le b$ and $a_n(x) \ne 0$ on $a \le x \le b$, then either $W(y_1, \ldots, y_n; x)$ vanishes identically on $a \le x \le b$ or $W(y_1, \ldots, y_n; x)$ is never zero on $a \le x \le b$.*

**Proof**   The proof is for the case $n = 2$. For brevity we set $W = W(y_1, y_2; x)$, $p(x) = a_1(x)/a_2(x)$, and $q(x) = a_0(x)/a_2(x)$. Note that both $p$ and $q$ are continuous functions since $a_2(x) \ne 0$. Now

$$W = \begin{vmatrix} y_1 & y_2 \\ y_1' & y_2' \end{vmatrix} = y_1 y_2' - y_1' y_2.$$

Therefore,

$$\begin{aligned}
W' &= y_1 y_2'' + y_1' y_2' - (y_1' y_2' + y_1'' y_2) = y_1 y_2'' - y_1'' y_2 \\
&= y_1(-py_2' - qy_2) - (-py_1' - qy_1)y_2 = -p(y_1 y_2' - y_1' y_2) \\
&= -pW.
\end{aligned}$$

Using Eq. (5) of Section 1.4, with $a(x) = p(x)$ and $b(x) = 0$, we find that the differential equation $W' = -pW$ has the solution

$$W(x) = W(x_0) \exp\left[ -\int_{x_0}^{x} p(s)ds \right]. \tag{8}$$

Since the exponential function is never zero, it follows from Eq. (8) that if $W$ is zero at a point $x_0$ on the interval $a \le x \le b$, then it is identically zero; and if $W$ is different from zero at a point $x_0$, then it is never zero.

### DEFINITION 3

*Suppose that $y_1, y_2, \ldots, y_n$ are n solutions of the differential equation*

$$a_n(x)y^{(n)} + a_{n-1}(x)y^{(n-1)} + \cdots + a_1(x)y' + a_0(x)y = 0.$$

*Suppose also that these functions are linearly independent on the interval of definition of this differential equation. We then say that these functions form a fundamental set (or fundamental system) of solutions for the differential equation.*

For example, the functions $\cos x$ and $\sin x$ constitute a fundamental set of solutions for the differential equation $y'' + y = 0$. Also, the functions $e^x$ and

$e^{-x}$ form a fundamental set of solutions for the differential equation $y'' - y = 0$.

Discovering a fundamental set of solutions for a homogeneous linear differential equation is of great importance because, as we shall prove in Section 2.3, any solution of a homogeneous linear differential equation is a linear combination of the solutions in the fundamental set. More precisely, if $y_1, y_2, \ldots, y_n$ form a fundamental set of solutions for a linear homogeneous differential equation of order $n$, then the expression

$$y = c_1 y_1 + c_2 y_2 + \cdots + c_n y_n,$$

where $c_1, c_2, \ldots, c_n$ are arbitrary constants, is the general solution of the differential equation.

**REMARK 2**  Our use of Wronskians is primarily as a test to determine whether or not a given collection of solutions to a differential equation are linearly independent. Eggan and Insell[6] present a different use of the Wronskian concept. Their results are embodied in the following theorem.

**THEOREM 4**

*If f is a real-valued function defined on the real line R and if there exists a positive integer n such that f has 2n continuous derivatives on R and $W(f, f', \ldots, f^{(n)}) = 0$ on R, then f is a solution of an nth-order homogeneous linear differential equation with constant coefficients, not all of which are zero.*

**EXERCISES**

In Exercises 1 through 15, compute the Wronskian of the set of functions.

**1.** $1, x, x^2$

**2.** $\sin x^2, \cos x^2$

**3.** $\sin 3x, \cos 3x$

**4.** $1 + x, 2 - 3x, 4 - x$

**5.** $\ln x, e^x$

**6.** $x^{1/2}, x^{3/2}$

**7.** $e^{3x}, e^{2x}$

**8.** $6x^2, 14x^2, 2x$

**9.** $x^2, x^2 \ln x$

**10.** $e^{7x}, e^{7x}$

**11.** $e^{7x}, xe^{7x}$

**12.** $e^x, xe^x, x^2 e^x$

**13.** $x^{\lambda_1}, x^{\lambda_2}; \lambda_1 \neq \lambda_2$

**14.** $e^{\lambda_1 x}, e^{\lambda_2 x}; \lambda_1 \neq \lambda_2$

**15.** $f(x), cf(x); c$ a constant

In Exercises 16 through 23, integrate the differential equation directly to find the general solution.

**16.** $y' = e^{2x}$     **17.** $y''' = 0$     **18.** $y''' = x^3$     **19.** $y'' = \sin x$

[6] L. C. Eggan and A. J. Insell, *Amer. Math. Monthly* 80, no. 3 (1973).

**20.** $y'' = 3x$      **21.** $y^{(4)} = 0$      **22.** $y''' = x^2$      **23.** $y^{(5)} = 0$

In Exercises 24 through 33, show that the functions form a fundamental set of solutions for the differential equation. Write down the general solution of the differential equation.

**24.** $\sin 3x$, $\cos 3x$; $y'' + 9y = 0$

**25.** $e^{2x}$, $e^{-2x}$; $y'' - 4y = 0$

**26.** $e^{iat}$, $e^{-iat}$; $\ddot{y} + a^2 y = 0$ $(i^2 = -1)$, $a \neq 0$

**27.** $\dfrac{1}{x}$; $xy' + y = 0, x < 0$

**28.** $e^{3x}$, $e^x$; $y'' - 4y' + 3y = 0$

**29.** $e^{-2x}$, $xe^{-2x}$; $y'' + 4y' + 4y = 0$

**30.** $e^x \cos 2x$, $e^x \sin 2x$; $y'' - 2y' + 5y = 0$

**31.** $\cos x + \sin x$, $\cos x - \sin x$; $y'' + y = 0$

**32.** $\dfrac{1}{x} \cos(\ln x)$, $\dfrac{1}{x} \sin(\ln x)$; $x^2 y'' + 3xy' + 2y = 0$, $x > 0$

**33.** $1$, $\cos x$, $\sin x$; $y''' + y' = 0$

In Exercises 34 through 37, a second-order differential equation is given. Write an expression for the Wronskian of two linearly independent solutions without finding the solutions themselves. [*Hint:* Refer to Eq. (8).]

**34.** $2y'' - 3y' + y = 0$      **35.** $xy'' - 3y' - 5y = 0$, $x \neq 0$

**36.** $y'' + (\cos x)y' + e^x y = 0$      **37.** $y'' + y = 0$

In Exercises 38 through 41, the functions are linearly independent solutions of the differential equation. In each case show that the corresponding Wronskian satisfies Eq. (8).

**38.** $y'' + 4y = 0$, $y_1 = \sin 2x$, $y_2 = \cos 2x$

**39.** $y'' - 7y' + 6y = 0$, $y_1 = e^x$, $y_2 = e^{6x}$

**40.** $xy'' + y' = 0$, $y_1 = \ln x$, $y_2 = 1$, $x > 0$

**41.** $3y'' + 48y' + 192y = 0$, $y_1 = e^{-8x}$, $y_2 = xe^{-8x}$

**42. Astronomy** Kopal[7] obtained a differential equation of the form

$$v \frac{d^2 \psi}{dv^2} + 4 \frac{d\psi}{dv} = 0, \qquad v > 0.$$

Show that $\psi = 1$ and $\psi = v^{-3}$ are linearly independent solutions of this differential equation and obtain the general solution.

[7]Z. Kopal, "Stress History of the Moon and of Terrestial Planets," *Icarus* 2 (1963): 381.

**43. Flows** Barton and Raynor, in their study of peristaltic flow in tubes,[8] obtained the following linear homogeneous differential equation with variable coefficients.

$$\frac{d^2p}{dv^2} + \frac{1}{v}\frac{dp}{dv} = 0, \quad v > 0.$$

Show that $p(v) = 1$ and $p(v) = \ln v$ are linearly independent solutions of this differential equation, and obtain the general solution.

## 2.3  EXISTENCE AND UNIQUENESS OF SOLUTIONS

As we pointed out in Chapter 1, it is reassuring to know that *there is* (existence) a solution to the differential equation that one hopes to solve. The situation is even better if, in addition, there is *only one* solution (uniqueness) to the differential equation. The main result of this section is Theorem 1 below, which states conditions on the coefficient functions that guarantee the existence and uniqueness of solutions to the IVP containing an $n$th-order linear differential equation. This *existence and uniqueness theorem* is, from one point of view, the most important theorem associated with $n$th-order linear differential equations. Although we choose not to present the proof, it is possible to prove this theorem using arguments based on a knowledge of calculus only.[9]

**THEOREM 1**   EXISTENCE AND UNIQUENESS

*Let $a_0(x)$, $a_1(x)$, . . . , $a_n(x)$ and $f(x)$ be defined and continuous on $a \le x \le b$ with $a_n(x) \neq 0$ for $a \le x \le b$. Let $x_0$ be such that $a \le x_0 \le b$ and let $\beta_0$, $\beta_1$, . . . , $\beta_{n-1}$ be any constants. Then there exists a unique solution $y$ satisfying the initial value problem*

$$a_n(x)y^{(n)} + a_{n-1}(x)y^{(n-1)} + \cdots + a_1(x)y' + a_0(x)y = f(x) \tag{1}$$

$$y(x_0) = \beta_0, y'(x_0) = \beta_1, \ldots, y^{(n-1)}(x_0) = \beta_{n-1}. \tag{2}$$

*Furthermore, the solution $y$ is defined in the entire interval $a \le x \le b$.*

Using Theorem 1 it is easy to *construct a fundamental set* of solutions for the *homogeneous* differential equation

$$a_n(x)y^{(n)} + a_{n-1}(x)y^{(n-1)} + \cdots + a_1(x)y' + a_0(x)y = 0. \tag{3}$$

We define $y_1$ to be the unique solution of Eq. (3) satisfying the initial conditions $y(x_0) = 1, y'(x_0) = 0, y''(x_0) = 0, \ldots, y^{(n-1)}(x_0) = 0$. Define $y_2$ to be the unique

[8]C. Barton and S. Raynor, *Bull. Math. Biophys.* 30 (1968): 663–80.
[9]See E. A. Coddington, *An Introduction to Ordinary Differential Equations* (Englewood Cliffs, N.J.: Prentice-Hall, 1961), or D. Willett, "The Existence-Uniqueness Theorem for an $n$th-Order Linear Ordinary Differential Equation," *Amer. Math. Monthly* 5, no. 2 (1968).

solution of Eq. (3) satisfying the initial conditions $y(x_0) = 0$, $y'(x_0) = 1$, $y''(x_0) = 0$, ..., $y^{(n-1)}(x_0) = 0$. Define $y_3$ to be the unique solution of Eq. (3) satisfying the initial conditions $y(x_0) = 0$, $y'(x_0) = 0$, $y''(x_0) = 1$, ..., $y^{(n-1)}(x_0) = 0$, and so on. Define $y_n$ to be the unique solution of Eq. (3) satisfying the initial conditions $y(x_0) = 0$, $y'(x_0) = 0$, ..., $y^{(n-1)}(x_0) = 1$. By Theorem 1 each of these functions exists and is uniquely determined. It remains to show that these functions are linearly independent. In this case

$$W(y_1, y_2, \ldots, y_n; x_0) = \begin{vmatrix} 1 & 0 & 0 & \cdots & 0 \\ 0 & 1 & 0 & \cdots & 0 \\ 0 & 0 & 1 & \cdots & 0 \\ & \cdot & \cdot & & \cdot \\ & \cdot & & & \cdot \\ & \cdot & \cdot & & \cdot \\ 0 & 0 & 0 & \cdots & 1 \end{vmatrix} = 1 \neq 0.$$

Thus, these functions constitute a fundamental set of solutions.

The observations of the preceding paragraph emphasize the fact that for any linear differential equation whose coefficients satisfy the hypotheses of Theorem 1, there always exists a fundamental set of solutions. These solutions, although guaranteed by the theory, are not always easy to construct in practice. Nevertheless, there are some forms of linear differential equations for which systematic construction of a fundamental set of solutions can be described. These forms of equations are discussed in Sections 2.4 through 2.12.

**REMARK 1**   The existence and uniqueness theorem for the IVP (1)–(2) is something that should be expected to be true when the IVP is a "good" mathematical model of a real-life situation. For example, in the case of the vibrating spring discussed in Section 2.1.1, the existence and uniqueness of the IVP (10)–(11) of that section simply means that the mass $m$ moves (existence) and, in fact, moves in only one way (uniqueness).

**THEOREM 2**

If $y_1, y_2, \ldots, y_n$ constitute a fundamental set of solutions of Eq. (3), then given any other solution, $y$, satisfying the initial conditions $y(x_0) = \beta_0$, $y'(x_0) = \beta_1, \ldots, y^{(n-1)}(x_0) = \beta_{n-1}$, there exist unique constants $c_1, c_2, \ldots, c_n$ such that

$$y = c_1 y_1 + c_2 y_2 + \cdots + c_n y_n.$$

That is, this other solution, $y$, can be expressed uniquely as a linear combination of the fundamental set of solutions.

**Proof**   Set $Y = c_1 y_1 + c_2 y_2 + \cdots + c_n y_n$. Thus, by Theorem 1 of Section 2.2, it follows that the function $Y$ is a solution of the differential equation (3). We calculate $Y', Y'', \ldots, Y^{(n-1)}$ and then evaluate $Y$ and each derivative at $x = x_0$. This process leads to the system

$$c_1 y_1(x_0) + c_2 y_2(x_0) + \cdots + c_n y_n(x_0) = \beta_0$$

$$c_1 y_1'(x_0) + c_2 y_2'(x_0) + \cdots + c_n y_n'(x_0) = \beta_1 \qquad (4)$$

$$\cdots \cdots \cdots \cdots \cdots \cdots \cdots \cdots \cdots \cdots \cdots$$

$$c_1 y_1^{(n-1)}(x_0) + c_2 y_2^{(n-1)}(x_0) + \cdots + c_n y_n^{(n-1)}(x_0) = \beta_{n-1}.$$

Equations (4) constitute a nonhomogeneous system of linear algebraic equations in the unknowns $c_1, c_2, \ldots, c_n$ whose determinant of coefficients is $W(y_1, \ldots, y_n; x_0)$.

Since these functions form a fundamental set of solutions, their Wronskian is not zero. By Cramer's rule (for details, see Appendix A) there is a unique solution for the unknowns $c_i$. Now the functions $Y$ and $y$ are both solutions of the differential equation (3), with the same initial conditions, and, by the uniqueness part of Theorem 1, $Y = y$. The proof is complete.

On the basis of Theorem 2, the following result is well justified.

**COROLLARY 1**

*If* $y_1, y_2, \ldots, y_n$ *form a fundamental set of solutions for Eq. (3), then the expression*

$$y = c_1 y_1 + c_2 y_2 + \cdots + c_n y_n,$$

*where the* $c_i$ *are arbitrary constants is the general solution of Eq. (3).*

For example, $y = c_1 \cos x + c_2 \sin x$ is the general solution of the differential equation $y'' + y = 0$, and $y = c_1 e^x + c_2 e^{-x}$ is the general solution of the differential equation $y'' - y = 0$.

At this point we realize that the problem of finding all solutions of Eq. (3) boils down to finding a fundamental set of solutions. Sections 2.4 through 2.8 are devoted to discussing methods associated with the problem of finding a fundamental set of solutions.

**EXAMPLE 1**

Consider the IVP

$$y'' - y' - 2y = 0 \qquad (5)$$

$$y(0) = 1, \, y'(0) = 8. \qquad (6)$$

(a) Show that the functions $y_1(x) = e^{-x}$ and $y_2(x) = e^{2x}$ form a fundamental set of solutions for Eq. (5).

(b) Find the general solution of Eq. (5).

(c) Find the unique solution of the IVP (5)–(6).

**Solution**   (a) By direct substitution of $y_1$ and $y_2$ into Eq. (5), we find

$$(e^{-x})'' - (e^{-x})' - 2(e^{-x}) = e^{-x} + e^{-x} - 2e^{-x} = 0$$

and

$$(e^{2x})'' - (e^{2x})' - 2(e^{2x}) = 4e^{2x} - 2e^{2x} - 2e^{2x} = 0.$$

Therefore $y_1$ and $y_2$ are solutions of (5). Their Wronskian

$$\begin{vmatrix} e^{-x} & e^{2x} \\ -e^{-x} & 2e^{2x} \end{vmatrix} = 2e^x + e^x = 3e^x$$

is never zero. Hence $y_1$ and $y_2$ are linearly independent solutions of Eq. (5). Since Eq. (5) is second order, it follows that $y_1$ and $y_2$ form a fundamental set of solutions for the equation.

(b) By Corollary 1 the general solution of Eq. (5) is

$$y(x) = c_1 e^{-x} + c_2 e^{2x}. \tag{7}$$

(c) To find the unique solution of the IVP (5)–(6), we must use the initial conditions (6) in (7) and

$$y'(x) = -c_1 e^{-x} + 2c_2 e^{2x}. \tag{8}$$

Setting $x = 0$ in (7) and (8) we find,

$$\left. \begin{array}{r} c_1 + c_2 = 1 \\ -c_1 + 2c_2 = 8 \end{array} \right\} \Rightarrow c_1 = -2, c_2 = 3.$$

The unique solution of the IVP (5)–(6) is therefore

$$y(x) = -2e^{-x} + 3e^{2x}. \tag{9}$$

If we want to check that (9) is indeed *the* solution of the IVP (5)–(6), we must check three things: First, (9) is a solution of Eq (5). Second, $y(0) = 1$. Third, $y'(0) = 8$. In fact,

$$(-2e^{-x} + 3e^{2x})'' - (-2e^{-x} + 3e^{2x})' - 2(-2e^{-x} + 3e^{2x}) =$$

$$(-2e^{-x} + 12e^{2x}) - (2e^{-x} + 6e^{2x}) - 2(-2e^{-x} + 3e^{2x}) =$$

$$-2e^x + 12e^{2x} - 2e^{-x} - 6e^{2x} + 4e^{-x} - 6e^{2x} = 0;$$

$$y(0) = -2e^0 + 3e^0 = 1; y'(x) = 2e^{-x} + 6e^{2x} \Rightarrow y'(0) = 2e^0 + 6e^0 = 8.$$

**EXAMPLE 2**   In Section 2.1 we encountered the differential equation

$$\ddot{y} + a^2 y = 0, \qquad a \neq 0. \tag{10}$$

Show that $y_1 = \cos at$ and $y_2 = \sin at$ form a fundamental set of solutions for Eq. (10), and find its general solution.

**Solution**  In Section 2.1 we demonstrated that $y_1$ and $y_2$ are solutions of Eq. (10). To show that they are linearly independent, we compute their Wronskian:

$$W(y_1, y_2; t) = \begin{vmatrix} \cos at & \sin at \\ -a\sin at & a\cos at \end{vmatrix} = a\cos^2 at + a\sin^2 at$$

$$= a(\cos^2 at + \sin^2 at) \qquad = a \neq 0.$$

Since $W(y_1, y_2; t) \neq 0$ and $y_1$ and $y_2$ are solutions, we conclude that $y_1$ and $y_2$ constitute a fundamental system of solutions for Eq. (10), and every solution of Eq. (10) can be expressed as a linear combination of $y_1$ and $y_2$. That is, if $y(t)$ is *any* solution of Eq. (10), then $y = c_1 y_1 + c_2 y_2$ for some choice of constants $c_1, c_2$. Since $e^{iat}$ and $e^{-iat}$ are also solutions of Eq. (10), these functions must be expressible as linear combinations of $y_1$ and $y_2$. These linear combinations are readily obtained if one uses the *Euler identity* of trigonometry, that is, $e^{i\theta} = \cos\theta + i\sin\theta$ $(\Rightarrow e^{-i\theta} = \cos\theta - i\sin\theta)$. Thus, $e^{iat} = c_1 y_1 + c_2 y_2$, with $c_1 = 1$, $c_2 = i$; and $e^{-iat} = c_1 y_1 + c_2 y_2$, with $c_1 = 1$, $c_2 = -i$. The functions $e^{iat}$ and $e^{-iat}$ also form a fundamental set of solutions for Eq. (10). (See Exercise 21.)

### EXERCISES

For each of the initial value problems in Exercises 1 through 10, indicate whether or not the existence and uniqueness theorem, Theorem 1, applies.

**1.** $y'' + xy = 0$, $y(0) = 1$, $y'(0) = 0$, $-5 \leq x \leq 5$

**2.** $xy'' + y = 0$, $y(0) = 1$, $y'(0) = 0$, $-5 \leq x \leq 5$

**3.** $xy'' + y = 0$, $y(1) = 1$, $y'(1) = 1$, $1 \leq x \leq 5$

**4.** $(1 - x)y'' - xy' + e^x y = 0$, $y(0) = 0$, $y'(0) = 4$, $-\frac{1}{2} \leq x \leq \frac{1}{2}$

**5.** $(\sin x)y'' + xy' + y = 2$, $y\left(\frac{3\pi}{4}\right) = 1$, $y'\left(\frac{3\pi}{4}\right) = 0$, $\frac{3\pi}{4} \leq x \leq \frac{5\pi}{4}$

**6.** $x^2 y'' + (x\cos x)y' - 3y = 1$, $y\left(\frac{3\pi}{4}\right) = 1$, $y'\left(\frac{3\pi}{4}\right) = 1$, $\frac{3\pi}{4} \leq x \leq \frac{5\pi}{4}$

**7.** $(x^3 - 1)y''' - 3y'' + 4xy = 0$, $y(-1) = 0$, $y'(-1) = 2$, $y''(-1) = 0$, $-1 \leq x \leq 0$

**8.** $y''' + yy' = 2$, $y(0) = 0$, $y'(0) = -1$, $0 \leq x \leq 3$

**9.** $3y'' + y' - 2y = 0$, $y(0) = 2$, $y'(0) = 0$, $0 \leq x < \infty$

**10.** $(\cos x)y'' + 3y = 1$, $y(1) = 0$, $y'(1) = 0$, $1 \leq x < \infty$

**11.** Solve the IVP

$$y'' + (\sin x)\,y' + e^x y = 0$$

$$y(0) = 0, \qquad y'(0) = 0.$$

(*Hint:* Note that $y = 0$ is a solution and then use Theorem 1.)

For each of Exercises 12 through 17, determine an interval in which there exists a unique solution determined by the conditions $y(x_0) = y_0$, $y'(x_0) = y_1$, where $x_0$ is to be a point of the interval.

**12.** $(x - 1)y'' + 3y' = 0$                **13.** $2xy'' - 7(\cos x)y' + y = e^{-x}$

**14.** $y'' + 4(\tan x)y' - xy = 0$         **15.** $(\cos x)y'' + y = \sin x$

**16.** $(x^2 - 4)y'' + 3x^3y' + \dfrac{4}{x - 1}y = 0$

**17.** $(x^2 + 1)y'' + (2x - 1)\,y' + 3y = 0$

**18.** As we explained in Section 2.1.1, the IVP

$$m\ddot{y} + Ky = 0 \qquad\qquad (11)$$

$$y(0) = y_0, \ \dot{y}(0) = v_0 \qquad\qquad (12)$$

describes the motion of a spring of mass $m$ and spring constant $K$. Initially, at time $t = 0$, the mass is located at the position $y_0$ and has initial velocity $v_0$.

(a) Show that

$$y_1(t) = \cos \sqrt{\frac{K}{m}}\, t, \qquad y_2(t) = \sin \sqrt{\frac{K}{m}}\, t$$

form a fundamental set of solutions of the differential equation (11).

(b) Find the general solution of the differential equation.

(c) Find the unique solution of the IVP (11)–(12).

**19.** Give a physical interpretation of the IVP

$$\ddot{y} + 4y = 0$$

$$y(0) = \dot{y}(0) = 0.$$

Find the solution to the IVP first by physical intuition and then by applying Theorem 1.

**20. Mechanics** A projectile of mass $m$ leaves the earth with initial velocity $v_0$ in a direction that makes an angle $\theta$ with the horizontal. Assume that the motion takes place on the $xy$ plane, that the projectile was fired at the origin 0, and that the $x$ axis is the horizontal (Figure 2.3). Use Newton's second law of motion to show that the position $(x(t), y(t))$ of the mass $m$ at time $t$ satisfies the two initial value problems

$$
\begin{array}{ccc}
m\ddot{x} = 0 & & m\ddot{y} = -mg \\
x(0) = 0 & \text{and} & y(0) = 0 \\
\dot{x}(0) = v_0 \cos \theta & & \dot{y}(0) = v_0 \sin \theta.
\end{array}
$$

Show (by direct integrations) that the solutions of these initial value problems are given by $x(t) = (v_0 \cos \theta)t$ and $y(t) = -\frac{1}{2}gt^2 + (v_0 \sin \theta)t$.

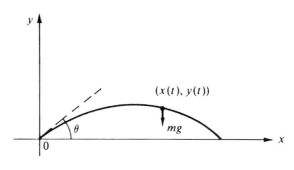

**Figure 2.3**

**21.** Show that the functions $y_1 = e^{iat}$ and $y_2 = e^{-iat}$ form a fundamental set of solutions for the differential equation

$$\ddot{y} + a^2y = 0, \qquad a \neq 0.$$

**22.** Show that the functions $y_1 = \cosh at$ and $y_2 = \sinh at$ form a fundamental set of solutions for the differential equation

$$\ddot{y} - a^2y = 0, \qquad a \neq 0.$$

**23.** Find the general solution of the differential equation

$$(x - a)(x - b)y'' + 2(2x - a - b)y' + 2y = 0$$

[*Hint:* Compute[10] the second derivative of the expression $(x - a)(x - b)y$.]

---

## 2.4 HOMOGENEOUS DIFFERENTIAL EQUATIONS WITH CONSTANT COEFFICIENTS—THE CHARACTERISTIC EQUATION

The linear homogeneous differential equation with constant coefficients has the form

$$a_ny^{(n)} + a_{n-1}y^{(n-1)} + \cdots + a_1y' + a_0y = 0, \tag{1}$$

where each coefficient $a_i$ is a constant and the leading coefficient $a_n \neq 0$. To find the general solution of Eq. (1), we need to find $n$ linearly independent solutions.

As we examine Eq. (1) carefully (especially in the case $n = 1$), we see that in a sense any solution of this equation must have the property that derivatives of this solution merely produce constant multiples of themselves. This observation motivates us to try $e^{\lambda x}$, with $\lambda$ a constant, as a solution of Eq. (1). If $y = e^{\lambda x}$, then $y' = \lambda e^{\lambda x}$, $y'' = \lambda^2 e^{\lambda x}, \ldots, y^{(n)} = \lambda^n e^{\lambda x}$. Substitution of $y = e^{\lambda x}$ in Eq. (1) yields

$$a_n\lambda^n e^{\lambda x} + a_{n-1}\lambda^{n-1}e^{\lambda x} + \cdots + a_1\lambda e^{\lambda x} + a_0e^{\lambda x} = 0$$

or

$$e^{\lambda x}(a_n\lambda^n + a_{n-1}\lambda^{n-1} + \cdots + a_1\lambda + a_0) = 0. \tag{2}$$

[10]See also *Math. Magazine* 44 (1971): 18, 56, 297.

The exponential function never takes the value zero. Thus, the only way that $y = e^{\lambda x}$ can be a solution of the differential equation is that $\lambda$ be a root of the polynomial equation

$$a_n\lambda^n + a_{n-1}\lambda^{n-1} + \cdots + a_1\lambda + a_0 = 0. \qquad (3)$$

We conclude, then, that if $\lambda$ is a root of Eq. (3), $y = e^{\lambda x}$ is a solution of the differential equation (1).

**DEFINITION 1**

*The polynomial $f(\lambda) \equiv a_n\lambda^n + a_{n-1}\lambda^{n-1} + \cdots + a_1\lambda + a_0$ is called the characteristic polynomial for Eq. (1), and the equation $f(\lambda) = 0$ is called the characteristic equation for Eq. (1). The roots of the characteristic equation are called characteristic roots.*

In Section 2.5 we will demonstrate that knowledge of the roots of the characteristic equation is all that is needed to construct the general solution of homogeneous linear differential equations with constant coefficients.

Consequently, the problem of solving homogeneous linear differential equations with constant coefficients can be reduced to the problem of finding the roots of the characteristic equation for the differential equation. The characteristic polynomial is of degree $n$ and the *fundamental theorem of algebra* states that every polynomial of degree $n$ $(n \geq 1)$ has at least one root and, therefore, counting multiplicities, exactly $n$ roots. However, if $n$ is large, indeed, when $n \geq 5$, the problem of finding the roots of the characteristic equation may be far from trivial. Some elementary methods for finding roots of polynomial equations are given in Appendix C.

**REMARK 1**  If $n = 1$, the characteristic equation is $a_1\lambda + a_0 = 0$ and has the root $\lambda = -a_0/a_1$. If $n = 2$, the characteristic equation is the quadratic equation $a_2\lambda^2 + a_1\lambda + a_0 = 0$ and its two roots are given by the quadratic formula

$$\lambda = \frac{-a_1 \pm \sqrt{a_1^2 - 4a_2a_0}}{2a_2}. \qquad (4)$$

If $n = 3$, the characteristic equation is a cubic equation, and the formulas for its solutions (called *Cardan's formulas*) were discovered by Scipione del Ferro in 1515. If $n = 4$, the characteristic equation is a quartic equation, and its solution was discovered by Ludovico Ferrari in the sixteenth century.[11] For $n \geq 5$, the characteristic equation cannot be solved (by radicals) in general unless it is of some special form. This fact was proved by Abel and Galois early in the nineteenth century.

In spite of the possibility of algebraic difficulties surrounding the solution of the characteristic equation, we can develop some information about linearly independent solutions to Eq. (1).

---

[11]For the solutions of the cubic and quartic equations, the interested reader is referred to the text by E. A. Grove and G. Ladas, *Introduction to Complex Variables* (Boston: Houghton Mifflin, 1974).

We emphasize once more that if $\lambda$ is a characteristic root, then $e^{\lambda x}$ is a solution of the linear homogeneous differential equation with constant coefficients. In the special case that $\lambda$ is a complex number, it is conventional to write the corresponding solution in an alternative form. First; to say that $\lambda$ is a complex number means that $\lambda$ is of the form $\alpha + i\beta$, where $\alpha$ and $\beta$ are real numbers and $i^2 = -1$. The real number $\alpha$ is called the *real part* of $\lambda$, and the real number $\beta$ is called the *imaginary part* of $\lambda$. We then write

$$e^{\lambda x} = e^{(\alpha + i\beta)x} = e^{\alpha x + i\beta x} = e^{\alpha x}e^{i\beta x}$$
$$= e^{\alpha x}(\cos \beta x + i \sin \beta x). \tag{5}$$

The last form follows from the Euler identity, $e^{i\theta} = \cos \theta + i \sin \theta$ ($\Rightarrow e^{-i\theta} = \cos \theta - i \sin \theta$). Suppose now that the coefficients $a_i$ of the differential equation are real numbers. Then the only way that a characteristic root can be a complex number is that the characteristic polynomial contain a quadratic factor of the form $b_2\lambda^2 + b_1\lambda + b_0$ for which $b_1^2 < 4b_2b_0$. In this case the complex characteristic roots occur in *conjugate pairs* ($\alpha - i\beta$ is called the *conjugate* of the complex number $\alpha + i\beta$). Thus, two characteristic roots would be $\alpha + i\beta$ and $\alpha - i\beta$, where [refer to Eq. (4) and recall that $\sqrt{-K} = i \sqrt{K}$]

$$\alpha = -\frac{b_1}{2b_2}, \qquad \beta = \frac{\sqrt{4b_0b_2 - b_1^2}}{2b_2}.$$

In addition to the solution (5) there would also be the solution $e^{\alpha x}(\cos \beta x - i \sin \beta x)$. For further discussion of this case, see Example 2 below, Exercises 21 through 29, and Remark 1 in Section 2.5.

**EXAMPLE 1**   Solve the differential equation $y'' - y = 0$.

**Solution**   Set $y = e^{\lambda x}$; then (since $a_2 = 1$, $a_1 = 0$, $a_0 = -1$) $\lambda$ must satisfy the characteristic equation

$$\lambda^2 - 1 = 0 \Rightarrow (\lambda + 1)(\lambda - 1) = 0 \Rightarrow \lambda = \pm 1.$$

We have two solutions, $y_1 = e^x$ and $y_2 = e^{-x}$. These two solutions are linearly independent, since their Wronskian has the value $-2(\neq 0)$. Thus, any solution $y$ of this differential equation is of the form

$$y = c_1 e^x + c_2 e^{-x},$$

where $c_1$ and $c_2$ are arbitrary constants.

**EXAMPLE 2**   Solve the differential equation of the vibrating spring example of Section 2.1.1:

$$\ddot{y} + a^2 y = 0,$$

with $a^2 = K/m$.

**Solution**　Set $y = e^{\lambda t}$; then (since $a_2 = 1$, $a_1 = 0$, $a_0 = a^2$) $\lambda$ must satisfy the equation

$$\lambda^2 + a^2 = 0, \qquad \lambda = \pm ia.$$

We have two solutions, $y_1 = e^{iat}$ and $y_2 = e^{-iat}$. These two solutions form a fundamental set of solutions, since their Wronskian has the value $-2ia \neq 0$. Hence, any solution to this differential equation is of the form

$$y(t) = c_1 e^{iat} + c_2 e^{-iat},$$

or (using the Euler identity $e^{i\theta} = \cos\theta + i\sin\theta$) we find

$$y(t) = \alpha_1 \cos at + \alpha_2 \sin at,$$

with $\alpha_1 = c_1 + c_2$ and $\alpha_2 = i(c_1 - c_2)$.

## EXERCISES

In Exercises 1 through 10, write out the characteristic equation for the differential equation.

**1.** $4y''' - 2y'' + 6y' - 7y = 0$

**2.** $2y^{iv} + 3y''' - y'' + 2y' - y = 0$

**3.** $y''' - y' + 2y = 0$

**4.** $y'' - y' + 6y = 0$

**5.** $5y''' - 5y'' + y' - 2y = 0$

**6.** $6y''' - 4iy'' + (3 + i)y' - 2y = 0$

**7.** $3y''' + 4y'' = 0$

**8.** $6y^{iv} - 3y''' + y'' - 7y' - 6y = 0$

**9.** $3y^v - 2y^{iv} + y'' - 2y' = 0$

**10.** $2y''' - 2y'' - y' + 3y = 0$

In Exercises 11 through 20, determine whether the associated characteristic equation has complex roots.

**11.** $y'' - 5y' + 6y = 0$

**12.** $2y'' + 3y' + y = 0$

**13.** $2y'' + 3y' - 2y = 0$

**14.** $y'' + 9y = 0$

**15.** $3y'' - 5y' + 3y = 0$

**16.** $y'' + y' + y = 0$

**17.** $2y'' - 4y' - y = 0$

**18.** $4y'' - 3y' + y = 0$

**19.** $y'' + 3y' + 4y = 0$

**20.** $2y'' + y = 0$

In Exercises 21 through 29, the second-order homogeneous linear differential equation has characteristic roots which occur in conjugate pairs. In each case determine the characteristic roots $\lambda_1$ and $\lambda_2$. Write the general solution in the form $y = c_1 e^{\lambda_1 x} + c_2 e^{\lambda_2 x}$. Next, following the method of Example 2, write the general solution so that the complex unit $i$ is absorbed into the constants.

**21.** $y'' + 16y = 0$

**22.** $2y'' + 14y' + 25y = 0$

**23.** $y'' + y' + y = 0$

**24.** $y'' + 9y = 0$

**25.** $4y'' - 8y' + 5y = 0$

**26.** $2y'' - 6y' + 5y = 0$

**27.** $y'' + 4y = 0$

**28.** $y'' - 3y' + \tfrac{5}{2}y = 0$

**29.** $y'' + 25y = 0$

**30.** The definitions of the hyperbolic cosine and hyperbolic sine are

$$\cosh x = \frac{e^x + e^{-x}}{2}$$

and

$$\sinh x = \frac{e^x - e^{-x}}{2}.$$

   (a) Show that these definitions can be manipulated algebraically to yield $e^x = \cosh x + \sinh x$ and $e^{-x} = \cosh x - \sinh x$.
   (b) Show that the general solution of the differential equation $y'' - y = 0$ can be written in the form $y = c_1 \cosh x + c_2 \sinh x$.

In Exercises 31 through 37 write two equivalent versions of the general solution. [*Hint:* See Example 1 and Exercise 30.]

**31.** $y'' - 4y = 0$               **32.** $y'' - 16y = 0$

**33.** $y'' - 25y = 0$              **34.** $y' - 49y = 0$

**35.** $y'' - 81y = 0$              **36.** $y'' - 121y = 0$

**37.** $y'' - 9y = 0.$

**38.** Apply the method of Example 2 to write the general solution of

$$y^{iv} + 13y'' + 36y = 0$$

in terms of trigonometric functions. [*Hint:* $\gamma^2 + 13\gamma + 36 = (\gamma + 4) \cdot (\gamma + 9).$]

**39. Economics** Second-order linear differential equations with constant coefficients of the form

$$\ddot{D} - \beta D = 0$$

occur in the Domar burden-of-debt model, where the income grows at a constant relative rate $\beta$ $(0 < \beta < 1)$ and $D$ is the total public debt.[12] Find the characteristic roots associated with this differential equation. Write the general solution of this differential equation in the form $y(t) = c_1 e^{\lambda_1 t} + c_2 e^{\lambda_2 t}$ and also in terms of the hyperbolic sine and hyperbolic cosine.

**40. The Vibrating Spring with Damping** In practice, a vibrating spring is most often subjected to frictional forces and other forces (for example, air resistance) which act to retard (dampen) the motion and eventually cause the system to come to rest. Thus, it is realistic to assume that the spring is subjected to a *damping force*. Generally speaking, the damping force is difficult to formulate precisely; however, experiments have shown that the magnitude of the damping force is approximately proportional to the velocity of the mass, provided that the velocity of the mass is small. Naturally,

[12]See E. D. Domar, "The Burden of the Debt and the National Income," *Amer. Econ. Rev.* (Dec. 1944): 798–827.

the damping force, as described, acts in a direction opposite to that of the mass. Thus, the damping force is negative when $dy/dt$ is positive and positive when $dy/dt$ is negative. Consequently, we can express the damping force as $-\alpha(dy/dt)$, where $\alpha > 0$ is called the *damping constant*.

Using Newton's second law, show that the equation of motion in this case is

$$m\ddot{y} + \alpha\dot{y} + Ky = 0. \tag{6}$$

Write the characteristic equation for this differential equation. Describe conditions on $\alpha$, $K$, and $m$ which guarantee that the characteristic roots will not be complex numbers. What conditions will guarantee that the characteristic roots are complex?

## 2.5  HOMOGENEOUS DIFFERENTIAL EQUATIONS WITH CONSTANT COEFFICIENTS—THE GENERAL SOLUTION

In Sections 2.2 and 2.3 we demonstrated that for linear differential equations, the basic problem is to find a fundamental set of solutions. In Section 2.4 we demonstrated that linear homogeneous differential equations with constant coefficients will have solutions of the form $e^{\lambda x}$ provided that $\lambda$ is a root of the characteristic equation associated with the differential equation. The loose end that still persists is to relate the solutions $e^{\lambda x}$ with a fundamental set of solutions of the differential equation. This we accomplish in the present section via theorems and a number of illustrative examples.

### THEOREM 1

*If all the roots of the characteristic equation are distinct, say $\lambda = \lambda_i$, $i = 1$, $2, \ldots, n$, then the $n$ functions $y_i = e^{\lambda_i x}$, $i = 1, 2, \ldots, n$ constitute a fundamental set of solutions.*

**Proof**  The proof is for $n = 2$.

$$\begin{aligned} W(y_1, y_2; x) &= \begin{vmatrix} e^{\lambda_1 x} & e^{\lambda_2 x} \\ \lambda_1 e^{\lambda_1 x} & \lambda_2 e^{\lambda_2 x} \end{vmatrix} \\ &= \lambda_2 e^{(\lambda_1 + \lambda_2)x} - \lambda_1 e^{(\lambda_1 + \lambda_2)x} \\ &= (\lambda_2 - \lambda_1)e^{(\lambda_1 + \lambda_2)x} \neq 0. \end{aligned}$$

Thus, $y_1$ and $y_2$ form a fundamental set of solutions. The proof for the general case would follow along similar lines of reasoning. However, the evaluation of the corresponding Wronskian would not be trivial. In fact, the Wronskian reduces to a determinant commonly called the *Vandermonde determinant*.[13]

---

[13]For the statement and evaluation of the van der Monde determinant, see C. W. Curtis, *Linear Algebra*, 2nd ed. (Boston: Allyn and Bacon, 1968), p. 140.

**EXAMPLE 1**   Find the general solution of the differential equation

$$y''' - 6y'' + 11y' - 6y = 0.$$

**Solution**   For this differential equation the characteristic equation is

$$\lambda^3 - 6\lambda^2 + 11\lambda - 6 = 0.$$

Note that the coefficients in this polynomial add up to zero. Whenever this happens it is always true that $\lambda = 1$ is a root. Thus we obtain (after long division or synthetic division)

$$\lambda^3 - 6\lambda^2 + 11\lambda - 6 = (\lambda - 1)(\lambda^2 - 5\lambda + 6) = (\lambda - 1)(\lambda - 2)(\lambda - 3).$$

Hence, $\lambda = 1, 2, 3$ and the fundamental set of solutions is $y_1 = e^x$, $y_2 = e^{2x}$, $y_3 = e^{3x}$. The general solution is

$$y = c_1 e^x + c_2 e^{2x} + c_3 e^{3x}.$$

If the roots of the characteristic equation are not all distinct, that is, if the characteristic equation has at least one repeated root, then we cannot generate a fundamental system of solutions by considering just the exponentials $e^{\lambda_i x}$. To try to obtain a feeling for the form of the general solution in this case, we first consider some special examples.

**EXAMPLE 2**   Solve the differential equation

$$y'' - 2y' + y = 0. \tag{1}$$

**Solution**   For the differential equation (1) the characteristic polynomial is $\lambda^2 - 2\lambda + 1 = (\lambda - 1)^2$. Thus $\lambda = 1$ is a root of multiplicity 2. We introduce a new unknown function $v$ by the relation $v = y' - y$. Then $v' = y'' - y'$, which, using the differential equation (1), takes the form

$$v' = (2y' - y) - y' = y' - y = v \Rightarrow v' = v \Rightarrow v = c_1 e^x.$$

Therefore,

$$v = y' - y = c_1 e^x.$$

This last differential equation is a first-order linear differential equation and can be solved by the methods of Section 1.4 to yield

$$y = e^x(\int e^{-x} c_1 e^x \, dx + c_2) = e^x(c_1 x + c_2) = c_1(xe^x) + c_2 e^x.$$

That is,

$$y = c_1 y_1 + c_2 y_2,$$

where $y_1 = xe^x$ and $y_2 = e^x$. The reader can verify that both $y_1$ and $y_2$ are solutions of Eq. (1). Furthermore,

$$W(y_1, y_2; x) = \begin{vmatrix} xe^x & e^x \\ (x+1)e^x & e^x \end{vmatrix} = e^x \cdot e^x \begin{vmatrix} x & 1 \\ x+1 & 1 \end{vmatrix} = e^{2x}(-1) \neq 0.$$

Our fundamental set of solutions is therefore $y_1 = xe^x$ and $y_2 = e^x$, and the general solution is

$$y = c_1xe^x + c_2e^x.$$

In this example we obtained a simpler problem by making the substitution $v = y' - y$. For an alternative approach, see Exercises 46 and 47.

**EXAMPLE 3**  Solve the differential equation

$$y''' - 3y'' + 3y' - y = 0. \tag{2}$$

**Solution**  The characteristic polynomial for the differential equation (2) is $\lambda^3 - 3\lambda^2 + 3\lambda - 1 = (\lambda - 1)^3$. Hence, $\lambda = 1$ is a root of the characteristic equation of multiplicity 3. *Taking a clue from Example 2, we guess that the fundamental system of solutions for Eq. (2) is $y_1 = e^x$, $y_2 = xe^x$, and $y_3 = x^2e^x$.* Certainly, $y_1$ is a solution of Eq. (2). As for $y_2$, we have

$$y_2 = xe^x \Rightarrow y_2' = (x + 1)e^x \Rightarrow y_2'' = (x + 2)e^x \Rightarrow y_2''' = (x + 3)e^x.$$

Substituting these results in Eq. (2) yields

$$(x + 3)e^x - 3(x + 2)e^x + 3(x + 1)e^x - xe^x = e^x(x + 3 - 3x - 6 + 3x + 3 - x)$$
$$= e^x(0) = 0.$$

Hence, $y_2$ is also a solution. Similarly, for $y_3$ we obtain

$$y_3 = x^2e^x \Rightarrow y_3' = (x^2 + 2x)e^x \Rightarrow y_3'' = (x^2 + 4x + 2)e^x$$
$$\Rightarrow y_3''' = (x^2 + 6x + 6)e^x.$$

Substitution of these results in Eq. (2) yields

$$e^x[x^2 + 6x + 6 - 3(x^2 + 4x + 2) + 3(x^2 + 2x) - x^2] = e^x(0) = 0.$$

Therefore, $y_3$ is a solution. To show that these solutions are linearly independent, we compute their Wronskian.

$$W(y_1, y_2, y_3; x) = \begin{vmatrix} e^x & xe^x & x^2e^x \\ e^x & (x + 1)e^x & (x^2 + 2x)e^x \\ e^x & (x + 2)e^x & (x^2 + 4x + 2)e^x \end{vmatrix}$$

$$= e^x \cdot e^x \cdot e^x \begin{vmatrix} 1 & x & x^2 \\ 1 & x + 1 & x^2 + 2x \\ 1 & x + 2 & x^2 + 4x + 2 \end{vmatrix}$$

$$= e^{3x}(2) \neq 0.$$

We conclude, then, that $y_1$, $y_2$, and $y_3$ form a fundamental set of solutions. The general solution is

$$y = c_1e^x + c_2xe^x + c_3x^2e^x.$$

The results embodied in Examples 2 and 3 are capable of generalization, and this generalization is stated without proof in the following theorem.[14]

### THEOREM 2

*If $\lambda = \lambda_i$ is a root of multiplicity $m$ of the characteristic equation associated with the differential equation $a_n y^{(n)} + a_{n-1} y^{(n-1)} + \cdots + a_1 y' + a_0 y = 0$, then there are $m$ linearly independent solutions associated with $\lambda = \lambda_i$. These solutions are as follows: $e^{\lambda_i x}, xe^{\lambda_i x}, x^2 e^{\lambda_i x}, \ldots, x^{m-1} e^{\lambda_i x}$.*

**REMARK 1**  When the coefficients $a_i$ of the differential equation are all real numbers, we know from the theory of polynomial equations that if the characteristic equation has a complex root, such as $\lambda = \alpha + i\beta$, of multiplicity $m$, then the characteristic equation also has the complex conjugate $\alpha - i\beta$ as a root again of multiplicity $m$. Applying Theorem 2 first to the root $\alpha + i\beta$ and then to $\alpha - i\beta$ we find $2m$ complex valued solutions. It can be shown however, that the real and imaginary parts of the solutions

$$e^{(\alpha + i\beta)x}, xe^{(\alpha + i\beta)x}, \ldots, x^{m-1} e^{(\alpha + i\beta)x}$$

and

$$e^{(\alpha - i\beta)x}, xe^{(\alpha - i\beta)x}, \ldots, x^{m-1} e^{(\alpha - i\beta)x},$$

namely the $2m$ functions

$$e^{\alpha x} \cos \beta x, e^{\alpha x} \sin \beta x, xe^{\alpha x} \cos \beta x, xe^{\alpha x} \sin \beta x, \ldots, x^{m-1} e^{\alpha x} \cos \beta x, x^{m-1} e^{\alpha x} \sin \beta x,$$

are linearly independent solutions. From this and the previous results it follows that *the general solution of a linear differential equation with real and constant coefficients can always be expressed as a linear combination of $n$ real valued functions.*

### COROLLARY 1

*Consider the second-order linear differential equation with real coefficients*

$$a_2 y'' + a_1 y' + a_0 y = 0, \, a_2 \neq 0. \tag{3}$$

*Assume that $\lambda_1$ and $\lambda_2$ are roots of the characteristic equation $a_2 \lambda^2 + a_1 \lambda + a_0 = 0$. Then the form of the general solution $y(x)$ of Eq. (3) is described by the following cases:*

### CASE 1  *Real and Distinct Roots*  $\lambda_1 \neq \lambda_2$

$$y(x) = c_1 e^{\lambda_1 x} + c_2 e^{\lambda_2 x},$$

*where $c_1$ and $c_2$ are arbitrary constants.*

[14]For a proof, see Coddington, *Ordinary Differential Equation* (Englewood Cliffs, N.J. Prentice-Hall, 1961)

**CASE 2**   *Equal Roots*   $\lambda_1 = \lambda_2 = \lambda$

$$y(x) = c_1 e^{\lambda x} + c_2 x e^{\lambda x}.$$

**CASE 3**   *Complex Conjugate Roots*   $\lambda_{1,2} = k \pm i\ell$

$$y(x) = c_1 e^{kx} \cos \ell x + c_2 e^{kx} \sin \ell x.$$

**EXAMPLE 4**   Solve the differential equation

$$y' - (2 + 3i)y = 0. \tag{4}$$

**Solution**   The characteristic equation for the differential equation (4) is $\lambda - (2 + 3i) = 0$, hence the solution is

$$y = c_1 e^{(2+3i)x} = c_1(e^{2x} \cos 3x + i e^{2x} \sin 3x).$$

Note that the functions $e^{2x} \cos 3x$ and $e^{2x} \sin 3x$ are not, by themselves, solutions of the differential equation (4), rather it is the specific combination $e^{2x} \cos 3x + i e^{2x} \sin 3x$ that is a solution. Naturally, the difference between this example and Case 3 of Corollary 1 is that the constants in the differential equation (4) are *complex* and that the complex characteristic root $\lambda = 2 + 3i$ *does not* occur in a conjugate pair of characteristic roots.

**EXAMPLE 5**   Solve the differential equation

$$y^{(5)} - 3y^{(4)} + 4y''' - 4y'' + 3y' - y = 0. \tag{5}$$

**Solution**   The characteristic equation for the differential equation (5) is

$$\lambda^5 - 3\lambda^4 + 4\lambda^3 - 4\lambda^2 + 3\lambda - 1 = 0.$$

Since the coefficients add to zero, it follows that $\lambda = 1$ is a root and we can write (see Appendix C)

$$\begin{aligned}
\lambda^5 - 3\lambda^4 + 4\lambda^3 - 4\lambda^2 + 3\lambda - 1 &= (\lambda - 1)(\lambda^4 - 2\lambda^3 + 2\lambda^2 - 2\lambda + 1) \\
&= (\lambda - 1)(\lambda - 1)(\lambda^3 - \lambda^2 + \lambda - 1) \\
&= (\lambda - 1)(\lambda - 1)(\lambda - 1)(\lambda^2 + 1).
\end{aligned}$$

Thus, $\lambda = 1, 1, 1, i, -i$, and the general solution has the form

$$y = c_1 e^x + c_2 x e^x + c_3 x^2 e^x + c_4 \cos x + c_5 \sin x.$$

**EXAMPLE 6**   The characteristic equation of a certain differential equation has the following roots: 1 as a simple root, $-2$ as a triple root, $2 \pm 3i$ as simple roots, and $-1 \pm 2i$ as double roots; that is, the characteristic polynomial can be factored as

$$a_{10}(\lambda - 1)(\lambda + 2)^3(\lambda^2 - 4\lambda + 13)(\lambda^2 + 2\lambda + 5)^2,$$

where $a_{10}$ is the coefficient of $y^{(10)}$ in the original differential equation. Write down the general solution of this differential equation.

**Solution**   It follows from Theorems 1 and 2 and Remark 1 that the general solution of the differential equation in question is

$$y = c_1 e^x + c_2 e^{-2x} + c_3 x e^{-2x} + c_4 x^2 e^{-2x} + c_5 e^{2x} \cos 3x + c_6 e^{2x} \sin 3x$$
$$+ c_7 e^{-x} \cos 2x + c_8 e^{-x} \sin 2x + c_9 x e^{-x} \cos 2x + c_{10} x e^{-x} \sin 2x.$$

## APPLICATION 2.5.1

It is suggested that the reader review Exercise 40 of Section 2.4 before studying this application.

■ The motion of the vibrating spring subjected to damping (retarding) forces is governed by the differential equation [see Eq. (6), Exercise 40, Section 2.4]

**Mechanics**
The Vibrating
Spring with
Damping

$$\ddot{y} + \frac{\alpha}{m}\dot{y} + \frac{K}{m}y = 0,$$

where $\alpha > 0$ is the damping constant, $K > 0$ is the spring constant, and $m > 0$ is the mass. To avoid fractions, it is convenient to set $2a = \alpha/m$ and $b^2 = K/m$, $b > 0$. Thus, the equation of motion has the form

$$\ddot{y} + 2a\dot{y} + b^2 y = 0. \tag{6}$$

The characteristic equation associated with Eq. (6) is

$$\lambda^2 + 2a\lambda + b^2 = 0.$$

Consequently, the characteristic roots are

$$\lambda_1 = -a + \sqrt{a^2 - b^2}, \qquad \lambda_2 = -a - \sqrt{a^2 - b^2}.$$

We realize that the form of the general solution of Eq. (6) depends very heavily upon the nature of the characteristic roots. The three possible cases are discussed separately.

**CASE 1**   $b^2 > a^2 (\Rightarrow b > a)$   The general solution of Eq. (6) is

$$y(t) = e^{-at}(c_1 \cos\sqrt{b^2 - a^2}\,t + c_2 \sin\sqrt{b^2 - a^2}\,t).$$

Graphically, the solution looks as shown in Figure 2.4. We observe that the motion [that is $y(t)$] oscillates about the $t$ axis, and for this reason we term the motion *oscillatory*. We note, however, that the "amplitude" of the motion[15] decreases as time goes on. This situation is referred to as being *underdamped*

---

[15] As in Exercise 21, Section 2.1, the solution in this case can be written in the form

$$y(t) = Ae^{-at}\cos\left[\sqrt{b^2 - a^2}\,t - \phi\right]$$
$$A = \sqrt{c_1^2 + c_2^2} \quad \text{and} \quad \phi = \cos^{-1}\left[c_1/(c_1^2 + c_2^2)^{1/2}\right].$$

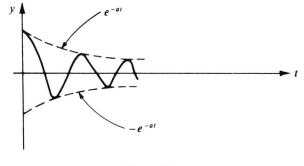

**Figure 2.4**

in the sense that the amount of damping (or the damping constant) is less than the spring stiffness, and hence there is not enough damping to overcome the oscillations.

**CASE 2**  $b^2 = a^2 \ (\Rightarrow b = a)$   The general solution of Eq. (6) is

$$y(t) = e^{-at}(c_1 + c_2 t).$$

Since $y(t)$ contains no sine or cosine terms, we note that there are no vibrations (oscillations). However, in reference to Case 1, we realize that even the slightest decrease in the damping will produce vibrations. For this reason, this case is called *critically damped.*

**CASE 3**  $b^2 < a^2 \ (\Rightarrow b < a)$   Here the general solution of Eq. (6) is

$$y(t) = c_1 e^{\lambda_1 t} + c_2 e^{\lambda_2 t},$$

with both $\lambda_1$ and $\lambda_2$ negative real numbers. In this case there are no oscillations and the graph of the solution approaches the $t$ axis very quickly as time goes on. That is, the damping force is so strong that it causes the system to "slow down" very quickly. This case is referred to as being *overdamped.*

**EXAMPLE 7**   It is known that the motion of a certain damped vibrating spring is governed by the differential equation

$$\ddot{y} + 4\dot{y} + 4y = 0.$$

Classify this case as being underdamped, critically damped, or overdamped.

**Solution**   Comparing this differential equation with Eq. (6), we note that $2a = 4$ and $b^2 = 4$; hence $a = b = 2$. This is Case 2, and therefore the motion is classified as critically damped.

## EXERCISES

In Exercises 1 through 10, write the general solution of the differential equation.

**1.** $2y'' + 3y' + y = 0$                          **2.** $8y'' - 6y' + y = 0$

**3.** $y'' + y' + y = 0$                            **4.** $9y'' - 6y' + y = 0$

**5.** $y'' + 6y = 0$                                **6.** $y'' - 9y = 0$

**7.** $y' - 3y = 0$                                 **8.** $y''' - 7y'' + 5y' + y = 0$

**9.** $y^{iv} + y''' - 3y'' - y' + 2y = 0$          **10.** $y''' - 6y'' + 12y' - 8y = 0$

In Exercises 11 through 19, the factored form of the characteristic equation for certain differential equations are given. In each case state the order of the differential equation and give the form of the general solution.

**11.** $(\lambda - 3)(\lambda + 2)^2(\lambda^2 - 2\lambda + 5)(\lambda^2 + 4\lambda + 5)^2$

**12.** $(\lambda + 1)(\lambda - 1)(\lambda^2 + 1)(\lambda^2 + \lambda + 1)$

**13.** $(2\lambda - 3)(5\lambda + 4)(\lambda^2 + 4)^2(\lambda^2 - \lambda - 1)$

**14.** $(\lambda - 3)^3(\lambda + 2)^4$

**15.** $(\lambda^2 + 4)(\lambda^2 + 1)^2$

**16.** $(\lambda + 1)(\lambda + 3)(\lambda + 2)^2(\lambda^2 - 2\lambda + 5)^2(\lambda^2 + 9)$

**17.** $(\lambda - 7)(2\lambda - 7)(\lambda^2 + \lambda + 1)(\lambda + 4)^3$

**18.** $(\lambda^2 + 3)(\lambda^2 + 1)^3(\lambda + 1)^3$

**19.** $(\lambda + 2)(\lambda - 5)^2(\lambda - 4)^3(\lambda^2 + 4)^3$

Solve the initial value problems in Exercises 20 and 21.

**20.** $y'' - 2y' + y = 0$            **21.** $y'' + y' + y = 0$
$\qquad y(0) = 1$                      $\qquad y(0) = 1$
$\qquad y'(0) = -1$                    $\qquad y'(0) = \sqrt{3}$

**22.** Solve the differential equation

$$y'' - iy' + 12y = 0.$$

Do the roots of the characteristic equation occur in conjugate pairs? Determine whether or not the real or imaginary part of either of the fundamental solutions satisfies the differential equation.

In Exercises 23 through 29, find the solution of the initial value problem.

**23.** $y'' + 3y = 0,$          $\quad y(0) = 3,$      $\quad y'(0) = -6\sqrt{3}$

**24.** $y'' - 4y = 0,$          $\quad y(0) = 0,$      $\quad y'(0) = 4$

**25.** $y''' - y'' = 0,$         $\quad y(0) = 1,$      $\quad y'(0) = 3,$      $\quad y''(0) = 2$

**26.** $y''' - 4y'' + 4y' = 0,$  $\quad y(0) = 2,$      $\quad y'(0) = -1$      $\quad y''(0) = 0$

**27.** $y''' - y'' - y' + y = 0,$　　$y(0) = 0,$　　$y'(0) = 5$　　$y''(0) = 2$

**28.** $y'' + 4y = 0,$　　　　　$y\left(\dfrac{\pi}{4}\right) = 1,$　　$y'\left(\dfrac{\pi}{4}\right) = 0$

**29.** $y''' - 2y'' + 2y' = 0,$　　$y(0) = -2,$　$y'(0) = 0,$　　$y''(0) = 4$

Find the solution of each boundary value problem in Exercises 30 and 31.

**30.** $y'' + 3y' + 2y = 0,$　$y(0) = 0,$　$y(-1) = e$

**31.** $y'' + y = 0,$　　　　$y(0) = 3,$　$y\left(\dfrac{\pi}{2}\right) = 2$

**32.** A spring is stretched 1.5 cm by a force of 4 dyn. A mass of 4 g is attached to the end of the spring. The system is then set into motion by pulling the mass 6 cm above the point of equilibrium and releasing it, at time $t = 0$, with an initial velocity of $5\sqrt{6}$ cm/sec directed downward. Find the position of the mass at any time $t$.

**33.** In Exercise 32, assume that, in addition, air resistance acts upon the mass. This air resistance force equals $\frac{20}{3}$ times the velocity of the mass at time $t$. Find the position of the mass as a function of time. Discuss the behavior of the motion as $t \to \infty$. [*Hint:* See Exercise 40, Section 2.4.]

Assume that each of the differential equations in Exercises 34 through 43 governs the motion of a certain damped vibrating spring. Classify the resulting motion as being underdamped, critically damped, or overdamped.

**34.** $\ddot{y} + 6\dot{y} + 12y = 0$　　　　　　**35.** $\ddot{y} + 20\dot{y} + 64y = 0$

**36.** $\ddot{y} + 9\dot{y} + 4y = 0$　　　　　　**37.** $5\ddot{y} + 10\dot{y} + 20y = 0$

**38.** $\ddot{y} + 2\dot{y} + y = 0$　　　　　　**39.** $6\ddot{y} + 4\dot{y} + y = 0$

**40.** $\ddot{y} + 5\dot{y} + y = 0$　　　　　　**41.** $\ddot{y} + 8\dot{y} + 16y = 0$

**42.** $4\ddot{y} + 8\dot{y} + 4y = 0$　　　　　　**43.** $\ddot{y} + \dot{y} + y = 0$

**44.** Show[16] that the graphs in the $xy$ plane of all solutions of the system of differential equations

$$\ddot{x} + \dot{y} + 6x = 0, \qquad \ddot{y} - \dot{x} + 6y = 0$$

which satisfy $\dot{x}(0) = \dot{y}(0) = 0$ are hypocycloids. (A hypocycloid is the path described by a fixed point on the circumference of a circle which rolls on the inside of a given fixed circle.) The problem that is proposed here is the following: Verify that the substitution $z = x + iy$ enables one to collapse the given system into the single differential equation

$$\ddot{z} - i\dot{z} + 6z = 0$$

with initial condition $\dot{z}(0) = 0$. Solve this differential equation.

[16]Problem 44 appeared on the 1971 William Lowell Putnam Mathematical Competition, *Amer. Math. Monthly* 80, no. 2 (1973).

**45.** Find[17] all twice differentiable functions $f$ such that for all $x$

$$f'(x) = f(-x).$$

Solve this problem. (*Hint:* Differentiate the expression, use the expression itself, and the result will be a differential equation that you can solve.)

**46.** One can motivate the form of the general solution in the case of equal roots of the characteristic equation in the following manner.[18]

(a) Show that the solution of the IVP

$$y'' - 2ry' + \left(r^2 - \frac{\alpha^2}{4}\right) y = 0$$

$$y(0) = 0, \qquad y'(0) = 1$$

is

$$y_\alpha(x) = \frac{1}{\alpha}\left[e^{(r+\alpha/2)x} - e^{(r-\alpha/2)x}\right].$$

(b) Show that $\lim_{\alpha \to 0} y_\alpha(x) = xe^{rx}$.

**47.** (a) Show that the solution of the IVP

$$y'' - 2(r + \beta)y' + r^2 y = 0$$

$$y(0) = 0, \qquad y'(0) = 1$$

is

$$y_\beta(x) = \frac{1}{2\sqrt{\beta(2r + \beta)}}\left[e^{[r+\beta+\sqrt{\beta(2r+\beta)}]x} - e^{[r+\beta-\sqrt{\beta(2r+\beta)}]x}\right].$$

(b) Show that $\lim_{\beta \to 0} y_\beta(x) = xe^{rx}$.

It is often possible to extract information about solutions to a differential equation without explicitly solving the differential equation. Perhaps one of the best examples of this idea[19] is the trigonometric differential equation $y'' + y = 0$.

*Define* $S(x)$ to be *the* solution of the IVP

$$y'' + y = 0$$

$$S(0) = 0, \qquad S'(0) = 1$$

*Define* $C(x)$ to be *the* solution of the IVP

$$y'' + y = 0$$

$$C(0) = 1, \qquad C'(0) = 0.$$

---

[17]Problem 45 appears in *Math. Magazine* 47, no. 5 (1974).
[18]See R. S. Baslaw and H. M. Hastings, "On the Critically Damped Oscillator," *Math Magazine* 48, no. 2 (1975). For further reference, see R. D. Larsson, "General Solutions of Linear Ordinary Differential Equations," *Amer. Math. Monthly* 65, no. 7 (1958).
[19]The outline presented in Exercises 48–53 is extracted from D. A. Kearns, "An Analytic Approach to Trigonometric Function," *Amer. Math. Monthly* 65, no. 8 (1958). See also G. Birkhoff and G–C. Rota, *Ordinary Differential Equations* (Boston: Ginn & Company, 1962), p. 53, and W. Leighton, *An Introduction to the Theory of Differential Equations* (New York: McGraw-Hill, 1952), p. 41.

**48.** Show that $C(x)$ and $S(x)$ are linearly independent.

**49.** Show that $S'(x) = C(x)$. (*Hint:* Use the differential equation, the initial conditions, and the uniqueness theorem.)

**50.** Show that $C'(x) = -S(x)$.

**51.** Show that $[S(x)]^2 + [C(x)]^2 \equiv 1$. (*Hint:* Multiply $S'' + S = 0$ by $S'$; and integrate.)

**52.** Show that for arbitrary $a$,

$$S(x + a) = S(x)C(a) + S(a)C(x).$$

(*Hint:* First show that $S(x + a)$ is a solution, then use the fact that $S(x)$ and $C(x)$ form a fundamental set.)

**53.** Show that for arbitrary $a$,

$$C(x + a) = C(x)C(a) - S(x)S(x).$$

---

## 2.6   HOMOGENEOUS EQUATIONS WITH VARIABLE COEFFICIENTS— OVERVIEW

Throughout this book we strive to find solutions of differential equations. In Chapter 1 we learned how to solve a few types of differential equations of order 1 (such as separable, linear, exact, and homogeneous). In the last two sections we learned that (subject to some algebraic difficulties only) we can solve any linear homogeneous differential equation with constant coefficients. In Sections 2.11 and 2.12 we will see that the nonhomogeneous linear differential equation with constant coefficients can be handled in a straightforward manner.

Unfortunately, matters are not so simple when we try to solve linear differential equations with variable coefficients of order 2 or higher. In general there is *no way* to solve, *explicitly,* a linear homogeneous differential equation with variable coefficients unless the differential equation is of a very special form; for example, the Euler-type differential equation and differential equations of order 2, for which we know already one solution. The following two sections deal with these two special cases. Once we solve the homogeneous differential equation the nonhomogeneous equation presents no difficulties (at least in theory), as we will see in Sections 2.10, 2.11, and 2.12. Apparently, then, it is difficult if not impossible to solve linear differential equations with variable coefficients. But what do we do if we have such a differential equation and we desire information about its solution(s)? Perhaps the best we can do with differential equations having variable coefficients, at this level, is to try to *approximate* their solutions by assuming that the solutions are power series. The supposition that the solution is a power series is valid only if the coefficient functions $a_i(x)$ satisfy certain differentiability requirements. When these requirements are met, the supposition that the solution is a power series is nothing more than looking at the Taylor-series representation of the solution. A direct substitution of the power-series solution into the differential equation enables us, in many cases, to compute as many coefficients of the power series as we please for our approximation. We have devoted an entire chapter (Chapter 5) to this approach. We include a brief discussion of power series here to round out the treatment within this chapter.

# 2.7 EULER DIFFERENTIAL EQUATION

An Euler differential equation is a differential equation of the form

$$a_n x^n y^{(n)} + a_{n-1} x^{n-1} y^{(n-1)} + \cdots + a_1 x y' + a_0 y = 0, \tag{1}$$

where $a_n, a_{n-1}, \ldots, a_1, a_0$ are constants and $a_n \neq 0$. Since the leading coefficient $a_n x^n$ should never be zero, the interval of definition of the differential equation (1) is either the open interval $(0, \infty)$ or the open interval $(-\infty, 0)$. That is, the differential equation should be solved for either $x > 0$ or $x < 0$. The Euler differential equation is probably the simplest type of linear differential equation with variable coefficients. The reason for this is that the change of independent variable

$$x = \begin{cases} e^t & \text{if} & x > 0 \\ -e^t & \text{if} & x < 0 \end{cases}$$

produces a differential equation with constant coefficients. We illustrate this fact for the second-order case.

**EXAMPLE 1** Show by means of the change of independent variables above that the Euler differential equation of second order,

$$a_2 x^2 y'' + a_1 x y' + a_0 y = 0, \tag{2}$$

with $a_0$, $a_1$, and $a_2$ given constants, is reduced to the differential equation

$$a_2 \ddot{y} + (a_1 - a_2) \dot{y} + a_0 y = 0, \tag{3}$$

where dots denote derivatives with respect to $t$.

**Proof** First, assume that $x > 0$. Then the transformation $x = e^t$ implies that $dx/dt = e^t$, and so $dt/dx = e^{-t}$. By the chain rule for differentiation, we have

$$y' = \frac{dy}{dx} = \frac{dy}{dt} \cdot \frac{dt}{dx} = \dot{y} e^{-t} = \dot{y} \frac{1}{x} \Rightarrow x y' = \dot{y}$$

and

$$y'' = \frac{dy'}{dt} \cdot \frac{dt}{dx} = \frac{d}{dt} (\dot{y} e^{-t}) e^{-t}$$

$$= (\ddot{y} e^{-t} - \dot{y} e^{-t}) e^{-t} = (\ddot{y} - \dot{y}) e^{-2t} = (\ddot{y} - \dot{y}) \frac{1}{x^2} \Rightarrow x^2 y'' = \ddot{y} - \dot{y}.$$

Thus, substituting $xy'$ and $x^2y''$ into the differential equation (2), we obtain

$$a_2x^2y'' + a_1xy' + a_0y = a_2(\ddot{y} - \dot{y}) + a_1\dot{y} + a_0y$$
$$= a_2\ddot{y} + (a_1 - a_2)\dot{y} + a_0y,$$

proving that Eq. (3) holds. Now, if we assume that $x < 0$, we set $x = -e^t$, so $dx/dt = -e^t$ and $dt/dx = -e^{-t}$. Again, by the chain rule we have $y' = -\dot{y}e^{-t}$ and $y'' = (\ddot{y} - \dot{y})e^{-2t}$. Thus,

$$a_2x^2y'' + a_1xy' + a_0y = a_2(-e^t)^2(\ddot{y} - \dot{y})e^{-2t} + a_1(-e^t)(-\dot{y}e^{-t}) + a_0y$$
$$= a_2\ddot{y} + (a_1 - a_2)\dot{y} + a_0y,$$

and Eq. (3) holds once again. The proof is complete.

**EXAMPLE 2**   Solve the Euler differential equation

$$2x^2y'' - 3xy' - 3y = 0,$$

(a) for $x > 0$, and (b) for $x < 0$.

**Solution**   (a) Using the transformation $x = e^t$, the differential equation becomes [see Eq. (3)]

$$2\ddot{y} + (-3 - 2)\dot{y} - 3y = 0$$
$$2\ddot{y} - 5\dot{y} - 3y = 0.$$

The characteristic equation for this latter differential equation is $2\lambda^2 - 5\lambda - 3 = 0$, and so the characteristic roots are $\lambda_1 = -\frac{1}{2}, \lambda_2 = 3$. Thus,

$$y(t) = c_1e^{-(1/2)t} + c_2e^{3t} = c_1(e^t)^{-1/2} + c_2(e^t)^3.$$

But $x = e^t$, so

$$y(x) = c_1x^{-1/2} + c_2x^3.$$

(b) Using the transformation $x = -e^t$, we obtain the same differential equation as in part (a), with the same characteristic roots. Thus,

$$y(t) = c_1e^{-(1/2)t} + c_2e^{3t} = c_1(e^t)^{-1/2} + c_2(e^t)^3.$$

In this case $e^t = -x$, so

$$y(x) = c_1(-x)^{-1/2} + c_2(-x)^3.$$

**EXAMPLE 3**   Discuss the form of the solution to the Euler differential equation of order 2.

**Solution**   From Example 1 we see that the nature of the solutions depends on the roots $\lambda_1$ and $\lambda_2$ of the characteristic equation for the differential equation (3),

$$a_2\lambda^2 + (a_1 - a_2)\lambda + a_0 = 0. \tag{4}$$

Thus, we have

$$y(t) = \begin{cases} c_1 e^{\lambda_1 t} + c_2 e^{\lambda_2 t} & \text{if } \lambda_1, \lambda_2 \text{ real } \lambda_1 \neq \lambda_2 \\ (c_1 + c_2 t)e^{\lambda_1 t} & \text{if } \lambda_1, \lambda_2 \text{ real, } \lambda_1 = \lambda_2 \\ c_1 e^{\alpha t} \cos \beta t + c_2 e^{\alpha t} \sin \beta t & \text{if } \lambda_1 = \alpha + i\beta, \lambda_2 = \alpha - i\beta. \end{cases}$$

Now, for $x > 0$, $e^{\lambda t} = (e^t)^\lambda = x^\lambda$ and $t = \ln x$. Thus, the general solution of the Euler differential equation has the form

$$y(x) = \begin{cases} c_1 x^{\lambda_1} + c_2 x^{\lambda_2} & \text{if } \lambda_1, \lambda_2 \text{ are real, } \lambda_1 \neq \lambda_2 \\ (c_1 + c_2 \ln x)x^{\lambda_1} & \text{if } \lambda_1, \lambda_2 \text{ are real, } \lambda_1 = \lambda_2 \\ c_1 x^\alpha \cos (\beta \ln x) + c_2 x^\alpha \sin (\beta \ln x) & \text{if } \lambda_1 = \alpha + i\beta, \lambda_2 = \alpha - i\beta. \end{cases}$$

If $x < 0$, then $x$ should be replaced by $-x$ in the formula for $y(x)$. For the last expression we used the relations $x^{\alpha + i\beta} = x^\alpha x^{i\beta} = x^\alpha e^{\ln(x^{i\beta})} = x^\alpha e^{i(\beta \ln x)} = x^\alpha[\cos (\beta \ln x) + i \sin (\beta \ln x)]$.

The results of Examples 1, 2, and 3 associated with the Euler differential equation of order 2 prompt us to conclude that in order to solve any Euler differential equation, we try a substitution of the form $y = x^\lambda$, where $\lambda$ is to be a root of a polynomial equation.

**EXAMPLE 4**   Solve the Euler differential equation

$$x^3 y''' - x^2 y'' - 2xy' - 4y = 0 \qquad \text{for } x > 0. \tag{5}$$

**Solution**   We substitute $y = x^\lambda$ in the differential equation (5) to obtain

$$\lambda(\lambda - 1)(\lambda - 2)x^\lambda - \lambda(\lambda - 1)x^\lambda - 2\lambda x^\lambda - 4x^\lambda = 0.$$

Thus,

$$x^\lambda[\lambda(\lambda - 1)(\lambda - 2) - \lambda(\lambda - 1) - 2\lambda - 4] = 0.$$

Since $x^\lambda \neq 0$, we require that

$$\lambda(\lambda - 1)(\lambda - 2) - \lambda(\lambda - 1) - 2\lambda - 4 = 0$$

or

$$\lambda^3 - 4\lambda^2 + \lambda - 4 = 0$$

$$\Rightarrow (\lambda^2 + 1)(\lambda - 4) = 0 \Rightarrow \lambda = 4, \pm i.$$

Therefore, the solution is

$$y = c_1 x^4 + c_2 \cos (\ln x) + c_3 \sin (\ln x).$$

**EXAMPLE 5**   Substitute $y = x^\lambda$ in Eq. (1) to find the characteristic equation for the general $n$th-order Euler differential equation.

**Solution**   Equation (1) is the general $n$th-order Euler differential equation

$$a_n x^n y^{(n)} + a_{n-1} x^{n-1} y^{(n-1)} + \cdots + a_1 xy' + a_0 y = 0$$

Set $y = x^\lambda \Rightarrow y' = \lambda x^{\lambda-1}$, $y'' = \lambda(\lambda - 1) x^{\lambda-2}, \ldots, y^{(n)} = \lambda(\lambda - 1) \cdots$
$(\lambda - n + 1) x^{\lambda-n}$. Substituting these results into the differential equation, we
obtain

$$a_n\lambda(\lambda - 1) \cdots (\lambda - n + 1)x^\lambda + a_{n-1}\lambda(\lambda - 1) \cdots (\lambda - n + 2)x^\lambda$$
$$+ \cdots + a_2\lambda(\lambda - 1)x^\lambda + a_1\lambda x^\lambda + a_0 x^\lambda = 0.$$

Since $x^\lambda \neq 0$ we find

$$a_n\lambda(\lambda - 1) \cdots (\lambda - n + 1) + a_{n-1}\lambda(\lambda - 1) \cdots (\lambda - n + 2)$$
$$+ \cdots + a_2\lambda(\lambda - 1) + a_1\lambda + a_0 = 0. \qquad (6)$$

Equation (6) is the characteristic equation for the general $n$th-order Euler differential equation.


### EXERCISES

Find the general solution of the Euler equations in Exercises 1 through 15.

**1.** $5x^2y'' - 3xy' + 3y = 0$, $x > 0$

**2.** $x^2y'' - xy' + y = 0$, $x > 0$

**3.** $3x^2y'' + 4xy' + y = 0$, $x < 0$

**4.** $x^2y'' + 3xy' + 2y = 0$, $x < 0$

**5.** $x^2y'' + 4xy' - 4y = 0$, $x < 0$

**6.** $(x - 1)^2y'' + 5(x - 1)y' + 4y = 0$, $x - 1 > 0$ [*Hint:* Try $y = (x - 1)^\lambda$.]

**7.** $2x^2y'' + xy' - 3y = 0$, $x < 0$

**8.** $x^2y'' + 3xy' + 5y = 0$, $x > 0$

**9.** $x^3y''' + 3x^2y'' - 2xy' + 2y = 0$, $x > 0$

**10.** $x^2y'' + \frac{7}{2}xy' - \frac{3}{2}y = 0$, $x > 0$

**11.** $(x + 3)^2y'' + 3(x + 3)y' + 5y = 0$, $x + 3 < 0$

**12.** $(x - 2)^2y'' - (x - 2)y' + y = 0$, $x > 2$

**13.** $x^2y'' - 6y = 0$, $x < 0$

**14.** $x^3y''' + 4x^2y'' - 8xy' + 8y = 0$, $x < 0$

**15.** $x^3y''' + 4x^2y'' - 8xy' + 8y = 0$, $x > 0$

Solve the initial value problems in Exercises 16 through 20.

**16.** $x^2y'' - xy' + y = 0$, $y(1) = 1$, $y'(1) = 0$

**17.** $x^2y'' - xy' + y = 0$, $y(-1) = 1$, $y'(-1) = 0$

**18.** $x^2y'' + 3xy' + 2y = 0$, $y(1) = 0$, $y'(1) = 1$

**19.** $x^2y'' + \frac{7}{2}xy' - \frac{3}{2}y = 0$, $y(-4) = 1$, $y'(-4) = 0$

**20.** $x^3y''' + 4x^2y'' - 8xy' + 8y = 0$, $y(1) = 0$, $y'(1) = 1$, $y''(1) = 0$

**21.** Just as in the case of linear differential equations with constant coefficients (see Exercises 46 and 47 of Section 2.5), the form of the solution in the case of equal characteristic roots can be obtained via a limit process.[20]
(a) If $\lambda_1$ and $\lambda_2$ are roots of the characteristic equation (4), then

$$y(x) = \frac{x^{\lambda_1} - x^{\lambda_2}}{\lambda_1 - \lambda_2}$$

is a solution of Eq. (2). (Why?)
(b) Show that

$$\lim_{\lambda_1 \to \lambda_2} y(x) = x^{\lambda_2} \ln x.$$

**22. Astronomy** Kopal[21] obtained a differential equation of the form

$$r\frac{d^2\psi}{dr^2} + 4\frac{d\psi}{dr} = 0, \quad r > 0.$$

Solve the differential equation. [*Hint:* Multiplication of the differential equation by $r$ produces an Euler differential equation.]

**23. Flows** Barton and Raynor[22] in their study of peristaltic flow in tubes, obtained the following linear homogeneous differential equation with variable coefficients:

$$\frac{d^2p}{dr^2} + \frac{1}{r}\frac{dp}{dr} = 0, \quad r > 0.$$

Solve this differential equation. [*Hint:* Multiplication by $r^2$ produces an Euler differential equation.]

In Exercises 24 through 32, write the characteristic equation associated with the Euler differential equation. Find the characteristic roots and obtain the general solution.

**24.** $3x^3y''' - x^2y'' + 4xy' - 4y = 0$, $x > 0$

**25.** $x^4y^{iv} - 5x^3y''' + 3x^2y'' - 6xy' + 6y = 0$, $x > 0$

**26.** $2x^4y^{iv} + 3x^3y''' - 4x^2y'' + 8xy' - 8y = 0$, $x < 0$

**27.** $2x^3y''' + x^2y'' - 12xy' - 2y = 0$, $x < 0$

**28.** $x^5y^v - 2x^3y''' + 4x^2y'' = 0$, $x < 0$

**29.** $7x^4y^{iv} - 2x^3y''' + 3x^2y'' - 6xy' + 6y = 0$, $x > 0$

**30.** $x^5y^v + 3x^3y''' - 9x^2y'' + 18xy' - 18y = 0$, $x > 0$

**31.** $x^6y^{vi} - 12x^4y^{iv} = 0$, $x > 0$

---

[20]See B. Schweizer, "On the Euler-Cauchy Equation," *Amer. Math. Monthly* 68, no. 6 (1961).
[21]Kopal, "Stress History of the Moon." See footnote 8.
[22]Barton and Raynor, *Bull. Math. Biophys.* See footnote 7.

**32.** $x^3y''' + 3x^2y'' - 2xy' - 2y = 0, x > 0$

Solve each of the differential equations in Exercises 33 through 40.

**33.** $y'' - \dfrac{5}{x}y' + \dfrac{5}{x^2}y = 0, x > 0$

**34.** $xy''' - \dfrac{6}{x^2}y = 0, x < 0$

**35.** $x^2y^{iv} - xy''' = 0, x < 0$

**36.** $3xy'' - 4y' + \dfrac{5}{x}y = 0, x > 0$

**37.** $(x - 4)y'' + 4y' - \dfrac{4}{x - 4}y = 0, x > 4$

**38.** $(x + 2)y'' - y' + \dfrac{1}{x + 2}y = 0, x + 2 > 0$

**39.** $y'' + \dfrac{5}{x - 1}y' + \dfrac{4}{(x - 1)^2}y = 0, x < 1$

**40.** $5y'' - \dfrac{3}{x - 3}y' + \dfrac{3}{(x - 3)^2}y = 0, x < 3$

---

## 2.8   REDUCTION OF ORDER

---

As the terminology implies, *reduction of order* is a device whereby the problem of solving a (linear) differential equation of certain order is replaced by a problem of solving a (linear) differential equation of lower order. The procedure associated with reduction of order is somewhat systematic and hinges upon our having specific knowledge of at least one solution to the original differential equation. Perhaps the most interesting application of the reduction of order is that *we can find a second linearly independent solution of a differential equation of order 2 with variable coefficients provided that we know one nontrivial (not identically zero) solution of it.* The method is embodied in the following theorem.

### THEOREM 1

*If $y_1$ is a known solution of the linear differential equation*

$$a_n(x)y^{(n)} + a_{n-1}(x)y^{(n-1)} + \cdots + a_1(x)y' + a_0(x)y = 0, \tag{1}$$

*which has the property that it is never zero in the interval of definition of the differential equation, then the change of dependent variable*

$$y = y_1u \tag{2}$$

*produces a linear differential equation of order $n - 1$ for $u'$.*

**Proof**   As we have done many times before, we consider a special case rather than confuse the reader with the cumbersome symbolism that the general case requires. Thus, we restrict our attention to the case $n = 2$. Now

$$y = y_1 u$$

$$y' = y_1 u' + y_1' u$$

$$y'' = y_1 u'' + 2y_1' u' + y_1'' u.$$

Substituting these results into Eq. (1) (when $n = 2$) yields

$$a_2(x)[y_1 u'' + 2y_1' u' + y_1'' u] + a_1(x)[y_1 u' + y_1' u] + a_0(x)[y_1 u] = 0$$

$$[a_2(x)y_1]u'' + [2a_2(x)y_1' + a_1(x)y_1]u' + [a_2(x)y_1'' + a_1(x)y_1' + a_0(x)y_1]u = 0$$

Since $y_1$ is a solution of the differential equation it follows that

$$a_2(x)y_1'' + a_1(x)y_1' + a_0(x)y_1 = 0,$$

and so

$$[a_2(x)y_1]u'' + [2a_2(x)y_1' + a_1(x)y_1]u' = 0. \tag{3}$$

Setting $u' = v$, Eq. (3) becomes a first-order linear differential equation which we can solve by the method shown in Section 1.4.

**EXAMPLE 1**   Given that $y_1 = x \sin x$ is a solution of the differential equation

$$xy'' - 2y' + \frac{x^2 + 2}{x} y = 0, \tag{4}$$

for $0 < x < \pi$, find another (linearly independent) solution.

**Solution**   We leave it to the reader to verify that $y_1$ is a solution of the differential equation (4) and proceed directly to Eq. (3). Thus,

$$[x(x \sin x)]u'' + [2x(x \cos x + \sin x) - 2(x \sin x)]u' = 0$$

$$\Rightarrow \quad (x^2 \sin x)u'' + (2x^2 \cos x)u' = 0$$

Setting $u' = v$ we obtain a first-order linear differential equation, which therefore we can solve explicitly:

$$v = e^{-2 \ln |\sin x|} = \csc^2 x.$$

Hence, $u' = v = \csc^2 x \Rightarrow u = -\cot x$, and so a second solution is

$$y_2 = y_1 u = x \sin x(-\cot x) = -x \cos x.$$

Furthermore, $W(y_1, y_2; x) = x^2 \neq 0$, since $0 < x < \pi$, and so $y_1$ and $y_2$ are linearly independent. Therefore, we conclude that the general solution of Eq. (4) is

$$y = c_1 y_1 + c_2 y_2$$
$$= x(c_1 \sin x - c_2 \cos x).$$

**EXAMPLE 2**   Apply the reduction-of-order technique to the differential equation

$$y''' - 4y'' + 5y' - 2y = 0, \tag{5}$$

given that $y_1 = e^x$ is a solution.

**Solution**   As in Example 1, we leave it to the reader to verify that $y_1$ is a solution of the differential equation (5). Now

$$y = y_1 u = e^x u$$

$$\Rightarrow \qquad y' = e^x u + e^x u'$$

$$\Rightarrow \qquad y'' = e^x u + 2e^x u' + e^x u''$$

$$\Rightarrow \qquad y''' = e^x u + 3e^x u' + 3e^x u'' + e^x u'''.$$

Substitution of these results into Eq. (5) yields (after simplification)

$$e^x u''' - e^x u'' = 0,$$

and setting $v = u'$ we find

$$e^x v'' - e^x v' = 0. \tag{6}$$

Equation (6) has variable coefficients. However, since $e^x$ is never zero, the differential equation is equivalent to $v'' - v' = 0$. This latter differential equation has $v_1 = 1$ and $v_2 = e^x$ as its solutions. Since $u' = v$, it follows that $u_1 = x$ and $u_2 = e^x$. Finally, $y_2 = u_1 y_1 = xe^x$ and $y_3 = u_2 y_1 = e^{2x}$. The general solution to Eq. (5) is

$$y = c_1 y_1 + c_2 y_2 + c_3 y_3$$
$$= c_1 e^x + c_2 xe^x + c_3 e^{2x}.$$

**REMARK 1**   Equation (5) has constant coefficients and can be solved by the method of Section 2.5. We solved it by reduction of order in order to illustrate that method with a relatively simple example.

**EXAMPLE 3**   Use the reduction-of-order method to find the general solution of the differential equation

$$(\tfrac{4}{3}x^3 + 1) y''' - 4x^2 y'' + 8xy' - 8y = 0, \qquad x > 0,$$

given that $y_1 = x$ is a solution.

**Solution**   Many of the details of the solution are left as exercises (see Exercises 30 through 34). Substituting $y_1 = x$, $y_1' = 1$, $y_1'' = y_1''' = 0$ into the differential equation yields

$$8x(1) - 8x = 0.$$

Thus, $y_1$ is indeed a solution. Setting $y = uy_1$ into the differential equation leads to the following differential equation for $u$:

$$[(\tfrac{4}{3}x^3 + 1)y_1]u''' + [3(\tfrac{4}{3}x^3 + 1)y_1' - 4x^2y_1]u''$$
$$+ [3(\tfrac{4}{3}x^3 + 1)y_1'' - 8x^2y_1' + 8xy_1]u'$$
$$+ [(\tfrac{4}{3}x^3 + 1)y_1''' - 4x^2y_1'' + 8xy_1' - 8y_1]u = 0. \qquad (7)$$

Since $y_1$ is a solution of the differential equation, the coefficient of $u$ is zero. Set $u' = v$; then the differential equation (7) takes the form

$$[(\tfrac{4}{3}x^3 + 1)y_1]v'' + [3(\tfrac{4}{3}x^3 + 1)y_1' - 4x^2y_1]v'$$
$$+ [3(\tfrac{4}{3}x^3 + 1)y_1'' - 8x^2y_1' + 8xy_1]v = 0.$$

Substituting $y_1 = x$, $y_1' = 1$, $y_1'' = 0$, we have

$$[(\tfrac{4}{3}x^3 + 1)x]v'' + [3(\tfrac{4}{3}x^3 + 1) - 4x^3]v' = 0. \qquad (8)$$

This differential equation has the solutions

$$v_1 = \tfrac{4}{3}x - \tfrac{1}{2}x^{-2} \qquad (9)$$

$$v_2 = 1. \qquad (10)$$

These two solutions give rise to

$$y_2 = u_1y_1 = \tfrac{2}{3}x^3 + \tfrac{1}{2} \qquad (11)$$

and

$$y_3 = u_2y_1 = x^2. \qquad (12)$$

Now

$$W(y_1, y_2, y_3; x) = -\tfrac{4}{3}x^3 - 1 \neq 0, \qquad (13)$$

because $x > 0$. Thus,

$$y = c_1x + c_2(\tfrac{2}{3}x^3 + \tfrac{1}{2}) + c_3x^2.$$

The substitution (2) always reduces the general differential equation (1) to an equation of lower order. In the special case $n = 2$, we can also use another approach. From the proof of Theorem 3 of Section 2.2 [Eq. (8)], we have the relation

$$W(y_1, y_2; x) = C \exp\left[-\int_{x_0}^{x} p(s)\,ds\right],$$

where $p(x) = a_1(x)/a_2(x)$. That is,

$$y_1 y_2' - y_1' y_2 = C \exp\left[-\int_{x_0}^x p(s)ds\right]$$

$$\Rightarrow \frac{y_1 y_2' - y_1' y_2}{y_1^2} = \frac{C \exp\left[-\int_{x_0}^x p(s)ds\right]}{y_1^2}$$

$$\Rightarrow \frac{d}{dx}\left(\frac{y_2}{y_1}\right) = \frac{C}{y_1^2} \exp\left[-\int_{x_0}^x p(s)ds\right]$$

$$\Rightarrow \frac{y_2}{y_1} = C\int_{x_0}^x \frac{\exp\left[-\int_{x_0}^t p(s)ds\right]}{[y_1(t)]^2} dt$$

$$\Rightarrow y_2 = Cy_1 \int_{x_0}^x \frac{\exp\left[-\int_{x_0}^t p(s)ds\right]}{[y_1(t)]^2} dt. \tag{14}$$

Equation (14) is called *Abel's formula* and gives a second linearly independent solution of a second-order linear differential equation when one solution is known. Actually, the constant $C$ in Eq. (14) can be omitted since in the general solution we consider $c_1 y_1 + c_2 y_2$ anyway. Also, the evaluation of the integral at $x_0$ can be neglected for the same reason.

**EXAMPLE 4**   Using Abel's formula, find a second linearly independent solution of the differential equation

$$xy'' - 2y' + \frac{x^2 + 2}{x}y = 0, \qquad 0 < x < \pi,$$

given that $y_1 = x \sin x$ is a solution.

**Solution**   Observe that $a_2(x) = x$, $a_1(x) = -2$ and hence that $p(x) = -2/x$. Substituting into Abel's formula, Eq. (14), we obtain

$$y_2 = (x \sin x) \int \frac{\exp\left[-\int^t -2/s\, ds\right]}{(t \sin t)^2} dt = x \sin x \int^x \frac{e^{2 \ln t}}{t^2 \sin^2 t} dt$$

$$= x \sin x \int^x \frac{t^2}{t^2 \sin^2 t} dt = x \sin x \int^x \csc^2 t\, dt = x \sin x(-\cot x)$$

$$= -x \cos x.$$

## APPLICATIONS 2.8.1

Second-order linear ordinary differential equations with variable coefficients occur rather frequently in theoretical physics. They also play an indirect role in such mathematical studies as approximation theory and probability theory. Most often at least one of the linearly independent solutions (very often both) of these equations is an infinite series. It is not uncommon in these circumstances to find one solution and then express the second solution in integral form by means of Abel's formula. How one obtains the first solution is another question altogether. A partial answer to this question is presented in Section 2.9. A more elaborate and detailed treatment of the question is presented in Chapter 5.

■ In theoretical physics a basic equation of quantum mechanics is the Schrödinger wave equation

**Quantum Mechanics**

$$\frac{ih}{2\pi}\frac{\partial \phi}{\partial t} = -\frac{h^2}{8m\pi^2}\left(\frac{\partial^2 \phi}{\partial x^2} + \frac{\partial^2 \phi}{\partial y^2} + \frac{\partial^2 \phi}{\partial z^2}\right) + V(x, y, z)\phi,$$

where $h$ is a constant known as Planck's constant, $V$ is the potential energy, and $m$ is the mass of the particle whose wave function is $\phi$ ($\phi$ is related to the probability that the particle will be found in a differential element of volume at any particular time). The search for acceptable solutions of Schrödinger's equation for the special case of a harmonic oscillator leads to the differential equation

$$y'' - 2xy' + 2py = 0, \tag{15}$$

where $p$ is an arbitrary constant. Equation (15), known as *Hermite's equation*, will be studied in detail in Section 5.4.1.

**EXAMPLE 5** Let $p = 2$ in Eq. (15). Given that $y_1 = 4x^2 - 2$ is a solution of Eq. (15) (with $p = 2$), use Abel's formula to obtain a second linearly independent solution.

**Solution** We leave it to the reader to verify that $y_1 = 4x^2 - 2$ is a solution of $y'' - 2xy' + 4y = 0$. Applying Abel's formula [Eq. (14)] we have

$$y_2 = (4x^2 - 2)\int^x \frac{\exp[-\int'(-2s)\,ds]}{(4t^2 - 2)^2}dt = (4x^2 - 2)\int^x \frac{e^{t^2}}{(4t^2 - 2)^2}dt.$$

Since the integral cannot be evaluated in terms of elementary functions, we content ourselves with having $y_2$ in integral form.

■ Many theoretical studies in physics utilize *Laplace's equation*,

**Mathematical Physics**

$$\frac{\partial^2 u}{\partial x^2} + \frac{\partial^2 u}{\partial y^2} + \frac{\partial^2 u}{\partial z^2} = 0.$$

There are various techniques associated with the problem of obtaining solutions to this equation, and there are a number of considerations that govern which technique is most efficient for a given situation. It is not our intention to discuss these considerations but rather to consider a special instance.

If Laplace's equation is expressed in terms of spherical coordinates and then solved by the method of separation of variables for partial differential equations, the result is *Legendre's differential equation* (for details, see Section 5.4.1),

$$(1 - x^2)y'' - 2xy' + p(p + 1)y = 0, \qquad x < 1, \tag{16}$$

where $p$ is a constant.

**EXAMPLE 6**  If $p = 2$, then $y_1 = \frac{3}{2}x^2 - \frac{1}{2}$ is a solution of Legendre's differential equation. Find an integral form for a second linearly independent solution.

**Solution**  We leave it to the reader to verify that $y_1$ is a solution of the differential equation $(1 - x^2)y'' - 2xy' + 6y = 0$. Applying Abel's formula, we have

$$y_2 = (\tfrac{3}{2}x^2 - \tfrac{1}{2}) \int^x \frac{\exp\left[-\int^t \frac{-2s}{(1 - s^2)}\,ds\right]}{(\tfrac{3}{2}t^2 - \tfrac{1}{2})^2}\,dt = (\tfrac{3}{2}x^2 - \tfrac{1}{2}) \int^x \frac{e^{-\ln(1 - t^2)}}{(\tfrac{3}{2}t^2 - \tfrac{1}{2})^2}\,dt$$

$$= (\tfrac{3}{2}x^2 - \tfrac{1}{2}) \int^x \frac{dt}{(1 - t^2)(\tfrac{3}{2}t^2 - \tfrac{1}{2})^2}. \tag{17}$$

The integral can be evaluated in this case (partial fractions) and is left as an exercise (Exercise 35).

### EXERCISES

In Exercises 1 through 12, a differential equation and one of its solutions is given. Apply the reduction-of-order method (as in Examples 1 and 2) to obtain another linearly independent solution.

**1.** $x^2y'' + xy' - y = 0, x > 0, y_1 = x$

**2.** $x^2y'' + (2x^2 - x)y' - 2xy = 0, x > 0, y_1 = e^{-2x}$

**3.** $x^3y'' + (5x^3 - x^2)y' + 2(3x^3 - x^2)y = 0, x > 0, y_1 = e^{-2x}$

**4.** $x^4y'' + 2x^3y' - y = 0, x > 0, y_1 = e^{1/x}$

**5.** $x^2y'' + x^2y' - (x + 2)y = 0, x > 0, y_1 = x^{-1}e^{-x}$

**6.** $xy'' + (x - 1)y' + (3 - 12x)y = 0, x > 0, y_1 = e^{3x}$

**7.** $x^2(1 - \ln x)y'' + xy' - y = 0, x > e, y_1 = x$

**8.** $y'' + \dfrac{2}{x}y' + \dfrac{9}{x^4}y = 0, x > 0, y_1 = \cos\dfrac{3}{x}$

**9.** $x^2y'' + xy' + (x^2 - \tfrac{1}{4})y = 0, x > 0, y_1 = x^{-1/2}\sin x$

**10.** $(1 - x^2)y'' - 7xy' + 7y = 0, x > 1, y_1 = x$

**11.** $y'' + x^{-1/3}y' + (\tfrac{1}{4}x^{-2/3} - \tfrac{1}{6}x^{-4/3} - 6x^{-2})y = 0, x > 0, y_1 = x^3 e^{-(3/4)x^{2/3}}$

**12.** $x(x - 2)y'' - 2(x^2 - 3x + 3)y' + (x^2 - 4x + 6)y = 0, x > 2,$
$y_1 = e^x(x^{-1} - x^{-2})$

In Exercises 13 through 19, show that $y_1$ is a solution of the differential equation. Use Abel's formula to obtain an expression for a second linearly independent solution.

**13.** $x(1 - 3x \ln x)y'' + (1 + 9x^2 \ln x)y' - (3 + 9x)y = 0$, $x > e$, $y_1 = e^{3x}$

**14.** $y'' - \left(1 + \dfrac{3}{2x}\right)y' + \dfrac{3}{2x^2}y = 0$, $x > 0$, $y_1 = x^{3/2}e^x$

**15.** $x(1 - 2x \ln x)y'' + (1 + 4x^2 \ln x)y' - (2 + 4x)y = 0$, $x > e$, $y_1 = \ln x$

**16.** $y'' - 2xy' + 2y = 0$ (Hermite $p = 1$), $y_1 = 2x$

**17.** $y'' - 2xy' + 6y = 0$ (Hermite $p = 3$), $y_1 = 8x^3 - 12x$

**18.** $(1 - x^2)y'' - 2xy' + 2y = 0$ (Legendre $p = 1$), $x < 1$, $y_1 = x$

**19.** $(1 - x^2)y'' - 2xy' + 12y = 0$ (Legendre $p = 3$), $|x| < 1$, $y_1 = \frac{5}{2}x^3 - \frac{3}{2}x$

**Quantum Mechanics**  The study of solutions to the Schrödinger wave equation for the hydrogen atom leads to *Laguerre's differential equation*,

$$xy'' + (1 - x)y' + py = 0, \qquad x > 0,$$

where $p$ is a constant. In Exercises 20, 21, and 22, verify that $y_1$ as given is a solution of Laguerre's differential equation for the given $p$. Using Abel's formula, obtain an integral form for a second linearly independent solution.

**20.** $p = 1$, $y_1 = 1 - x$ \qquad\qquad **21.** $p = 2$, $y_1 = 2 - 4x + x^2$

**22.** $p = 3$, $y_1 = 6 - 18x + 9x^2 - x^3$

**Approximation Theory**  One aspect of approximation theory (from an oversimplified point of view) is the problem of approximating a nonpolynomial function by a polynomial. An important class of polynomials for this purpose is the class of Chebyshev polynomials. *Chebyshev polynomials* are solutions of the differential equation (called Chebyshev's differential equation)

$$(1 - x^2)y'' - xy' + p^2y = 0, \qquad |x| < 1,$$

in the special case that $p$ is a nonnegative integer. In Exercises 23, 24, and 25, verify that $y_1$ is a solution of Chebyshev's differential equation for the given $p$. In each case use Abel's formula to obtain in integral form a second linearly independent solution.

**23.** $p = 1$, $y_1 = x$ \quad **24.** $p = 2$, $y_1 = 2x^2 - 1$ \quad **25.** $p = 3$, $y_1 = 4x^3 - 3x$

**26.** Find the general solution of the differential equation

$$(x^2 + 1)y'' - 2xy' + 2y = 0,$$

given that $y_1(x) = -x$ is a solution.

**27.** Using Abel's formula, find a second linearly independent solution of the differential equation

$$xy'' + (1 - 2x)y' + (x - 1)y = 0 \qquad \text{for} \qquad x > 0,$$

given that $y_1(x) = e^x$ is a solution.

**28.** Solve the IVP

$$xy'' + (1 - 2x)y' + (x - 1)y = 0$$
$$y(1) = 2e$$
$$y'(1) = -3e,$$

given that $y_1(x) = e^x$ is a solution of the differential equation.

**29.** Show that $y_1 = xe^x$ is a solution of the differential equation

$$y''' - 4y'' + 5y' - 2y = 0.$$

Show that the reduction-of-order method leads to the differential equation

$$xv'' + (3 - x)v' - 2v = 0.$$

**30.** Verify Eq. (7) of this section.

**31.** Verify Eq. (8) of this section.

**32.** Set $w = v'$ in Eq. (8) to obtain

$$[(\tfrac{4}{3}x^3 + 1)x]w' + [3(\tfrac{4}{3}x^3 + 1) - 4x^3]w = 0.$$

Solve this differential equation by separation of variables and then integrate the result to obtain Eq. (9).

**33.** Verify that $v_2 = 1$ is a solution of the differential equation (8).

**34.** Verify Eq. (13) of this section.

**35.** Perform the integrations required to simplify Eq. (17).

**36.** Let $a_2 y'' + a_1 y' + a_0 y = 0$ be a differential equation with constant coefficients. Assume that $y_1$ is a nontrivial solution of the differential equation. Show that the reduction-of-order method leads to a (first-order) differential equation with constant coefficients if and only if $y_1'/y_1$ is equal to a constant.

---

## 2.9   SOLUTIONS OF LINEAR HOMOGENEOUS DIFFERENTIAL EQUATIONS BY THE METHOD OF TAYLOR SERIES

---

In this section we provide a brief introduction to the method of finding Taylor-series representations for solutions of differential equations. This method is perhaps the most widely applicable technique for solving (or approximating the solutions of) differential equations with variable coefficients. There are many ramifications associated with the Taylor-series technique. Discussion of these ramifications is postponed to Chapter 5, where the method is discussed in detail. We include the section here simply to round out our treatment of linear differential equations.

We reiterate the assumption made earlier in this chapter that all the coefficient functions are continuous and that the leading coefficient function $a_n(x)$ does not vanish in the interval of definition of the differential equation.

Consider the initial value problem

$$y'' + xy = 0 \tag{1}$$
$$y(0) = 1, \qquad y'(0) = 0.$$

Theorem 1 of Section 2.3 guarantees us that this IVP possesses a unique solution. None of the methods that we have discussed can be applied to this differential equation. First, recall from calculus that a Taylor-series representation for a function $f$ is of the form

$$f(x) = f(x_0) + f'(x_0)(x - x_0) + \frac{f''(x_0)}{2!}(x - x_0)^2 + \frac{f'''(x_0)}{3!}(x - x_0)^3 + \cdots$$

or, compactly,

$$f(x) = \sum_{n=0}^{\infty} \frac{f^{(n)}(x_0)}{n!}(x - x_0)^n.$$

We will seek a Taylor expansion for the function $y$ *defined* by the IVP (1). There are essentially two approaches that one can follow, and we illustrate each approach via an example.

**EXAMPLE 1**   Using a Taylor-series approach, solve the IVP (1).

**Solution**   The Taylor (Maclaurin) expansion has the form

$$y(x) = y(0) + y'(0)x + \frac{y''(0)}{2!}x^2 + \frac{y'''(0)}{3!}x^3 + \cdots \tag{2}$$

Now, $y(0) = 1$, $y'(0) = 0$, and from (1) we have

$$y'' = -xy \tag{3}$$
$$\Rightarrow \quad y''(0) = -(0)\cdot(1) = 0.$$

Differentiating (3), we obtain

$$y''' = -y - xy' \tag{4}$$
$$\Rightarrow \quad y'''(0) = -1 - (0)(0) = -1.$$

Differentiating (4), we have

$$y^{\mathrm{iv}} = -2y' - xy''$$
$$\Rightarrow \quad y^{\mathrm{iv}}(0) = 0.$$

Similarly,

$$y^{(5)} = -3y'' - xy''' \quad \Rightarrow \quad y^{(5)}(0) = 0$$
$$y^{(6)} = -4y''' - xy^{\mathrm{iv}} \quad \Rightarrow \quad y^{(6)}(0) = 4.$$

We could proceed in this fashion to generate as many terms of the Taylor expansion as we want. (In general, we cannot find them all in this manner, since there are an infinity of terms; therefore, the number of terms to be computed would have to be prescribed at the onset.) Substituting our results into Eq. (2) yields

$$y = 1 - \frac{1}{3!}x^3 + \frac{4}{6!}x^6 + \cdots .$$

**EXAMPLE 2**  Using power series, solve the differential equation

$$(2 - x)y'' + y = 0. \tag{5}$$

**Solution**  In this example we are not given the initial values; therefore, we cannot compute the coefficients in the Taylor expansion in the manner of Example 1. Here we assume that the solution $y$ of Eq. (5) has an expansion of the form

$$y = a_0 + a_1x + a_2x^2 + a_3x^3 + a_4x^4 + a_5x^5 + \cdots , \tag{6}$$

where the coefficients $a_i$ are constants to be determined. From (6), we obtain

$$y' = a_1 + 2a_2x + 3a_3x^2 + 4a_4x^3 + 5a_5x^4 + \cdots$$

and

$$y'' = 2a_2 + 6a_3x + 12a_4x^2 + 20a_5x^3 + \cdots .$$

Substituting these results in the differential equation (5) yields

$$(2 - x)(2a_2 + 6a_3x + 12a_4x^2 + 20a_5x^3 + \cdots)$$
$$+ (a_0 + a_1x + a_2x^2 + a_3x^3 + a_4x^4 + a_5x^5 + \cdots) = 0.$$

Doing the multiplication and gathering together terms with like powers of $x$, we have

$$(4a_2 + a_0) + (12a_3 - 2a_2 + a_1)x + (24a_4 - 6a_3 + a_2)x^2$$
$$+ (40a_5 - 12a_4 + a_3)x^3 + \cdots = 0. \tag{7}$$

In order that Eq. (7) hold *for every x* in the interval of definition of the differential equation, we must have that the coefficient of each power of $x$ must vanish. Thus, we are led to the *infinite* system of equations

$$4a_2 + a_0 = 0$$

$$12a_3 - 2a_2 + a_1 = 0$$

$$24a_4 - 6a_3 + a_2 = 0 \tag{8}$$

$$40a_5 - 12a_4 + a_3 = 0,$$

and so on. The first of Eqs. (8) yields $a_2 = -\frac{1}{4}a_0$ and the second yields $a_3 = \frac{1}{6}a_2 - \frac{1}{12}a_1$, which in light of the result for $a_2$ reduces to $a_3 = -\frac{1}{24}a_0 - \frac{1}{12}a_1$.

Analogous manipulations apply to each subsequent equation, and we are led to the results

$$a_2 = -\tfrac{1}{4}a_0$$

$$a_3 = -\tfrac{1}{24}a_0 - \tfrac{1}{12}a_1 \tag{9}$$

$$a_4 = -\tfrac{1}{48}a_1$$

$$a_5 = \tfrac{1}{960}a_0 - \tfrac{1}{240}a_1,$$

and so on. Thus,

$$
\begin{aligned}
y &= a_0 + a_1 x + \left(-\tfrac{1}{4}a_0\right)x^2 + \left(-\tfrac{1}{24}a_0 - \tfrac{1}{12}a_1\right)x^3 + \left(-\tfrac{1}{48}a_1\right)x^4 \\
&\quad + \left(\tfrac{1}{960}a_0 - \tfrac{1}{240}a_1\right)x^5 + \cdots \\
&= a_0\left(1 - \tfrac{1}{4}x^2 - \tfrac{1}{24}x^3 + \tfrac{1}{960}x^5 + \cdots\right) \\
&\quad + a_1\left(x - \tfrac{1}{12}x^3 - \tfrac{1}{48}x^4 - \tfrac{1}{240}x^5 + \cdots\right). \tag{10}
\end{aligned}
$$

We note that the solution (10) depends on two arbitrary constants, $a_0$ and $a_1$, as it should, since the differential equation (5) is of second order. We also conjecture that the parenthetical expressions in Eq. (10) are the two linearly independent solutions to the differential equation (5). Furthermore, these expressions are infinite series and therefore their convergence must be considered. That is, do the expressions in parentheses in Eq. (10) converge for every $x$ in the interval of definition of the differential equation? These conjectures and observations will be discussed in Chapter 5.

Example 2 illustrates the Taylor-series method most often used. We assume a solution of the form $y = \sum_{n=0}^{\infty} a_n(x - x_0)^n$. We substitute this expression into the differential equation and then rearrange terms so that like powers of $(x - x_0)$ can be combined. The next step is to set the coefficient of each power of $(x - x_0)$ equal to zero, thus obtaining an infinite system of equations [for example, Eqs. (8)]. Finally, we obtain the coefficients $a_n$ from this infinite system.

## EXERCISES

In Exercises 1 through 10, use the Taylor-series method of Example 1 to solve the initial value problem. In each exercise, compute the coefficients out to the fifth power of the $x$ term.

**1.** $y'' + xy' + y = 0$, $y(0) = 1$, $y'(0) = 0$

**2.** $y''' - (\sin x)y = 0$, $y(0) = 0$, $y'(0) = 1$, $y''(0) = 0$

**3.** $y'' + y' + e^x y = 0$, $y(0) = 2$, $y'(0) = 1$

**4.** $y^{iv} - \ln(1 + x)y = 0$, $y(0) = 1$, $y'(0) = 1$, $y''(0) = 0$, $y'''(0) = 0$

**5.** $y'' + (3 + x)y = 0$, $y(0) = 1$, $y'(0) = 0$

**6.** $y'' + x^2 y = 0$, $y(0) = 0$, $y'(0) = 2$

**7.** $y'' - 2xy' + y = 0$, $y(0) = 2$, $y'(0) = 1$

**8.** $y''' - 3x^2y' + 2xy = 0$, $y(0) = 1$, $y'(0) = 1$, $y''(0) = 0$

**9.** $y''' + 4y'' + 2y' - x^3y = 0$, $y(0) = 1$, $y'(0) = 1$, $y''(0) = 0$

**10.** $y'' + (\cos x)y = 0$, $y(0) = 0$, $y'(0) = 1$

In Exercises 11 through 20, use the Taylor-series method of Example 2 to solve the differential equation. In each exercise, compute the coefficients out to the fifth power of the $x$ term.

**11.** $y'' + y = 0$                           **12.** $y'' + xy = 0$

**13.** $y'' + x^2y' + xy = 0$                  **14.** $y''' + y = 0$

**15.** $(x^2 + 2)y'' - 3xy' + 4y = 0$          **16.** $y'' + x^2y = 0$

**17.** $(2x^2 + 1)y'' + 3y' - xy = 0$          **18.** $y' + 3xy = 0$

**19.** $y''' - 2xy = 0$                        **20.** $(x + 1)y'' + 3y' - 2x^2y = 0$

**21. Electronics** The differential equation

$$(1 + a \cos 2x)y'' + \lambda y = 0, \qquad a \neq -1,$$

has been studied by Campi.[23] The constants $a$ and $\lambda$ are physical parameters. Recall that

$$\cos t = 1 - \frac{t^2}{2!} + \frac{t^4}{4!} - \cdots,$$

and find $a_2$, $a_3$, $a_4$, and $a_5$ in terms of $a$, $\lambda$, $a_0$, and $a_1$ if the solution $y$ is taken in the form

$$y = a_0 + a_1x + a_2x^2 + a_3x^3 + a_4x^4 + a_5x^5 + \cdots.$$

---

## 2.10  NONHOMOGENEOUS DIFFERENTIAL EQUATIONS

---

In the analysis of the deflection of a beam one encounters the differential equation $EIy^{(4)} = f(x)$ (see Exercises 59–64, Section 2.11), where $E$ and $I$ are physical constants and $f$ is a given function of the independent variable. This equation, although linear, is nonhomogeneous. For nonhomogeneous differential equations, in contrast to homogeneous differential equations, there is no guarantee that $c_1y_1 + c_2y_2$ is a solution merely because $y_1$ and $y_2$ happen to be solutions. In fact, it is generally the case for nonhomogeneous differential equations that $c_1y_1 + c_2y_2$ is not a solution. As a consequence, it requires a little ingenuity to describe the general solution for nonhomogeneous linear differential equations. However, this is always possible provided we know the general solution of the associated homogeneous differential equation.

---

[23]E. Campi, "Trigonometric Components of a Frequency-Modulated Wave," *Proc. IRE* (1948).

The general nonhomogeneous linear differential equation has the form

$$a_n(x)y^{(n)} + a_{n-1}(x)y^{(n-1)} + \cdots + a_1(x)y' + a_0(x)y = f(x) \qquad (1)$$

or, briefly,

$$\sum_{i=0}^{n} a_i(x)y^{(i)} = f(x). \qquad (2)$$

The function $f$ is called the *nonhomogeneous term* for the differential equation (1).

## DEFINITION 1

*With every nonhomogeneous differential equation (2), there is an associated homogeneous differential equation defined by*

$$\sum_{i=0}^{n} a_i(x)y^{(i)} = 0. \qquad (3)$$

## DEFINITION 2

*If the n functions $y_1, y_2, \ldots, y_n$ constitute a fundamental system of solutions for the homogeneous differential equation (3), then the function $y_h$ defined by*

$$y_h = c_1 y_1 + c_2 y_2 + \cdots + c_n y_n, \qquad (4)$$

*where the $c_i$ are arbitrary constants, is called the homogeneous solution for Eq. (2). [Note that the homogeneous solution is not an actual solution of Eq. (2). It is the general solution of the associated homogeneous differential equation. In some texts the homogeneous solution is referred to as the complementary solution.]*

**EXAMPLE 1**   Find the homogeneous solution of the differential equation

$$y'' - 3y' + 2y = \cos x. \qquad (5)$$

**Solution**   The homogeneous equation associated with the differential equation (5) is

$$y'' - 3y' + 2y = 0.$$

Using the methods of Section 2.5, the general solution of this equation is $c_1 e^x + c_2 e^{2x}$. Thus, the homogeneous solution of the differential equation (5) is

$$y_h = c_1 e^x + c_2 e^{2x}.$$

**EXAMPLE 2**   Find the homogeneous solution of the differential equation

$$2x^2 y'' - 3xy' - 3y = e^x, \qquad x > 0. \qquad (6)$$

**Solution**  The homogeneous equation associated with the differential equation (6) is

$$2x^2y'' - 3xy' - 3y = 0.$$

This latter differential equation is an Euler differential equation, and its solution (given in Example 2 of Section 2.7) is $c_1x^{-1/2} + c_2x^3$. Thus,

$$y_h = c_1x^{-1/2} + c_2x^3.$$

If, somehow or other, we find a function that satisfies Eq. (2), we refer to this function as a *particular solution* of Eq. (2) and denote it by $y_p$.

**THEOREM  1**

*If $y_1, y_2, \ldots, y_n$ form a fundamental system of solutions for Eq. (3), and if $y_p$ is any particular solution of Eq. (2), then the general solution of Eq. (2) can be written in the form*

$$y = y_h + y_p = c_1y_1 + c_2y_2 + \cdots c_ny_n + y_p. \tag{7}$$

**Proof**  It suffices to demonstrate that $y$ as given in the expression (7) satisfies the differential equation (2) and that any solution $Y$ of the differential equation (2) can be written in the form of Eq. (7). Substituting into the differential equation (2), we have

$$\sum_{i=0}^{n} a_i(x)y^{(i)} = \sum_{i=0}^{n} a_i(x)[c_1y_1 + c_2y_2 + \cdots + c_ny_n + y_p]^{(i)}$$

$$= \sum_{i=0}^{n} a_i(x)[c_1y_1^{(i)} + c_2y_2^{(i)} + \cdots + c_ny_n^{(i)} + y_p^{(i)}]$$

$$= c_1 \sum_{i=0}^{n} a_i(x)y_1^{(i)} + c_2 \sum_{i=0}^{n} a_i(x)y_2^{(i)}$$

$$+ \cdots + c_n \sum_{i=0}^{n} a_i(x)y_n^{(i)} + \sum_{i=0}^{n} a_i(x)y_p^{(i)}$$

$$\Rightarrow \quad \sum_{i=0}^{n} a_i(x)y^{(i)} = c_1 \cdot 0 + c_2 \cdot 0 + \cdots + c_n \cdot 0 + f(x)$$

$$= f(x).$$

That is, the function $y$ as defined by Eq. (7) satisfies the differential equation (2). Now let $Y$ be any solution of the differential equation (2). We should prove that $Y$ can be written in the form of Eq. (7). This is equivalent to proving that the function $Y - y_p$ satisfies the homogeneous differential equation (3). In fact,

$$\sum_{i=0}^{n} a_i(x)(Y - y_p)^{(i)} = \sum_{i=0}^{n} a_i(x)Y^{(i)} - \sum_{i=0}^{n} a_i(x)y_p^{(i)} = f(x) - f(x) = 0.$$

The proof is complete.

From this theorem and the fact that the last few sections provided us with information concerning the fundamental system of solutions of Eq. (3), we realize that we need only concentrate on methods associated with finding the particular solution $y_p$. There are essentially two popular methods for finding a particular solution: the *method of undetermined coefficients* and the *method of variation of parameters*. (See Sections 2.11 and 2.12.)

## EXERCISES

In Exercises 1 through 15, find the homogeneous solution of each nonhomogeneous differential equation.

**1.** $y'' - 5y' + 6y = x^2 + 3$      **2.** $y'' + 4y' + 4y = e^x + e^{-2x}$

**3.** $y'' - y' - 2y = \cos x$      **4.** $y'' - y = e^x$

**5.** $y'' + 9y = \cos 3x - \sin 3x$      **6.** $y''' + y' - 2y = x^3$

**7.** $y'' - 13y' + 36y = xe^{4x}$

**8.** $x^2y'' - 2xy' + 2y = \tan x,\ 0 < x < \pi/2$

**9.** $y'' - 10y' + 25y = x^2e^{5x}$      **10.** $y''' - y = 3 \ln x,\ x > 0$

**11.** $y^{iv} - y = x^2$

**12.** $x^2y'' + 4xy' - 10y = x \ln x,\ x > 0$

**13.** $3x^2y'' - 2xy' - 8y = 3x + 5,\ x > 0$

**14.** $y'' + y' = e^{-x}$

**15.** $xy'' - 2y' + \dfrac{x^2 + 2}{x}y = 4 + \tan x,\quad 0 < x < \pi/2$   (*Hint:* See Example 1, Section 2.8.)

Answer true or false.

**16.** $3xy''' - 4xy = \cos y$ is a nonhomogeneous differential equation.

**17.** $y'' + 5y' = \sin x$ has $c_1 + c_2e^{-5x}$ as its homogeneous solution.

**18.** $y'' + y = x$ has a particular solution $y_p = x$.

**19.** $y''' - 3xy'' + 4y = x^2$ has a particular solution $y_p = \frac{1}{4}x^2 + \frac{3}{8}x$.

**20.** $y'' - 3y = \cos x$ has a particular solution $y_p = \cos x$.

**21.** $3xy''' + 5y'' + 6y' - 3 \cos x = 0$ is a homogeneous differential equation.

**22.** $y'' + 2y = e^x$ has a particular solution $y_p = \frac{1}{3}e^x$.

**23.** $y'' - y = e^x$ has a particular solution $y_p = \frac{1}{2}xe^x$.

# 2.11   THE METHOD OF UNDETERMINED COEFFICIENTS

The *method of undetermined coefficients* is used when we want to compute a particular solution of the nonhomogeneous differential equation

$$a_n y^{(n)} + a_{n-1} y^{(n-1)} + \cdots + a_1 y' + a_0 y = f(x), \tag{1}$$

where the coefficients $a_0, a_1, \ldots, a_n$ are constants and $f(x)$ is a linear combination of functions of the following types:

1. $x^\alpha$, where $\alpha$ is a positive integer or zero.
2. $e^{\beta x}$, where $\beta$ is a nonzero constant.
3. $\cos \gamma x$, where $\gamma$ is a nonzero constant.
4. $\sin \delta x$, where $\delta$ is a nonzero constant.
5. A (finite) product of two or more functions of types 1–4.

For example, the function

$$f(x) = 3x^2 - 2 + 5e^{3x} - x(\sin x)e^{2x} + 5 \cos 2x + xe^x$$

is a linear combination of functions of types 1–5. On the other hand, the functions $1/x$ and $\log x$ are not of these types.

The key observation that makes the method of undetermined coefficients work is the fact that not only $f(x)$ but also *any derivative* (zeroth derivative included; recall that the zeroth derivative of $f$ is $f$ itself) of any term of $f(x)$ is a linear combination of functions of types 1–5. For example, any derivative of $3x^2$ is a linear combination of the functions $x^2$, $x$, and 1, which are all of type 1. We use the symbolism

$$3x^2 \to \{x^2, x, 1\}$$

to denote that any derivative of the function $3x^2$ on the left is a linear combination of the functions $x^2$, $x$, 1 on the right. We also say that the functions $x^2$, $x$, 1 *span* the derivatives of the function $3x^2$. As further illustrations, we have

$$2e^{3x} \to \{e^{3x}\}$$

$$-5 \to \{1\}$$

$$5 \cos 2x \to \{\cos 2x, \sin 2x\}$$

$$xe^x \to \{xe^x, e^x\}$$

$$-xe^{2x} \sin x \to \{xe^{2x} \sin x, xe^{2x} \cos x, e^{2x} \sin x, e^{2x} \cos x\}.$$

Roughly speaking, a particular solution of Eq. (1) is a linear combination of those functions of types 1–5 that span all the derivatives of $f(x)$. The coefficients of this linear combination are the undetermined coefficients (hence the name attached to the method) that are to be determined by substituting the assumed particular solution into the differential equation (1) and equating coefficients of similar terms.

The following example contains the typical features of the method of undetermined coefficients, and the reader is urged to study this example very carefully. (For motivation of the method, see Exercises 65 and 66.)

**EXAMPLE 1**   Compute a particular solution.of the differential equation

$$y'' - y = -2x^2 + 5 + 2e^x. \tag{2}$$

**Solution**   The method of undetermined coefficients is applicable to this example since the differential equation (2) is of the form of Eq. (1), and $f(x) = -2x^2 + 5 + 2e^x$ is a linear combination of the functions $x^2$, 1, and $e^x$ (which are of types 1, 1, and 2, respectively). We first compute the functions that span the derivatives of each one of the three terms of the function $f$. With our previous notation, we have

$$-2x^2 \rightarrow \{x^2, x, 1\} \tag{3}$$

$$5 \rightarrow \{1\} \tag{4}$$

$$2e^x \rightarrow \{e^x\}. \tag{5}$$

Therefore, the derivatives of $f$ are spanned by the functions in the sets $\{x^2, x, 1\}$ and $\{e^x\}$. The set $\{1\}$ in (4) is omitted because it is contained in the larger set (3). Now if *none* of the functions in the sets $\{x^2, x, 1\}$ and $\{e^x\}$ is a solution of the associated homogeneous equation for (2), that is, $y'' - y = 0$, then the particular solution of Eq. (2) is of the form

$$y_p = Ax^2 + Bx + C + De^x,$$

where $A$, $B$, $C$, and $D$ are the undetermined coefficients. On the other hand, if *any* one of the functions in *any* one of the sets $\{x^2, x, 1\}$ and $\{e^x\}$ is a solution of the associated homogeneous equation, then all the elements of that set should be multiplied by the lowest integral power of $x$, so that the resulting new set does not contain any function that is a solution of the associated homogeneous equation. Now, no function in the set $\{x^2, x, 1\}$ is a solution of $y'' - y = 0$, and so we leave this set as it is. On the other hand, $e^x$ in the set $\{e^x\}$ is a solution of $y'' - y = 0$, and so we must multiply the function in this set by the lowest integral power of $x$, so that the resulting function is not a solution of $y'' - y = 0$. Since $e^x$ is, but $xe^x$ is not, a solution of $y'' - y = 0$, we should multiply $e^x$ by $x$, *obtaining* $\{xe^x\}$. Thus, a particular solution of Eq. (2) is a linear combination of the functions in the sets $\{x^2, x, 1\}$ and $\{xe^x\}$. Hence,

$$y_p = Ax^2 + Bx + C + Dxe^x. \tag{6}$$

To obtain the undetermined coefficients $A$, $B$, $C$, and $D$, we compute

$$y_p' = 2Ax + B + De^x + Dxe^x$$

$$y_p'' = 2A + 2De^x + Dxe^x$$

and substitute these results into the differential equation (2), obtaining

$$(2A + 2De^x + Dxe^x) - (Ax^2 + Bx + C + Dxe^x) = -2x^2 + 5 + 2e^x.$$

Equating coefficients of similar terms, we obtain the following system of equations:

$$2A - C = 5$$
$$2D = 2$$
$$-B = 0$$
$$-A = -2.$$

Thus, $A = 2$, $B = 0$, $C = -1$, $D = 1$, and a particular solution of Eq. (2) is

$$y_p = 2x^2 - 1 + xe^x.$$

The crucial aspect of the method of undetermined coefficients is that we assumed the proper form for a particular solution (such as, $xe^x$ instead of $e^x$ in Example 1). Should we assume an inappropriate form for a particular solution, a contradiction will occur in the resulting system of equations when we attempt to compute the undetermined coefficients. Alternatively, it may happen that the coefficients of unnecessary terms will be found to be equal to zero.

To save space, we suggest that the explanations that lead us to the form (6) for a particular solution of Eq. (2) be abbreviated in accordance with the following format:

$$-2x^2 \rightarrow \{x^2, x, 1\}$$
$$5 \rightarrow \{1\}$$
$$2e^x \rightarrow \{e^x\} \rightarrow \{xe^x\}$$

($e^x$ is a solution of the associated homogeneous equation and $xe^x$ is not). So

$$y_p = Ax^2 + Bx + C + Dxe^x$$

is the correct form of a particular solution of Eq. (2).

**EXAMPLE 2**   Find the form of a particular solution of the differential equation

$$y'' + 2y' - 3y = 3x^2 e^x + e^{2x} + x \sin x + 2 + 3x.$$

**Solution**   Since $e^x$ is a solution of the associated homogeneous equation but $xe^x$ is not, we have

$$3x^2 e^x \rightarrow \{x^2 e^x, xe^x, e^x\} \rightarrow \{x^3 e^x, x^2 e^x, xe^x\}$$
$$e^{2x} \rightarrow \{e^{2x}\}$$
$$x \sin x \rightarrow \{x \sin x, x \cos x, \sin x, \cos x\}$$
$$2 \rightarrow \{1\}$$
$$3x \rightarrow \{x, 1\}.$$

Thus,

$$y_p = Ax^3e^x + Bx^2e^x + Cxe^x + De^{2x} + Ex \sin x$$
$$+ Fx \cos x + G \sin x + H \cos x + Ix + J.$$

**EXAMPLE 3**   Compute a particular solution of the differential equation

$$y'' - 2y' + y = 2e^x - 3e^{-x}. \tag{7}$$

**Solution**   We have $2e^x \rightarrow \{e^x\} \rightarrow \{x^2e^x\}$ ($e^x$ and $xe^x$ are solutions of the associated homogeneous equation and $x^2e^x$ is not) and $-3e^{-x} \rightarrow \{e^{-x}\}$. Thus,

$$y_p = Ax^2e^x + Be^{-x};$$

hence,

$$y_p' = 2Axe^x + Ax^2e^x - Be^{-x}$$

$$y_p'' = 2Ae^x + 4Axe^x + Ax^2e^x + Be^{-x}.$$

Substituting $y_p$, $y_p'$, and $y_p''$ into the differential equation (7), we have

$$(2Ae^x + 4Axe^x + Ax^2e^x + Be^{-x}) - 2(2Axe^x + Ax^2e^x - Be^{-x})$$
$$+ (Ax^2e^x + Be^{-x}) = 2e^x - 3e^{-x},$$
$$\Rightarrow \quad 2Ae^x + 4Be^{-x} = 2e^x - 3e^{-x}.$$

Equating the coefficients of $e^x$ and equating the coefficients of $e^{-x}$, we obtain $A = 1$ and $B = -\frac{3}{4}$; thus,

$$y_p = x^2e^x - \frac{3}{4}e^{-x}.$$

**EXAMPLE 4**   Compute the general solution of the differential equation

$$y'' - y' - 2y = e^{3x} \cos 2x. \tag{8}$$

**Solution**   The general solution of the nonhomogeneous differential equation is

$$y = y_h + y_p.$$

We first compute the homogeneous solution $y_h$, which is the general solution of the associated homogeneous equation $y'' - y' - 2y = 0$. Now the characteristic equation is $\lambda^2 - \lambda - 2 = 0$, and therefore the characteristic roots are $\lambda_1 = 2$ and $\lambda_2 = -1$. Hence,

$$y_h = c_1e^{2x} + c_2e^{-x}.$$

As for $y_p$, we observe that

$$e^{3x} \cos 2x \rightarrow \{e^{3x} \cos 2x, e^{3x} \sin 2x\}.$$

Since neither of the functions $e^{3x} \cos 2x$ and $e^{3x} \sin 2x$ is a solution of the associated homogeneous equation, it follows that a particular solution of Eq. (8) is of the form

$$y_p = Ae^{3x} \cos 2x + Be^{3x} \sin 2x = e^{3x}(A \cos 2x + B \sin 2x).$$

Substituting this $y_p$ into Eq. (8) and equating coefficients of similar terms, we find that $A = 0$ and $B = \frac{1}{10}$ and so

$$y_p = \tfrac{1}{10}e^{3x} \sin 2x.$$

Therefore, the general solution of the differential equation (8) is

$$y = c_1 e^{2x} + c_2 e^{-x} + \tfrac{1}{10}e^{3x} \sin 2x.$$

**EXAMPLE 5**  Solve the initial value problem

$$y''' + y' = 2 + \sin x \tag{9}$$
$$y(0) = 0, \qquad y'(0) = 1, \qquad y''(0) = -1.$$

**Solution**  We first compute the general solution of Eq. (9), which is the sum of the general solution $y_h$ of the associated homogeneous equation $y''' + y' = 0$, and a particular solution $y_p$ of Eq. (9). As for $y_h$, the characteristic equation is $\lambda^3 + \lambda = 0$, hence the characteristic roots are $\lambda_1 = 0, \lambda_2 = i, \lambda_3 = -i$. Thus,

$$y_h = c_1 + c_2 \cos x + c_3 \sin x.$$

From Eq. (9) the nonhomogeneous term $2 + \sin x$ gives rise to the sets

$$2 \rightarrow \{1\} \rightarrow \{x\}$$

$$\sin x \rightarrow \{\sin x, \cos x\} \rightarrow \{x \sin x, x \cos x\}.$$

Consequently, we assume $y_p$ to be of the form

$$y_p = Ax + Bx \sin x + Cx \cos x.$$

Therefore,

$$y_p' = A + B \sin x + Bx \cos x + C \cos x - Cx \sin x$$
$$y_p'' = 2B \cos x - Bx \sin x - 2C \sin x - Cx \cos x$$
$$y_p''' = -3B \sin x - Bx \cos x - 3C \cos x + Cx \sin x.$$

Substituting $y_p'$ and $y_p'''$ into the differential equation (9), we obtain

$$A - 2B \sin x - 2C \cos x = 2 + \sin x.$$

Therefore, $A = 2, B = -\frac{1}{2}, C = 0$, and $y_p = 2x - \frac{1}{2}x \sin x.$

So the general solution of Eq. (9) is

$$y = c_1 + c_2 \cos x + c_3 \sin x + 2x - \tfrac{1}{2}x \sin x. \tag{10}$$

Finally, we use the initial conditions to determine $c_1$, $c_2$, and $c_3$. Differentiating Eq. (10), we have

$$y' = -c_2 \sin x + c_3 \cos x + 2 - \tfrac{1}{2} \sin x - \tfrac{1}{2} x \cos x$$

$$\Rightarrow \quad y'' = -c_2 \cos x - c_3 \sin x - \cos x + \tfrac{1}{2} x \sin x.$$

Now

$$y(0) = 0 \Rightarrow c_1 + c_2 = 0$$

$$y'(0) = 1 \Rightarrow c_3 + 2 = 1$$

$$y''(0) = -1 \Rightarrow -c_2 - 1 = -1.$$

Thus, $c_1 = 0$, $c_2 = 0$, $c_3 = -1$, and the solution of the initial value problem is

$$y(x) = -\sin x + 2x - \tfrac{1}{2} x \sin x.$$

### APPLICATIONS 2.11.1

■ The electric circuit shown in Figure 2.5 is customarily called an *RLC*-series **Electric Circuits** circuit. Here a generator supplying a voltage of $V(t)$ volts is connected in series with an *R*-ohm resistor, an inductor of *L* henries, and a capacitor of *C* farads. When the switch is closed, a current of $I = I(t)$ amperes flows in the circuit. We would like to find the current *I* as a function of time. We also want to find the charge $Q = Q(t)$ coulombs in the capacitor at any time *t*. By definition

$$I = \frac{dQ}{dt}. \tag{11}$$

Thus, it suffices to compute *Q*. It is known from physics that the current *I* produces a voltage drop across the resistor equal to $RI$, a voltage drop across the inductor equal to $L(dI/dt)$, and a voltage drop across the capacitor equal to

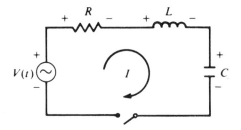

**Figure 2.5**

$(1/C)Q$. Kirchhoff's voltage law states that the voltage supplied is equal to the sum of the voltage drops in the circuit. Applying this law to the circuit of Figure 2.5 (with the switch closed), we have

$$L\frac{dI}{dt} + RI + \frac{1}{C}Q = V(t). \tag{12}$$

Using the relation (11), Eq. (12) can be written in the form

$$L\frac{d^2Q}{dt^2} + R\frac{dQ}{dt} + \frac{1}{C}Q = V(t). \tag{13}$$

This is a second-order linear nonhomogeneous differential equation. In order to find the charge $Q(t)$ in the capacitor, we must find the general solution of Eq. (13). This general solution involves two arbitrary constants. To determine these two constants, we impose two initial conditions $Q(0) = Q_0$ and $\dot{Q}(0) = 0$. $Q_0$ is the initial charge on the capacitor and $\dot{Q}(0) = I(0) = 0$, because at time $t = 0$ there is no current flow in the circuit. To find the current $I(t)$, we can use the relation (11) or the differential equation

$$L\frac{d^2I}{dt^2} + R\frac{dI}{dt} + \frac{1}{C}I = \frac{dV(t)}{dt}, \tag{14}$$

which is obtained by differentiating Eq. (12). Initial conditions associated with this differential equation are $I(0) = 0$ and $\dot{I}(0) = (1/L)[V(0) - (1/C)Q(0)]$, the latter condition being obtained from the differential equation (12).

**Electromechanical ■** If we compare Eq. (13) with Eq. (16) of Exercise 57, we observe the analogies
**Analogies** between mechanical and electrical systems given in Table 2.2. These analogies are very useful in applications. Often the study of a mechanical system is converted to the study of an equivalent electrical system, which is usually easier to study. For example, this is one of the basic ideas behind the design of analog computers.

| Mechanical | | Electrical |
|---|---|---|
| Mass | $m \leftrightarrow L$ | Inductance |
| Friction (damping) | $\alpha \leftrightarrow R$ | Resistance |
| Spring stiffness | $K \leftrightarrow \dfrac{1}{C}$ | Reciprocal of capacitance |
| Displacement | $y \leftrightarrow Q$ | Charge on capacitor |
| External force | $F(t) \leftrightarrow V(t)$ | Electromotive force |
| Velocity | $\dot{y} \leftrightarrow I$ | Current |

**Table 2.2**   Mechanical–Electrical Analogies

## EXERCISES

In Exercises 1 through 6, a nonhomogeneous differential equation is given. Verify that the expression for $y_p$ is the correct form.

**1.** $y'' + y = x + 2e^{-x}$; $y_p = Ax + B + Ce^{-x}$

**2.** $y'' - y = e^x + \sin x$; $y_p = Axe^x + B \sin x + C \cos x$

**3.** $y'' - 4y' + 4y = e^{2x}$; $y_p = Ax^2e^{2x}$

**4.** $y''' + y' = x$; $y_p = Ax^2 + Bx$

**5.** $y^{(6)} - 3y^{(4)} = 1$; $y_p = Ax^4$          **6.** $y'' - y = xe^x$; $y_p = Ax^2e^x + Bxe^x$

In Exercises 7 through 12, write out the assumed form of $y_p$ but do not carry out the calculation of the undetermined coefficients.

**7.** $y'' - y = x^3 + 3x$          **8.** $y'' - 3y' + 2y = e^{2x} + \sin x$

**9.** $y' + y = x^2 + \cos x$          **10.** $y'' + 4y' + 4y = e^{-2x} + xe^{-2x}$

**11.** $y''' - y' = x^5 + \cos x$          **12.** $y''' + y' = xe^x + 3x.$

In Exercises 13 through 26, compute $y_p$ for each differential equation.

**13.** $y'' + y = x + 2e^{-x}$          **14.** $y'' - y = e^x + \sin x$

**15.** $y'' - 4y' + 4y = e^{2x}$          **16.** $y''' + y' = x$

**17.** $y^{(6)} - 3y^{(4)} = 1$          **18.** $y'' - y = xe^x$

**19.** $y''' + 3y'' - 4y = e^{-2x}$          **20.** $y'' + 4y = 4x^3 - 8x^2 - 14x + 7$

**21.** $y''' - y' = e^x$          **22.** $y'' + y = e^x(x + 1)$

**23.** $y'' - y = x \sin x$          **24.** $y'' - 3y' + 2y = e^{-x} \cos x$

**25.** $2y'' + y' - y = e^x(x^2 - 1)$          **26.** $y'' - 2y' + y = xe^x$

In Exercises 27 through 38, compute the general solution of each nonhomogeneous equation.

**27.** $y'' + y = \sin x$          **28.** $y'' + 4y = \sin x$

**29.** $y'' - y' - 2y = 2xe^{-x} + x^2$          **30.** $y''' - 3y'' + 3y' - y = x^2 + 5e^x$

**31.** $y'' - y = 4 \cosh x$ [*Hint:* $\cosh x = (e^x + e^{-x})/2$]

**32.** $y'' = 3$          **33.** $x^2y'' + 4xy' - 4y = x^3, x > 0$

**34.** $x^2y'' - xy' + y = x^2, x > 0$          **35.** $y'' - 5y' + 6y = e^x \sin x$

**36.** $y'' - 7y' - 8y = e^x(x^2 + 2)$

**37.** $y'' - 5y' + 4y = e^{2x} \cos x + e^{2x} \sin x$

**38.** $y'' + 2y' - 3y = e^{2x}(x + 3)$

In Exercises 39 through 42, solve the initial value problems.

**39.** $y'' + y = x + 2e^{-x}$
$\qquad y(0) = 1$
$\qquad y'(0) = -2$

**40.** $y'' - y = xe^x$
$\qquad y(0) = 0$
$\qquad y'(0) = 1$

**41.** $y'' - 4y' + 4y = e^{2x}$
$\qquad y(0) = 0$
$\qquad y'(0) = 0$

**42.** $y''' + y' = x$
$\qquad y(0) = 0$
$\qquad y'(0) = 1$
$\qquad y''(0) = 0$

In Exercises 43 through 47, answer true or false.

**43.** A particular solution of $y'' + 3y' + 2y = e^x$ is of the form $Axe^x$.

**44.** A particular solution of $y'' - 3y' + 2y = e^x$ is of the form $Axe^x$.

**45.** A particular solution of $y'' + y = 1/x$ cannot be found by the method of undetermined coefficients.

**46.** A particular solution of $y'' + y = \cos x$ is of the form $Ax \cos x + Bx \sin x$.

**47.** A particular solution of $y'' - 3y = x \ln x$ can be found by the method of undetermined coefficients.

Exercises 48 through 50 refer to the *RLC*-series circuit shown in Figure 2.5. Assuming that initially $Q(0) = I(0) = 0$, find the charge and the current at any time $t$.

**48.** $R = 4$ ohms; $L = 1$ henry; $C = 1/13$ farads; $V(t) = 26$ volts.

**49.** $R = 10$ ohms; $L = 2$ henries; $C = 1/12$ farads; $V(t) = \cos 2t$ volts.

**50.** $R = 4$ ohms; $L = 2$ henries; $C = 0.5$ farads; $V(t) = 2e^{-t}$ volts.

**51. Learning Theory**  In 1962, Brady and Marmasse,[24] in an effort to describe experimental data recorded in a simple avoidance situation using rats, arrived at the following initial value problem:

$$\ddot{u} + \frac{1}{\tau}\dot{u} + bu = \frac{\beta}{\tau}$$

$$u(0) = \alpha, \qquad \dot{u}(0) = \beta,$$

where $u(t)$ represents the value of the learning curve of a rat at time $t$; $\alpha$ and $\beta$ are initial values; and $b$ and $\tau$ are positive constants characteristic of the experimental situation. Without any additional information about the constants $\tau$, $b$, and $\beta$, one would assume a particular solution of the form $u_p = A$. Show that such a $u_p$ can never satisfy the homogeneous equation if $b \neq 0$. Find $u_p$.

[24]J. P. Brady and C. Marmasse, *Psychol. Record* 12 (1962): 361–68; see also *Bull. Math. Biophys.* 26 (1964): 77–81.

52. Referring to Exercise 51, show that when $(1/\tau^2) - 4b < 0$, the solutions contain oscillatory terms (linear combinations of sines and cosines). (This phenomenon was observed in the actual experiment with rats.)

53. Solve the nonhomogeneous differential equation of Exercise 51 if $\tau = \frac{1}{5}$, $b = 6$, $\alpha = 1$, and $\beta = 1$. Discuss the behavior of the solution at $t \to \infty$.

54. **Medicine: Periodic Relapsing Catatonia**  In the development of a mathematical theory of thyroid–pituitary interactions, Danziger and Elmergreen[25] obtained the following third-order linear differential equations:

$$\alpha_3 \dddot{\theta} + \alpha_2 \ddot{\theta} + \alpha_1 \dot{\theta} + (1 + K)\theta = KC, \qquad \theta < C \qquad (15)$$

and

$$\alpha_3 \dddot{\theta} + \alpha_2 \ddot{\theta} + \alpha_1 \dot{\theta} + \theta = 0, \qquad \theta > C.$$

These equations describe the variation of thyroid hormone with time. Here $\theta = \theta(t)$ is the concentration of thyroid hormone at time $t$ and $\alpha_1$, $\alpha_2$, $\alpha_3$, $K$, and $C$ are constants. If $K \neq -1$, find $\theta_p$ for Eq. (15).

55. **Minerals**  In finding the optimum path for the exploitation of minerals in the simple situation where cost rises linearly with cumulated production, Herfindahl and Kneese[26] presented the following second-order differential equation:

$$\ddot{x} - r\dot{x} - \frac{r\beta}{b}x = -\frac{rK}{b},$$

where $r$, $\beta$, $b$, and $K$ are nonzero constants. Find $x_p$ for this differential equation.

56. **Social Behavior**  Second-order linear differential equations with constant coefficients of the form

$$\ddot{y} - Ay = B, \qquad A > 0,$$

were obtained by Rashevsky[27] in his study of riots by oppressed groups. Here $B$ is an arbitrary constant. Find the general solution of this differential equation.

57. **Forced Motion of the Vibrating Spring**  In addition to the damping force assumed in Exercise 40, Section 2.4, suppose that an external force equal to $f(t)$ acts upon the mass of the spring. Prove, in this case, that the differential equation of the vibrating spring is given by

$$m\ddot{y} + \alpha\dot{y} + Ky = f(t) \qquad (16)$$

Such a motion is called a *forced motion*. When the external force is identically equal to zero, the motion is called a *free motion*.

[25]L. Danziger and G. L. Elmergreen, *Bull. Math. Biophys.* 18 (1956): 1–13. See also N. Rashevsky, *Bull. Math. Biophys.* 30 (1968): 735–49.
[26]O. C. Herfindahl and A. V. Kneese, *Economic Theory of Natural Resources* (Columbus, Ohio: Charles E. Merrill, 1974), p. 182.
[27]N. Rashevsky, *Bull. Math. Biophys.* 30 (1968): 501–18.

**58.** For the forced motion of Exercise 57, assume that $m = 4$ g, $K = 3$ dyn/cm, and that air resistance acts upon the mass with a force that is 7 times its velocity at time $t$. The mass is also subjected to an external periodic force equal to $5 \cos t$. The mass is raised to a position 3 cm above the equilibrium position and released with a velocity of 5 cm/sec downward. Thus, the motion of the mass satisfies the IVP

$$4\ddot{y} + 7\dot{y} + 3y = 5 \cos t$$
$$y(0) = -3$$
$$\dot{y}(0) = 5.$$

Solve this IVP. Describe the limiting motion at $t \to \infty$.

**Elasticity: Static Deflection of a Beam**   In the consideration of the static deflection $u$ of a beam of uniform cross section (for example, an I beam) subjected to a transverse load $f(x)$, there arises the differential equation[28]

$$EI\frac{d^4u}{dx^4} = f(x), \tag{17}$$

where $E$ and $I$ are constants, $E$ being Young's modulus of elasticity and $I$ the moment of inertia of a cross section of the beam about some fixed axis.

In Exercises 59 through 64, assume that $EI = 1$ and solve Eq. (17) for the given $f$.

**59.** $f(x) = \cos x$      **60.** $f(x) = e^{-x}$      **61.** $f(x) = \sinh x$

**62.** $f(x) = 1$      **63.** $f(x) = x^2$      **64.** $f(x) = x^4$

The portion of the method of undetermined coefficients associated with the case wherein $f(x)$ has terms that are solutions of the homogeneous equation may seem unmotivated. Certainly, we did not give any indication where the proper form for $y_p$ came from. To this end, consider the following exercises.[29]

**65.** (a) Show that the general solution of the differential equation

$$y'' - 2y' + y = e^{ax}, \qquad a \neq 1,$$

is

$$y(x) = c_1e^x + c_2xe^x + \frac{e^{ax}}{(a-1)^2}.$$

(b) Given that $y(0) = y_0$ and $y'(0) = y_1$, show that $c_1$ and $c_2$ in the solution of part (a) are

$$c_1 = y_0 - \frac{1}{(a-1)^2} \qquad \text{and} \qquad c_2 = y_1 - y_0 - \frac{1}{a-1}$$

hence

$$y(x) = y_0e^x + (y_1 - y_0)xe^x + \frac{e^{ax} - e^x + (1-a)xe^x}{(a-1)^2}.$$

---

[28]S. Timoshenko, *Vibration Problems in Engineering* (Princeton, N.J.: Van Nostrand, 1928).
[29]These exercises are extracted from A. B. Farnell, "On the Solutions of Linear Differential Equations," *Amer. Math. Monthly* 54, no. 3 (1974).

(c) Show that

$$\lim_{a \to 1} y(x) = y_0 e^x + (y_1 - y_0) x e^x + \tfrac{1}{2} x^2 e^x.$$

**66.** (a) Show that the general solution of the differential equation

$$y'' + y = \sin ax$$

is

$$y(x) = c_1 \sin x + c_2 \cos x + \frac{\sin ax}{1 - a^2}, \qquad a \neq \pm 1.$$

(b) Given that $y(0) = y_0$ and $y'(0) = y_1$, show that $c_1$ and $c_2$ in the solution of part (a) are $c_1 = y_1 - a/(1 - a^2)$ and $c_2 = y_0$; hence

$$y(x) = y_1 \sin x + y_0 \cos x + \frac{\sin ax - a \sin x}{1 - a^2}.$$

(c) Show that

$$\lim_{a \to 1} y(x) = y_1 \sin x + y_0 \cos x + \tfrac{1}{2}(\sin x - x \cos x).$$

**67. Resonance** When an undamped vibrating spring is subjected to a periodic external force of the form $f(t) = M \cos wt$ the differential equation takes the form (see Eq. (16) of Ex. 57)

$$\ddot{y} + w_0^2 y = F \cos wt, \tag{18}$$

where $w_0^2 = K/m$ and $F = M/m$. If $f(t) \equiv 0 \ (\Rightarrow M = 0)$ the spring is said to be *free*. For this reason $w_0$ is called the *natural frequency* of the free vibration.

(a) If $w \neq w_0$, show that $y_p = \dfrac{F}{w_0^2 - w^2} \cos wt$.

(b) If $w$ is "close to" $w_0$, the solution in part (a) shows that the amplitude of $y_p$ is large and gets even larger if $w$ is chosen closer to $w_0$. This phenomenon associated with the unbounded growth of the amplitude of $y_p$ as $w \to w_0$ is referred to as *resonance*. That is, resonance is said to occur when the frequency of the external force matches the natural frequency of the system. Find $y_p$ for Eq. (18) in the case $w = w_0$.

## 2.12 VARIATION OF PARAMETERS

The method of variation of parameters, like the method of undetermined coefficients, is used to compute a particular solution of the nonhomogeneous differential equation

$$a_n(x)y^{(n)} + a_{n-1}(x)y^{(n-1)} + \cdots + a_1(x)y' + a_0(x)y = f(x). \tag{1}$$

The method of undetermined coefficients, however, requires that the coefficients

$a_0, a_1, \ldots,$ and $a_n$ of the differential equation be constants, and that the function $f$ be one of the types described in Section 2.11. When we encounter differential equations for which we cannot apply the method of undetermined coefficients, we use the *variation-of-parameters method*. This method, which in theory "always works," evolves from the following consideration. Suppose that we know the general solution

$$y_h = c_1 y_1 + c_2 y_2 + \cdots + c_n y_n$$

of the associated homogeneous differential equation

$$a_n(x)y^{(n)} + a_{n-1}(x)y^{(n-1)} + \cdots + a_1(x)y' + a_0(x)y = 0. \tag{2}$$

Is it possible to treat the constants $c_i$ as functions (variables) $u_i$ and to impose appropriate conditions on these functions so that the expression

$$u_1 y_1 + u_2 y_2 + \cdots + u_n y_n$$

is a solution of the nonhomogeneous equation (1)? Such an approach has been shown to lead to meaningful results, which we present in Theorem 1. This theorem is stated for the general $n$th-order equation, but for simplicity of presentation we prove it for the case $n = 2$.

**THEOREM 1**

*If $y_1, y_2, \ldots, y_n$ form a fundamental system of solution for Eq. (2), and if the functions $u_1, u_2, \ldots, u_n$ satisfy the system of equations*

$$
\begin{aligned}
y_1 u_1' + y_2 u_2' + \cdots + y_n u_n' &= 0, \\
y_1' u_1' + y_2' u_2' + \cdots + y_n' u_n' &= 0, \\
y_1'' u_1' + y_2'' u_2' + \cdots + y_n'' u_n' &= 0, \\
&\cdots\cdots\cdots\cdots\cdots\cdots\cdots\cdots\cdots \\
y_1^{(n-2)} u_1' + y_2^{(n-2)} u_2' + \cdots + y_n^{(n-2)} u_n' &= 0, \\
y_1^{(n-1)} u_1' + y_2^{(n-1)} u_2' + \cdots + y_n^{(n-1)} u_n' &= \frac{f(x)}{a_n(x)},
\end{aligned}
\tag{3}
$$

*then $y = u_1 y_1 + u_2 y_2 + \cdots u_n y_n$ is a particular solution of Eq. (1).*

**Proof**  (Special case $n = 2$) By hypothesis, $y_1$ and $y_2$ form a fundamental system of solutions for the equation $a_2(x)y'' + a_1(x)y' + a_0(x)y = 0$, and that $u_1$ and $u_2$ satisfy the system of algebraic equations

$$y_1 u_1' + y_2 u_2' = 0 \tag{4}$$

$$y_1' u_1' + y_2' u_2' = \frac{f(x)}{a_2(x)}.$$

The determinant of the coefficients of the nonhomogeneous system (4) is the Wronskian of the linearly independent solutions $y_1$ and $y_2$. Since this determinant is different from zero, the system has a unique solution $u_1', u_2'$. (See Appendix

A.) From $u_1'$ and $u_2'$ we determine $u_1$ and $u_2$ via integration. All that remains to be shown is that the function

$$y = u_1 y_1 + u_2 y_2 \;. \tag{5}$$

is a solution of the differential equation. Indeed, from (5) we have

$$y' = u_1 y_1' + u_1' y_1 + u_2 y_2' + u_2' y_2 = u_1 y_1' + u_2 y_2'.$$

The last expression is obtained by applying the first of Eqs. (4). Hence,

$$y'' = u_1 y_1'' + u_1' y_1' + u_2 y_2'' + u_2' y_2' = u_1 y_1'' + u_2 y_2'' + \frac{f(x)}{a_2(x)}.$$

Once again the last expression follows by applying the second of Eqs. (4). Now

$$a_2(x)y'' + a_1(x)y' + a_0(x)y = a_2(x)\left[u_1 y_1'' + u_2 y_2'' + \frac{f(x)}{a_2(x)}\right]$$

$$+ a_1(x)[u_1 y_1' + u_2 y_2'] + a_0(x)[u_1 y_1 + u_2 y_2]$$

$$= u_1[a_2(x)y_1'' + a_1(x)y_1' + a_0(x)y_1]$$

$$+ u_2[a_2(x)y_2'' + a_1(x)y_2' + a_0(x)y_2] + f(x)$$

$$= f(x).$$

Therefore, $y$ satisfies Eq. (1) (for $n = 2$) and the proof is complete. Other than the tedious notation, the proof in the general case follows in exactly the same manner.

The reader should observe that Theorem 1 and its proof show us how to construct the particular solution $y_p$. We first find the fundamental system of solutions for Eq. (2). Using these functions, we then solve Eqs. (3) for $u_1', u_2', \ldots, u_n'$. Next, we integrate each of the $u_1', u_2', \ldots, u_n'$ to find, respectively, $u_1, u_2, \ldots, u_n$. (In these integrations, there is no loss in generality if we set the constant of integration equal to zero, for otherwise the term corresponding to a nonzero constant of integration would be absorbed into the homogeneous solution when we wrote out the general solution). And, finally,

$$u_1 y_1 + u_2 y_2 + \cdots + u_n y_n$$

is the desired particular solution $y_p$.

We note that the variation-of-parameters method applies to any nonhomogeneous differential equation no matter what the coefficients and the function $f$ happen to be. The reason for introducing the method of undetermined coefficients is that it is sometimes quicker and easier to apply. The method of undetermined coefficients involves differentiation, while the method of variation of parameters involves integration, and in some cases differentiation is easier than integration.

**EXAMPLE 1**  Solve the differential equation

$$y'' + y = \csc x.$$

**Solution**  The homogeneous solution is $y_h = c_1 \sin x + c_2 \cos x$. The functions $u_1$ and $u_2$ are to be determined from the system of equations [see Eqs. (4)]

$$\sin x \, u_1' + \cos x \, u_2' = 0$$

$$\cos x \, u_1' - \sin x \, u_2' = \csc x.$$

Thus,

$$u_1' = \cos x \csc x \Rightarrow u_1 = \ln |\sin x|$$

$$u_2' = -1 \qquad \Rightarrow u_2 = -x.$$

Therefore, we have $y_p = [\ln |\sin x|] \sin x + (-x) \cos x$, and the general solution is

$$y = y_h + y_p = [c_1 + \ln |\sin x|] \sin x + (c_2 - x) \cos x.$$

**EXAMPLE 2**  Find $y_p$ for the beam-deflection problem [see Eq. (17), Section 2.11] subjected to the load $f(x) = \ln x$.

**Solution**  The differential equation in question is

$$EI \frac{d^4 y}{dx^4} = \ln x,$$

where $E$ and $I$ are constants. Therefore, the homogeneous solution is $y_h = c_1 + c_2 x + c_3 x^2 + c_4 x^3$, and $u_1, u_2, u_3, u_4$ are to be determined from the system of equations

$$u_1' + x u_2' + x^2 u_3' + x^3 u_4' = 0$$

$$u_2' + 2x u_3' + 3x^2 u_4' = 0 \qquad\qquad (6)$$

$$2u_3' + 6x u_4' = 0$$

$$6u_4' = \frac{\ln x}{EI}.$$

Equations (6) can be solved by starting with the last equation and working "upward." Thus,

$$u_4' = \frac{\ln x}{6EI} \qquad \Rightarrow u_4 = \frac{1}{6EI}(x \ln x - x)$$

$$u_3' = \frac{-1}{2EI}(x \ln x) \Rightarrow u_3 = \frac{-1}{2EI}\left(\frac{x^2}{2} \ln x - \frac{x^2}{4}\right)$$

$$u_2' = \frac{1}{2EI}(x^2 \ln x) \Rightarrow u_2 = \frac{1}{2EI}\left(\frac{x^3}{3} \ln x - \frac{x^3}{9}\right)$$

$$u_1' = \frac{-1}{6EI} x^3 \ln x \quad \Rightarrow u_1 = \frac{-1}{6EI}\left(\frac{x^4}{4} \ln x - \frac{x^4}{16}\right).$$

The particular solution $y_p$ is then $u_1 \cdot (1) + u_2 \cdot (x) + u_3 \cdot (x^2) + u_4 \cdot (x^3)$ which, after simplification, has the form

$$y_p = \frac{x^4}{288EI}(12 \ln x - 25).$$

We state the following theorem without proof.

## THEOREM 2

*Suppose that $y_1, y_2, \ldots, y_n$ constitutes a fundamental set of solutions for the equation*

$$a_n(x)y^{(n)} + a_{n-1}(x)y^{(n-1)} + \cdots + a_1(x)y' + a_0(x)y = 0.$$

*Then a particular solution of the equation*

$$a_n(x)y^{(n)} + a_{n-1}(x)y^{(n-1)} + \cdots + a_1(x)y' + a_0(x)y = f(x)$$

*is given by*

$$y_p(x) = \sum_{k=1}^{n} y_k(x) \int_{x_0}^{x} \frac{W_k(y_1, \ldots, y_n; s)}{W(y_1, \ldots, y_n; s)} \cdot \frac{f(s)}{a_n(s)} \, ds$$

*where $W(y_1, \ldots, y_n; s)$ is the Wronskian of $y_1, \ldots, y_n$ evaluated at $s$ and $W_k(y_1, \ldots, y_n; s)$ is the determinant obtained from $W(y_1, \ldots, y_n; s)$ by replacing the $k$th column by $(0, 0, \ldots, 0, 1)$. Furthermore, $y_p(x_0) = y_p'(x_0) = \cdots = y_p^{(n-1)}(x_0) = 0$. (For the important special case $n = 2$, see Exercise 30.)*

## EXERCISES

In Exercises 1 through 17, use the variation-of-parameters method to compute the general solution of the nonhomogeneous differential equations.

**1.** $y'' + y = \tan x,\ 0 < x < \pi/2$

**2.** $y'' + y = \sec x,\ 0 < x < \pi/2$

**3.** $y'' - 2y' + y = \dfrac{1}{x} e^x,\ x > 0$

**4.** $y'' + 10y' + 25y = e^{-5x} \ln x/x^2,\ x > 0$

**5.** $y'' + 6y' + 9y = \dfrac{1}{x^3} e^{-3x},\ x > 0$

**6.** $y'' + y = \csc x \cot x,\ 0 < x < \pi/2$

**7.** $y'' - 12y' + 36y = e^{6x} \ln x,\ x > 0$

**8.** $y'' + 4y' + 5y = e^{-2x} \sec x,\ 0 < x < \pi/2$

**9.** $y'' + y = \sec^3 x,\ 0 < x < \pi/2$

**10.** $y'' - 4y' + 4y = e^{2x} x^{-4},\ x > 0$

**11.** $y'' + 2y' + y = x^{-2}e^{-x}\ln x, \ x > 0$

**12.** $y'' - 2y' + y = e^{2x}/(e^x + 1)^2$

**13.** $5x^2y'' - 3xy' + 3y = x^{1/2}, \ x > 0$

**14.** $x^2y'' + 4xy' - 4y = x^{1/4}\ln x, \ x > 0$

**15.** $2x^2y'' + xy' - 3y = x^{-3}, \ x > 0$

**16.** $2x^2y'' + 7xy' - 3y = x^{-2}\ln x, \ x > 0$

**17.** $x^2y'' + 5xy' + 4y = (x^{-3} + x^{-5})\ln x, \ x > 0$

In Exercises 18 through 25, solve the initial value problems.

**18.** $y'' + y = \csc x$
$y(\pi/2) = 0$
$y'(\pi/2) = 1$

**19.** $y'' + y = \tan x$
$y(0) = 1$
$y'(0) = 0$

**20.** $y'' - 2y' + y = (1/x)e^x$
$y(1) = e$
$y'(1) = 0$

**21.** $y'' + 6y' + 9y = x^{-3}e^{-3x}$
$y(1) = 4e^{-3}$
$y'(1) = -2e^{-3}$

**22.** $y'' + y = \sec^3 x$
$y(0) = 1$
$y'(0) = 1$

**23.** $y'' - 4y' + 4y = x^{-4}e^{2x}$
$y(1) = 0$
$y'(1) = e^2$

**24.** $y'' - 2y' + y = e^{2x}/(e^x + 1)^2$
$y(0) = 3$
$y'(0) = \frac{5}{2}$

**25.** $2x^2y'' + xy' - 3y = x^{-3}$
$y(\frac{1}{4}) = 0$
$y'(\frac{1}{4}) = \frac{14}{9}$

**26.** Find the solution of the IVP

$$(x^2 + 1)y'' - 2xy' + 2y = (x^2 + 1)^2$$
$$y(0) = 0, \qquad y'(0) = 1$$

given that $y_1 = x$ is a solution of the corresponding homogeneous differential equation.

**27.** Solve the differential equation

$$y^{(5)} - y' - \frac{4}{x}y = 0, \qquad x > 0.$$

[*Hint:* Setting $y = vx$ and then $v^{(4)} - v = w$, we obtain the differential equation $xw' + 5w = 0$. Thus, $v^{(4)} - v = cx^{-5}$, where $c$ is an arbitrary constant. Use the variation-of-parameters method to obtain an integral formula for $v$ (the resulting integrals cannot be expressed in terms of elementary functions).[30]]

---

[30]The preceding transformations (and a generalization to higher-order equations) are due to M. S. Klamkin, *Math. Magazine* 43 (1970): 272–75.

**28.** Find the solution of the IVP

$$x(1 - 2x \ln x)y'' + (1 + 4x^2 \ln x)y' - (2 + 4x)y$$
$$= e^{2x}(1 - 2x \ln x)^2, \qquad \tfrac{1}{2} \le x \le 1$$

$$y(\tfrac{1}{2}) = \frac{e}{2} ; \qquad y'(\tfrac{1}{2}) = e(2 + \ln 2)$$

given that $y_1 = \ln x$ is a solution of the corresponding homogeneous differential equation.

**29. Seismology** In recording an earthquake, the response of the accelerogram is governed by the equation

$$\ddot{x} + 2\eta_0\omega_0\dot{x} + \omega_0^2 x = -a(t),$$

where $x(t)$ is the instrument response recorded on the accelerogram, $a(t)$ the ground acceleration, $\eta_0$ the damping ratio of the instrument, and $\omega_0$ the undamped natural frequency of the instrument.[31] Take $\eta_0 = \omega_0 = 1$. Find the general solution of this differential equation when $a(t) = e^{-t}/(t + 1)^2$.

**30.** If $n = 2$ show that the formula of Theorem 2 for $y_p$ yields

$$y_p(x) = \int_{x_0}^{x} \frac{y_1(s)y_2(x) - y_2(s)y_1(x)}{y_1(s)y_2'(s) - y_1'(s)y_2(s)} \cdot \frac{f(s)}{a_2(s)} \, ds, \qquad \text{with } y_p(x_0) = y_p'(x_0) = 0.$$

## REVIEW EXERCISES

In Exercises 1 through 3, classify the differential equation as being either homogeneous or nonhomogeneous, with constant coefficients or variable coefficients; also state its order and indicate verbally a method of solution.

**1.** $x^2y'' + 8x^3y''' - 46y - 3 = 0$

**2.** $8y - 2y'' - 7x = 0$

**3.** $y'' + 8xy' = 8y$

In Exercises 4 and 5, write an expression for the Wronskian of two linearly independent solutions without finding the solutions themselves.

**4.** $2y'' + \dfrac{7}{x}y' - y = 0$

**5.** $y'' - 8y' + 8e^xy = 0$

**6.** Using an appropriate theorem from this chapter, verify that there exists a unique solution to the initial value problem

$$(2 - x^2)y'' + (\cos x)y' + (\tan x)y = 0, \qquad -1 < x < 1,$$
$$y(0) = 0, \qquad y'(0) = 1.$$

[31] *Bull. Seismol. Soc. Amer.* 59 (1969): 1591–98.

What do we know about the interval of existence of the solution?

**7.** A certain homogeneous linear differential equation with constant coefficients has the characteristic equation

$$(\lambda - 7)^3(\lambda^2 - 3)(\lambda^2 + 2\lambda + 2) = 0.$$

Write the general solution for this differential equation and state its order.

In Exercises 8 through 10, find the first five coefficients in the Taylor expansion of the solution about the point $x_0 = 1$.

**8.** $x^2y'' - \left(\sin \dfrac{\pi x}{2}\right)y' + y = 0, \qquad y(1) = -1, \qquad y'(1) = 1$

**9.** $(2 + x)y'' + xy' - (3 + x^2)y = 0, \qquad y(1) = 0, \qquad y'(1) = 3$

**10.** $(5 - x)y' - (\sec^2 \pi x)y = 0, \qquad y(1) = 1$

In Exercises 11 through 26, write the general solution of the given differential equation.

**11.** $y'' + 17y' + 16y = e^x + 4e^{-x}$

**12.** $(x^2 - 2x)y'' + (2 - x^2)y' + 2(x - 1)y = 0, \quad 0 < x < 1,$ given that $y_1 = x^2$ is a solution.

**13.** $3x^2y'' - 2xy' - 12y = 0, x > 0$

**14.** $y'' + y = \csc x + 1, 0 < x < 1$

**15.** $y''' - 9y'' + 6y' + 16y = 0$

**16.** $y'' - y = x^3e^x$

**17.** $(\tan x - \sec^2 x)y'' + \tan x(2 \sec^2 x - 1)y' + \sec^2 x(1 - 2 \tan x)y = 0,$ $0 < x < 1,$ given that $y_1 = \tan x$ is a solution.

**18.** $x^2y'' + 7xy' + 9y = 0, x < 0$

**19.** $y''' + 27y = x$

**20.** $9y'' + 48y' + 64y = \sin 2x$

**21.** $y'' + y = \cot x, 0 < x < \pi$

**22.** $(x \cos x - \sin x)y'' + (x \sin x)y' - (\sin x)y = 0, \pi/4 < x < \pi/2,$ given that $y_1 = \sin x$ is a solution.

**23.** $y'' + y = x \sin x$

**24.** $y^{iv} - 2y''' + 2y'' - 2y' + y = 0$

**25.** $x^2y'' + 9xy' + 17y = 0, x > 0$

**26.** $3x^2y'' - 2xy' - 12y = x^{-2}, x > 0$

**27.** A projectile is fired horizontally[32] from a gun which is located 144 ft (44m) above a horizontal plane and has a muzzle speed of 800 ft/sec (240 m/sec). (a) How long does the projectile remain in the air? (b) At what horizontal distance does it strike the ground? (c) What is the magnitude of the vertical component of its velocity as it strikes the ground? Take $g = 32.2$ ft/sec$^2$.

**28.** A rifle[33] with a muzzle velocity of 1500 ft/sec shoots a bullet at a small target 150 ft away. How high above the target must the gun be aimed so that the bullet will hit the target? Take $g = 32.2$ ft/sec$^2$.

**29.** A batter[34] hits a pitched ball at a height 4.0 ft above ground so that its angle of projection is 45° and its horizontal range is 350 ft. The ball is fair down the left field line where a 24 ft high fence is located 320 ft from home plate. Will the ball clear the fence? Take $g = 32.2$ ft/sec$^2$.

**30.** It has been determined experimentally[35] that the magnitude in ft/sec$^2$ of the deceleration due to air resistance of a projectile is $0.001 \, v^2$, where $v$ is expressed in ft/sec. If the projectile is released from rest and keeps pointing downward, determine its velocity after it has fallen 500 ft. [*Hint:* The total acceleration is $g - 0.001v^2$, where $g = 32.2$ ft/sec$^2$.]

**31.** The acceleration due to gravity[36] of a particle falling toward the earth is $a = -gR^2/r^2$, where $r$ is the distance from the *center* of the earth to the particle, $R$ is the radius of the earth, and $g$ is the acceleration due to gravity at the surface of the earth. Derive an expression for *escape velocity*, the minimum velocity with which a particle should be projected vertically upward from the surface of the earth if it is not to return to the earth. [*Hint:* $v = 0$ for $r = \infty$.]

**32.** Automobile $A$ starts[37] from 0 and accelerates at the constant rate of 4 ft/sec$^2$. A short time later it is passed by truck $B$, which is travelling in the opposite direction at a constant speed of 45 ft/sec. Knowing that truck $B$ passes point $O$ 25 sec after automobile $A$ started from there, determine when and where the vehicles pass each other.

**33.** In a vertical takeoff,[38] rocket engines exert a thrust that gives it a constant acceleration of 2.0 $g$ for 40 sec. What velocity is attained at this time and what is its height?

[32]This is Exercise 8 in David Halliday and Robert Resnik, *Physics, Part I*, 3rd ed. (New York: John Wiley & Sons, 1977), p. 68. Copyright © 1977 by John Wiley & Sons. Reprinted by permission of John Wiley & Sons, Inc.

[33]This is Exercise 6, ibid., p. 67. Copyright © 1977 by John Wiley & Sons. Reprinted by permission of John Wiley & Sons, Inc.

[34]This is Exercise 14, ibid., p. 68. Copyright © 1977 by John Wiley & Sons. Reprinted by permission of John Wiley & Sons, Inc.

[35]This is Exercise 11.16 in Ferdinand P. Beer and E. Russell Johnston, Jr., *Vector Mechanics for Engineers: Statics and Dynamics*, 3rd ed. (New York: McGraw-Hill, 1977), p. 446. Reprinted by permission of McGraw-Hill Book Company.

[36]This is Exercise 11.21, ibid., p. 447. Reprinted by permission of McGraw-Hill Book Company.

[37]This is Exercise 11.28, ibid., p. 454. Reprinted by permission of McGraw-Hill Book Company.

[38]This is Exercise 1.3-2 in Eduard C. Pestel and William T. Thompson, *Dynamics* (New York: McGraw-Hill, 1968), p. 17. Reprinted by permission of McGraw-Hill Book Company.

**Figure 2.6**

**34.** A train[39] can attain its maximum speed of 60 mph in 4 min., then stop within a distance of 1/2 mile. Assuming that acceleration and deceleration are uniform, find the least time required for the train to go between two stations $8\frac{1}{2}$ miles apart. Draw a velocity-time diagram.

**35.** It is estimated[40] that an athlete, in putting a shot, releases the shot at a height of 7 ft, with a velocity of 34 ft/sec, at an angle of 40° above the horizontal. Determine the range, the maximum height attained by the shot, and the time of flight.

**36.** A telephone pole[41] is raised by backing a truck against it, as shown in Figure 2.6. If the angular velocity of the pole is to be a constant, show that the required speed of the truck is $\dot{x} = h\dot{\theta}/\sin^2 \theta$.

**37.** The displacement[42] of a simple harmonic oscillator is given by $x = a \sin \omega t$. If the values of this displacement $x$ and the velocity $\dot{x}$ are plotted on perpendicular axes, eliminate $t$ to show that the locus of points $(x, \dot{x})$ is an ellipse. Show that this ellipse represents a path of constant energy. [*Hint:* Energy of a simple harmonic oscillator equals $\frac{1}{2}$ (mass)(velocity)$^2$ $+ \frac{1}{2}$ (stiffness)·(displacement)$^2$. The basic differential equation for the motion is $\ddot{x} + \omega^2 x = 0$, where $\omega^2 = s/m$, $s$ is the stiffness, and $m$ is the mass.]

**38.** An overdamped spring mass system is displaced a distance $A$ from its equilibrium position and released from rest. Find the position of the mass at any time $t$. [*Hint:* The differential equation is $m\ddot{y} + \alpha\dot{y} + Ky = 0$.]

**39.** The equation $m\ddot{x} + sx = F_0 \sin \omega t$ describes[43] the motion of an undamped simple harmonic oscillator driven by a force of frequency $\omega$. Show that the steady-state solution (the particular solution) is given by

$$x = \frac{F_0 \sin \omega t}{m(\omega_0^2 - \omega^2)},$$

[39]This is Exercise 1.3-18, ibid., p. 19. Reprinted by permission of McGraw-Hill Book Company.
[40]This is Exercise 1.4-13, ibid., p. 34. Reprinted by permission of McGraw-Hill Book Company.
[41]This is Exercise 1.4-42, ibid., p. 39. Reprinted by permission of McGraw-Hill Book Company.
[42]This is Exercise 1.11 in H. J. Pain, *The Physics of Vibrations and Waves*, 2nd ed. (New York: John Wiley & Sons, 1976), p. 43. Reprinted by permission of John Wiley & Sons, Inc. Copyright © 1976 by John Wiley & Sons.
[43]This is Exercise 2.3, ibid., p. 68. Copyright © 1976 by John Wiley & Sons. Reprinted by permission of John Wiley & Sons, Inc.

**Figure 2.7**

where $\omega_0^2 = s/m$. Sketch the behavior of $x$ versus $\omega$ and note that the change of sign as $\omega$ passes through $\omega_0^2$ defines a phase change of $\pi$ radians in the displacement. Now show that the general solution for the displacement is given by

$$x = \frac{F_0 \sin \omega t}{m(\omega_0^2 - \omega^2)} + A \cos \omega_0 t + B \sin \omega_0 t,$$

where $A$ and $B$ are constant.

**40.** Consider[44] the nonlinear time-invariant subharmonic generating circuit shown in Figure 2.7. The inductor is linear and the capacitor has a characteristic

$$v_c = \frac{1}{18} q + \frac{2}{27} q^3.$$

(a) Verify that for an input $e_s = (1/54) \cos t$ volt, a response $q(t) = \cos (t/3)$ coulombs satisfies the differential equation. [*Note:* the charge oscillates at *one-third* of the frequency of the source.]
(b) Calculate the current through the source for the charge found in part (a).

**41.** The circuit[45] shown in Figure 2.8 is made of linear time-invariant elements. The input is $e_s$, and the response is $v_c$. Knowing that $e_s(t) = \sin 2t$ volts, and that at time $t = 0$ the state is $i_L = 2$ amp and $v_c = 1$ volt, calculate the complete response.

**Figure 2.8**

[44]This is Exercise 4 in Charles A. Desoer and Ernest S. Kuh, *Basic Circuit Theory* (New York: McGraw-Hill, 1969), p. 332. Reprinted by permission of McGraw-Hill Book Company.
[45]This is Exercise 8, ibid., p. 333. Reprinted by permission of McGraw-Hill Book Company.

**Figure 2.9**

42. Let[46] a sinusoidal voltage $e_s(t) = 3 \cos (10^6 t)$ volts be applied at time $t = 0$ to the linear time invariant $LC$-circuit shown in Figure 2.9. Given $i(0) = 1$ mA and $v(0) = 0$, calculate and sketch $i(t)$ for $t \geq 0$.

43. What[47] is the time of flight of a projectile fired with a muzzle velocity $v_0$ at an angle $\alpha$ above the horizontal if the launching site is located at height $h$ above the target?

44. Determine[48] the general solution for the motion of an underdamped harmonic oscillator having mass $m$, damping constant $b$, and spring constant $K$ when the driving force (the external force) is a constant $F_0$. Solve the nonhomogeneous equation by assuming a form for the particular solution and then determining the constant coefficients.

45. By differentiation,[49] confirm that $\Phi(\phi) = \dfrac{1}{2\pi} (\cos m\phi + i \sin m\phi)$ is a solution of

$$\frac{1}{\Phi} \frac{d^2 \Phi}{d\phi^2} = -m^2.$$

46. If $u$ and $v$ are linearly independent particular solutions of the differential equation

$$y''' + Py'' + Qy' + Ry = 0,$$

prove that the general solution is given by $y = Au + Bv$, where $A$ and $B$ satisfy the equations

$$\frac{A'}{v} = -\frac{B'}{u} = c \, \frac{e^{-\int P dx}}{(u'v - uv')^2}$$

and $c$ is an arbitrary constant.[50]

47. Solve[51] the differential equation

$$(y')^3 - 3y(y')^2 + 4y^3 = a.$$

[*Hint:* Differentiate both sides of the equation with respect to $x$ and factor.]

[46]This is Exercise 11a, ibid., p. 334. Reprinted by permission of McGraw-Hill Book Company.
[47]This is Exercise 2-2.2 in J. Norwood, Jr., *Intermediate Classical Mechanics* (Englewood Cliffs, N.J.: Prentice-Hall, 1979), p. 51. Reprinted by permission of Prentice-Hall, Inc.
[48]This is Exercise 3-3.6, ibid., p. 102. Reprinted by permission of Prentice-Hall, Inc.
[49]This is Exercise 14 in Bernard G. Harvey, *Introduction to Nuclear Physics and Chemistry*, 2nd ed. (Englewood Cliffs, N.J.: Prentice-Hall, 1969), p. 26. Reprinted by permission of Prentice-Hall, Inc.
[50]See *Amer. Math. Monthly* 36 (1929): 112–13.
[51]See *Amer. Math. Monthly* 60 (1953): 264.

# CHAPTER 3

# Linear Systems

---

## 3.1 INTRODUCTION AND BASIC THEORY

---

The differential equations that we studied in Chapters 1 and 2 are ordinary differential equations involving one unknown function. Consequently, in each case up to now, we had to solve *one* differential equation which involved *one* unknown function. For many reasons, including applications and generalizations, one is interested in studying systems of $n$ differential equations in $n$ unknown functions where $n$ is an integer $\geq 2$. Since in this book we are mainly interested in obtaining solutions, we shall restrict our attention to systems of linear differential equations, *linear systems. Nonlinear systems,* that is, systems of nonlinear differential equations, are usually (and successfully) studied in more advanced courses, by approximating each nonlinear equation of the system by a linear one. For some results on nonlinear systems, see Chapter 8.

In this chapter we mainly consider systems of *two* linear differential equations in *two* unknown functions of the form

$$\begin{aligned}
\dot{x}_1 &= a_{11}(t)x_1 + a_{12}(t)x_2 + f_1(t) \\
\dot{x}_2 &= a_{21}(t)x_1 + a_{22}(t)x_2 + f_2(t),
\end{aligned} \tag{1}$$

where the coefficients $a_{11}$, $a_{12}$, $a_{21}$, $a_{22}$, and the functions $f_1$, $f_2$ are all given continuous functions of $t$ on some interval $I$ and $x_1$, $x_2$ are unknown functions of $t$. As usual, the dots over $x_1$ and $x_2$ in (1) indicate differentiation with respect to the independent variable $t$.

**REMARK 1**　In the treatment of systems of differential equations, it is customary to use the symbols $x_1, x_2, \ldots, x_n$ to denote unknown functions and to use $t$ as the independent variable. This notation has obvious advantages when we are dealing with general systems. However, in special cases alternative notation may be used.

In this section we present a few definitions and state without proof some basic theorems about the solutions of system (1). These definitions and theorems extend easily to systems of $n$ linear differential equations in $n$ unknown functions of the form

$$\begin{aligned}
\dot{x}_1 &= a_{11}(t)x_1 + a_{12}(t)x_2 + \cdots + a_{1n}(t)x_n + f_1(t) \\
\dot{x}_2 &= a_{21}(t)x_1 + a_{22}(t)x_2 + \cdots + a_{2n}(t)x_n + f_2(t) \\
&\;\cdots\cdots\cdots\cdots\cdots\cdots\cdots\cdots\cdots\cdots\cdots\cdots\cdots\cdots\cdots\cdots \\
\dot{x}_n &= a_{n1}(t)x_1 + a_{n2}(t)x_2 + \cdots + a_{nn}(t)x_n + f_n(t),
\end{aligned} \tag{2}$$

or, more compactly,

$$\dot{x}_i = \sum_{j=1}^{n} a_{ij}(t)x_j + f_i(t), \qquad i = 1, 2, \ldots, n.$$

In subsequent sections we shall present two elementary methods for finding explicit solutions of system (1) when the coefficients $a_{11}$, $a_{12}$, $a_{21}$, and $a_{22}$ are all known constants. These methods, with varying degrees of technical difficulties, can be extended to solving system (2) when its coefficients $a_{ij}$ are all known constants. For the reader's sake we mention here that the *method of elimination* discussed in Section 3.2 is, perhaps, the easiest method to solve system (1). The *matrix method* outlined in Section 3.3 is recommended for system (2) when $n \geq 3$. This is an elegant method, but its detailed presentation and justification requires a good knowledge of matrix analysis and linear algebra.

### DEFINITION 1

*A solution of the system* (1) *is a pair of functions $x_1(t)$ and $x_2(t)$, each being differentiable on an interval I and which, when substituted into the two equations in* (1), *makes them identities in t for all t in I.*

For example,

$$x_1(t) = -\tfrac{3}{2}t - \tfrac{5}{4}, x_2(t) = -t - \tfrac{3}{2}$$

is a solution of the system

$$\dot{x}_1 = x_2 + t \tag{3}$$
$$\dot{x}_2 = -2x_1 + 3x_2 + 1$$

for all $t$, as the reader can easily verify.

Sometimes it is convenient (and suitable for studying more general systems) to denote a solution of (1) by the column vector

$$\begin{bmatrix} x_1(t) \\ x_2(t) \end{bmatrix}.$$

For example,

$$\begin{bmatrix} -\tfrac{3}{2}t - \tfrac{5}{4} \\ -t - \tfrac{3}{2} \end{bmatrix}$$

is a solution of system (3).

**REMARK 2**  The algebra of column vectors that is used in this chapter is embodied in the following statements:

(a) *Equality*

$$\begin{bmatrix} x_1(t) \\ x_2(t) \end{bmatrix} = \begin{bmatrix} y_1(t) \\ y_2(t) \end{bmatrix} \text{ if and only if } \quad x_1(t) = y_1(t) \text{ and } x_2(t) = y_2(t).$$

(b) *Linear Combination.* For any constants $c_1$ and $c_2$,

$$c_1\begin{bmatrix} x_1(t) \\ x_2(t) \end{bmatrix} + c_2\begin{bmatrix} y_1(t) \\ y_2(t) \end{bmatrix} = \begin{bmatrix} c_1x_1(t) + c_2y_1(t) \\ c_1x_2(t) + c_2y_2(t) \end{bmatrix}.$$

## DEFINITION 2

*When both functions $f_1$ and $f_2$ in (1) are equal to zero, the system is called homogeneous. Otherwise it is called nonhomogeneous.*

For example, system (3) is nonhomogeneous. On the other hand, the system

$$\begin{aligned} \dot{x}_1 &= x_2 \\ \dot{x}_2 &= -2x_1 + 3x_2 \end{aligned} \tag{4}$$

is homogeneous. The reader can verify that each of the two vectors

$$\begin{bmatrix} e^t \\ e^t \end{bmatrix} \quad \text{and} \quad \begin{bmatrix} e^{2t} \\ 2e^{2t} \end{bmatrix} \tag{5}$$

is a solution of system (4).

To a great extent, the theory of linear systems resembles the theory of linear differential equations. Let us consider the homogeneous system

$$\begin{aligned} \dot{x}_1 &= a_{11}(t)x_1 + a_{12}(t)x_2 \\ \dot{x}_2 &= a_{21}(t)x_1 + a_{22}(t)x_2, \end{aligned} \tag{6}$$

where the coefficients $a_{11}$, $a_{12}$, $a_{21}$, and $a_{22}$ are all continuous functions on some interval $I$. As in the case of homogeneous linear differential equations, we have the following.

## THEOREM 1

*Any linear combination of solutions of (6) is also a solution of (6).*

For example, each of the two column vectors in (5) is a solution of system (4); therefore, for any constants $c_1$ and $c_2$, the linear combination

$$c_1\begin{bmatrix} e^t \\ e^t \end{bmatrix} + c_2\begin{bmatrix} e^{2t} \\ 2e^{2t} \end{bmatrix} = \begin{bmatrix} c_1e^t + c_2e^{2t} \\ c_1e^t + 2c_2e^{2t} \end{bmatrix} \tag{7}$$

is also a solution of (4).

## DEFINITION 3

*The column vector*

$$\begin{bmatrix} 0 \\ 0 \end{bmatrix},$$

*that is, $x_1(t) \equiv 0$, $x_2(t) \equiv 0$, is a solution of (6) for any choice of the coefficients. This solution is called the trivial solution. Any other solution of (6) is called a nontrivial solution.*

**DEFINITION 4**

*Two solutions*

$$\begin{bmatrix} x_{11}(t) \\ x_{21}(t) \end{bmatrix} \quad and \quad \begin{bmatrix} x_{12}(t) \\ x_{22}(t) \end{bmatrix} \tag{8}$$

*of system* (6) *are called linearly independent on an interval I if the statement*

$$c_1 \begin{bmatrix} x_{11}(t) \\ x_{21}(t) \end{bmatrix} + c_2 \begin{bmatrix} x_{12}(t) \\ x_{22}(t) \end{bmatrix} = \begin{bmatrix} 0 \\ 0 \end{bmatrix} \quad for \ all \ t \ in \ I,$$

*that is,*

$$c_1 x_{11}(t) + c_2 x_{12}(t) = 0 \quad and \quad c_1 x_{21}(t) + c_2 x_{22}(t) = 0$$

*for all t in I implies that* $c_1 = c_2 = 0$. *Otherwise the solutions* (8) *are called linearly dependent.*

For example, we can demonstrate that the two solutions (5) are linearly independent solutions of the system (4). In fact, if

$$c_1 \begin{bmatrix} e^t \\ e^t \end{bmatrix} + c_2 \begin{bmatrix} e^{2t} \\ 2e^{2t} \end{bmatrix} = \begin{bmatrix} 0 \\ 0 \end{bmatrix},$$

then $c_1 e^t + c_2 e^{2t} = 0$ and $c_1 e^t + 2c_2 e^{2t} = 0$. Subtracting the first equation from the second, we find that $c_2 e^{2t} = 0$. Thus $c_2 = 0$. Consequently, the first equation becomes $c_1 e^t = 0$ and so $c_1 = 0$. Therefore, $c_1 = c_2 = 0$, which establishes our claim. The reader can verify that

$$\begin{bmatrix} e^{2t} \\ 2e^{2t} \end{bmatrix} \quad and \quad \begin{bmatrix} 3e^{2t} \\ 6e^{2t} \end{bmatrix} \tag{9}$$

are linearly dependent solutions of (4).

The following criterion, reminiscent of the Wronskian determinant, can be used to check the linear dependence or independence of solutions of system (6).

**THEOREM 2**

*The two solutions*

$$\begin{bmatrix} x_{11}(t) \\ x_{21}(t) \end{bmatrix} \quad and \quad \begin{bmatrix} x_{12}(t) \\ x_{22}(t) \end{bmatrix}$$

*of the system*

$$\dot{x}_1 = a_{11}(t)x_1 + a_{12}(t)x_2$$

$$\dot{x}_2 = a_{21}(t)x_1 + a_{22}(t)x_2$$

*are linearly independent on an interval I if and only if the Wronskian determinant*

$$\begin{vmatrix} x_{11}(t) & x_{12}(t) \\ x_{21}(t) & x_{22}(t) \end{vmatrix}$$

*is never zero on I.*

For example, the two solutions (5) of system (4) are linearly independent because

$$\begin{vmatrix} e^t & e^{2t} \\ e^t & 2e^{2t} \end{vmatrix} = 2e^{3t} - e^{3t} = e^{3t} \neq 0 \qquad \text{for all } t.$$

On the other hand, the two solutions (9) of system (4) are linearly dependent because

$$\begin{vmatrix} e^{2t} & 3e^{2t} \\ 2e^{2t} & 6e^{2t} \end{vmatrix} = 6e^{4t} - 6e^{4t} = 0.$$

The following basic *existence and uniqueness theorem* for linear systems is stated here without proof. The reader should recognize the strong resemblance between this theorem and the existence–uniqueness theorem of Chapter 2.

### THEOREM 3

*Assume that the coefficients $a_{11}$, $a_{12}$, $a_{21}$, $a_{22}$ and the functions $f_1$ and $f_2$ of the system*

$$\begin{aligned} \dot{x}_1 &= a_{11}(t)x_1 + a_{12}(t)x_2 + f_1(t) \\ \dot{x}_2 &= a_{21}(t)x_1 + a_{22}(t)x_2 + f_2(t) \end{aligned} \tag{1}$$

*are all continuous on an interval I. Let $t_0$ be a point in I and let $x_{10}$ and $x_{20}$ be two given constants. Then the IVP consisting of system (1) and the initial conditions*

$$x_1(t_0) = x_{10}, \qquad x_2(t_0) = x_{20}$$

*has a unique solution*

$$\begin{bmatrix} x_1(t) \\ x_2(t) \end{bmatrix}.$$

*Furthermore, this unique solution is valid throughout the interval I.*

Using this theorem, the reader can verify (by direct substitution) that the unique solution of the IVP

$$\begin{aligned} \dot{x}_1 &= x_2 + t \\ \dot{x}_2 &= -2x_1 + 3x_2 + 1 \\ x_1(0) &= -1 \\ x_2(0) &= -\tfrac{5}{4} \end{aligned} \tag{10}$$

is

$$\begin{aligned} x_1(t) &= \tfrac{1}{4}e^t - \tfrac{3}{2}t - \tfrac{5}{4} \\ x_2(t) &= \tfrac{1}{4}e^t - t - \tfrac{3}{2}. \end{aligned} \tag{11}$$

### THEOREM 4

*There exist two linearly independent solutions of the system*

$$\begin{aligned} \dot{x}_1 &= a_{11}(t)x_1 + a_{12}(t)x_2 \\ \dot{x}_2 &= a_{21}(t)x_1 + a_{22}(t)x_2. \end{aligned} \tag{6}$$

*Furthermore, if the two column vectors*

$$\begin{bmatrix} x_{11}(t) \\ x_{21}(t) \end{bmatrix} \quad and \quad \begin{bmatrix} x_{12}(t) \\ x_{22}(t) \end{bmatrix}$$

*are linearly independent solutions of (6), the general solution of system (6) is given by*

$$\begin{bmatrix} x_1(t) \\ x_2(t) \end{bmatrix} = c_1 \begin{bmatrix} x_{11}(t) \\ x_{21}(t) \end{bmatrix} + c_2 \begin{bmatrix} x_{12}(t) \\ x_{22}(t) \end{bmatrix};$$

*that is,*

$$x_1(t) = c_1 x_{11}(t) + c_2 x_{12}(t)$$

$$x_2(t) = c_1 x_{21}(t) + c_2 x_{22}(t),$$

*where $c_1$ and $c_2$ are arbitrary constants.*

For example, the general solution of (4) is given by

$$\begin{bmatrix} x_1(t) \\ x_2(t) \end{bmatrix} = c_1 \begin{bmatrix} e^t \\ e^t \end{bmatrix} + c_2 \begin{bmatrix} e^{2t} \\ 2e^{2t} \end{bmatrix};$$

that is,

$$x_1(t) = c_1 e^t + c_2 e^{2t} \quad and \quad x_2(t) = c_1 e^t + 2c_2 e^{2t}. \tag{12}$$

**THEOREM 5**

*If the two solutions*

$$\begin{bmatrix} x_{11}(t) \\ x_{21}(t) \end{bmatrix} \quad and \quad \begin{bmatrix} x_{12}(t) \\ x_{22}(t) \end{bmatrix}$$

*of the system*

$$\dot{x}_1 = a_{11}(t)x_1 + a_{12}(t)x_2$$

$$\dot{x}_2 = a_{21}(t)x_1 + a_{22}(t)x_2$$

*are linearly independent and if*

$$\begin{bmatrix} x_{1p}(t) \\ x_{2p}(t) \end{bmatrix}$$

*is a particular solution of the system*

$$\dot{x}_1 = a_{11}(t)x_1 + a_{12}(t)x_2 + f_1(t)$$

$$\dot{x}_2 = a_{21}(t)x_1 + a_{22}(t)x_2 + f_2(t), \tag{1}$$

*the general solution of (1) is given by*

$$\begin{bmatrix} x_1(t) \\ x_2(t) \end{bmatrix} = c_1 \begin{bmatrix} x_{11}(t) \\ x_{21}(t) \end{bmatrix} + c_2 \begin{bmatrix} x_{12}(t) \\ x_{22}(t) \end{bmatrix} + \begin{bmatrix} x_{1p}(t) \\ x_{2p}(t) \end{bmatrix};$$

*that is,*

$$x_1(t) = c_1 x_{11}(t) + c_2 x_{12}(t) + x_{1p}(t)$$

$$x_2(t) = c_1 x_{21}(t) + c_2 x_{22}(t) + x_{2p}(t).$$

For example, the general solution of system (3) is.

$$\begin{bmatrix} x_1(t) \\ x_2(t) \end{bmatrix} = c_1 \begin{bmatrix} e^t \\ e^t \end{bmatrix} + c_2 \begin{bmatrix} e^{2t} \\ 2e^{2t} \end{bmatrix} + \begin{bmatrix} -\frac{3}{2}t - \frac{5}{4} \\ -t - \frac{3}{2} \end{bmatrix};$$

that is,

$$x_1(t) = c_1 e^t + c_2 e^{2t} - \tfrac{3}{2}t - \tfrac{5}{4}$$

$$x_2(t) = c_1 e^t + 2c_2 e^{2t} - t - \tfrac{3}{2}. \tag{13}$$

## APPLICATIONS 3.1.1

Linear systems of differential equations are utilized in many interesting mathematical models of real-life situations. Here we present a sampling of these applications.

■ Consider two species which coexist in a given environment and which interact **Ecology** between themselves in a specific way. For example, one species eats the other, or both species help each other to survive. Let us denote by $N_1(t)$ and $N_2(t)$ the number, at time $t$, of the first and second species, respectively. If the two species were in isolation from each other, it would be reasonable to assume that they vary at a rate proportional to their number present. That is,

$$\dot{N}_1 = aN_1 \quad \text{and} \quad \dot{N}_2 = dN_2,$$

where the constants of proportionalities $a$ and $d$ are certain constants (positive, negative, or zero, depending on whether the population increases, decreases, or remains constant, respectively).

However, we have assumed that the two species interact between themselves. Let us assume for simplicity that the rate of change of one population due to interaction of the two species is proportional to the size of the other population. Then we obtain the following linear system of differential equations:

$$\dot{N}_1 = aN_1 + bN_2$$

$$\dot{N}_2 = cN_1 + dN_2,$$

where $b$ and $c$ are certain constants (positive, negative, or zero).

■ In Turing's theory of morphogenesis,[1] the state of a cell is represented by **Morphogenesis** the numerical values of a pair of substances $x$ and $y$ which are called *morphogens*. As an illustration, Turing proposed that these morphogens interact chemically according to the kinetic equations

$$\dot{x} = 5x - 6y + 1 \tag{14}$$

$$\dot{y} = 6x - 7y + 1.$$

The general solution of the system (14) is given by

$$x = c_1 e^{-t} + c_2 t e^{-t} + 1 \tag{15}$$

$$y = c_1 e^{-t} + c_2 e^{-t}(t - \tfrac{1}{6}) + 1.$$

[1]Turing, *Royal Soc. London Phil. Trans.* B237 (1952): 42. See also R. Rosen, *Foundations of Mathematical Biology*, vol. II (New York: Academic Press, 1972), pp. 61–66.

**Medicine**
Endocrinology

■ Systems of linear differential equations of the form

$$\dot{x}_i = a_{i0} + \sum_{j=1}^{n} a_{ij} x_j, \qquad i = 1, 2, \ldots, n, \tag{16}$$

have been used to describe periodic fluctuations of the concentrations of hormones in the bloodstream associated with the menstrual cycle.[2] In 1957, Danziger and Elmergreen[3] proposed mathematical models of endocrine systems that are linear systems of differential equations. For example, reactions of the pituitary and thyroid glands in the regulation of the metabolic rate and reactions of the pituitary and ovary glands in the control of the menstrual cycle can be represented by simple linear systems of the form

$$\dot{x}_1 = -a_{11}x_1 - a_{12}x_2 + a_{10} \tag{17}$$
$$\dot{x}_2 = a_{21}x_1 - a_{22}x_2 + a_{20},$$

where all the $a_{ij}$ are positive constants with the exception of $a_{20}$, which is equal to zero.

Recently, Ackerman and colleagues[4] obtained system (17) with $a_{20} > 0$ as a linear approximation of a mathematical model for the detection of diabetes.

**Circuit Theory**

■ Consider the electric circuits shown in Figures 3.1 and 3.2. Assume that at time $t$ the current flows as indicated in each closed path. Show in each case that the currents satisfy the given *systems* of linear differential equations.

**Figure 3.1**

$$\dot{I}_1 = -\frac{R_1}{L_1}I_1 + \frac{R_1}{L_1}I_2 + \frac{V(t)}{L_1}$$

$$\dot{I}_2 = \frac{R_1}{L_2}I_1 - \frac{R_1 + R_2}{L_2}I_2. \tag{18}$$

[2]A. Rapoport, *Bull. Math. Biophys.* 14 (1952): 171–83.
[3]L. Danziger and G. L. Elmergreen, *Bull. Math. Biophys.* 19 (1957): 9–18.
[4]E. Ackerman, L. C. Gatewood, J. W. Rosevear, and G. D. Molnal, "Blood Glucose Regulation and Diabetes," in *Concepts and Models of Biomathematics*, ed. F. Heinmets (New York: Marcel Dekker, 1969), chapter 4.

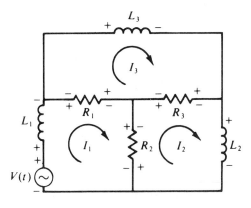

**Figure 3.2**

$$\dot{I}_1 = -\frac{R_1 + R_2}{L_1}I_1 + \frac{R_2}{L_1}I_2 + \frac{R_1}{L_1}I_3 + \frac{V(t)}{L_1}$$

$$\dot{I}_2 = \frac{R_2}{L_2}I_1 - \frac{R_2 + R_3}{L_2}I_2 + \frac{R_3}{L_2}I_3 \qquad (19)$$

$$\dot{I}_3 = \frac{R_1}{L_3}I_1 + \frac{R_3}{L_3}I_2 - \frac{R_3 + R_1}{L_3}I_3.$$

**Solution**  These systems are obtained by applying Kirchhoff's voltage law (see Section 1.1.1) to each closed path. For example, the first equation in (19) is obtained by applying Kirchhoff's voltage law to the closed path in Figure 3.2, where the current $I_1$ flows. In fact, we have

$$V(t) - L_1\dot{I}_1 - R_1I_1 + R_1I_3 - R_2I_1 + R_2I_2 = 0, \qquad (20)$$

which after some algebraic manipulations leads to the first differential equation in the system (19). Clearly, the sign in front of any term in Eq. (20) is $+$ or $-$ according to whether the term represents voltage increase or drop, respectively.

**Arms Races**
**The Richardson Model**

■ Lewis F. Richardson, in a classic monograph,[5] devised a mathematical model of arms races between two nations. In this model two very simplistic assumptions are made:

1. There are only two nations involved, denoted by $A$ and $B$.
2. There is only one kind of weapon or missile available.

Let $M_A(t)$ and $M_B(t)$ denote the number of missiles available to countries $A$ and $B$, respectively. Then $\dot{M}_A(t)$ and $\dot{M}_B(t)$ are the time rate of change of the missile stocks in the two countries. The Richardson model for a two-nation

[5]Lewis F. Richardson, "Generalized Foreign Politics," *British J. Psychol. Monograph Suppl.* 23 (1939).

armament race is given by the following system of nonhomogeneous linear differential equations

$$\dot{M}_A = -a_1 M_A + b_1 M_B + c_1$$

$$\dot{M}_B = a_2 M_A - b_2 M_B + c_2,$$

where the coefficients $a_1, b_1, a_2,$ and $b_2$ are nonnegative constants. The constants $b_1$ and $a_2$ are the "defense coefficients" of the respective nations; the constants $a_1$ and $b_2$ are the "fatigue and expense" coefficients, respectively; and the constants $c_1$ and $c_2$ are the "grievance coefficients" and indicate the effects of all other factors on the acquisition of missiles.[6]

**Economics** ■ The economics application described in Section 1.4.1 can be extended to interrelated markets. For simplicity we assume that we have two commodities. Assume that the rates of change of the prices $P_1$ and $P_2$ of these commodities are, respectively, proportional to the differences $D_1 - S_1$ and $D_2 - S_2$ of the demand and supply at any time $t$. Thus, Eq. (10) of Section 1.4.1 is replaced by the system of equations

$$\frac{dP_1}{dt} = \alpha_1(D_1 - S_1)$$

and

$$\frac{dP_2}{dt} = \alpha_2(D_2 - S_2).$$

In the simple case where $D_1, D_2, S_1,$ and $S_2$ are given as linear functions of the prices $P_1$ and $P_2$, the system above takes the form of a linear system in the unknowns $P_1$ and $P_2$. That is,[7]

$$\frac{dP_1}{dt} = a_{11}P_1 + a_{12}P_2 + b_1$$

$$\frac{dP_2}{dt} = a_{21}P_1 + a_{22}P_2 + b_2.$$

**EXERCISES**

**1.** (a) Show that the column vectors

$$\begin{bmatrix} e^{3t} \\ e^{3t} \end{bmatrix} \quad \text{and} \quad \begin{bmatrix} e^{5t} \\ 3e^{5t} \end{bmatrix}$$

are linearly independent solutions of the linear system

$$\dot{x}_1 = 2x_1 + x_2$$
$$\dot{x}_2 = -3x_1 + 6x_2.$$

---

[6]For more information on arms races, the interested reader is referred to M. D. Intriligator, *Proceedings of a Conference on the Application of Undergraduate Mathematics in the Engineering, Life, Managerial, and Social Sciences*, ed. P. J. Knopp and G. H. Meyer (Atlanta: Georgia Institute of Technology, 1973), p. 253.

[7]For more details, see R. G. D. Allen, *Mathematical Economics* (London: Macmillan, 1957), p. 480.

(b) Using Theorem 4 of Section 3.1, write down the general solution of the system above.

2. (a) Show that the column vectors

$$\begin{bmatrix} e^t \\ e^t \end{bmatrix} \quad \text{and} \quad \begin{bmatrix} e^{-t} \\ \frac{1}{2}e^{-t} \end{bmatrix}$$

are linearly independent solutions of the linear system

$$\dot{x}_1 = -3x_1 + 4x_2$$
$$\dot{x}_2 = -2x_1 + 3x_2.$$

(b) Using Theorem 4 of Section 3.1, write down the general solution of the system above.

(c) Using your answer in part (b), solve the IVP

$$\dot{x}_1 = -3x_1 + 4x_2$$

$$\dot{x}_2 = -2x_1 + 3x_2$$

$$x_1(0) = -1, \qquad x_2(0) = 3.$$

3. Use Theorem 3 of Section 3.1 to prove that the only solution of the IVP

$$\dot{x}_1 = (2 \sin t)x_1 + (\ln t)x_2$$

$$\dot{x}_2 = \frac{1}{t-2}x_1 + \frac{e^t}{t+1}x_2$$

$$x_1(3) = 0, \qquad x_2(3) = 0$$

is $x_1(t) \equiv 0$ and $x_2(t) \equiv 0$ in the interval $(2, \infty)$.

4. Show that the unique solution of the IVP

$$\dot{x} = 5x - 6y + 1$$
$$\dot{y} = 6x - 7y + 1$$
$$x(0) = 0, \qquad y(0) = 0$$

is $x(t) = 1 - e^{-t}, y(t) = 1 - e^{-t}$.

5. Show that the column vectors

$$\begin{bmatrix} e^{-t} \\ e^{-t} \end{bmatrix} \quad \text{and} \quad \begin{bmatrix} te^{-t} \\ (t - \frac{1}{6})e^{-t} \end{bmatrix}$$

are linearly independent solutions of the homogeneous system

$$\dot{x} = 5x - 6y$$
$$\dot{y} = 6x - 7y$$

and that the column vector

$$\begin{bmatrix} 1 \\ 1 \end{bmatrix}$$

is a particular solution of the nonhomogeneous system (14) of Section 3.1.1.

Use Theorem 5 of Section 3.1 to show that Eqs. (15) in Section 3.1.1 give the general solution of system (14).

In Exercises 6 through 11, answer true or false.

**6.** The column vectors

$$\begin{bmatrix} e^{-t} \\ e^{-3t} \end{bmatrix} \quad \text{and} \quad \begin{bmatrix} e^{-t} \\ -e^{-3t} \end{bmatrix}$$

are linearly independent solutions of the system

$$\dot{x} = -2x + y, \quad \dot{y} = x - 2y.$$

**7.** The column vectors

$$\begin{bmatrix} e^{-t} \\ e^{-t} \end{bmatrix} \quad \text{and} \quad \begin{bmatrix} e^{-3t} \\ -e^{-3t} \end{bmatrix}$$

are linearly independent solutions of the system

$$\dot{x} = -2x + y, \quad \dot{y} = x - 2y.$$

**8.** The general solution of the system

$$\dot{x} = 3x - 2y, \quad \dot{y} = 2x - 2y$$

is

$$x = c_1 e^{-t} + c_2 e^{2t}, \quad y = 2c_1 e^{-t} + \tfrac{1}{2} c_2 e^{2t}.$$

**9.** The column vectors

$$\begin{bmatrix} \cos t \\ -\sin t \end{bmatrix} \quad \text{and} \quad \begin{bmatrix} \sin t \\ \cos t \end{bmatrix}$$

are linearly independent solutions of the system

$$\dot{x} = x + y, \quad \dot{y} = x - y.$$

**10.** The column vector

$$\begin{bmatrix} \cos t \\ -\sin t \end{bmatrix}$$

is the unique solution of the IVP

$$\dot{x} = y, \quad \dot{y} = -x$$
$$x(0) = 1, \quad y(0) = 0.$$

**11.** The system

$$\dot{x} = -x + y, \quad \dot{y} = -x - y$$

has solutions of the form

$$\begin{bmatrix} Ae^{\lambda t} \\ Be^{\lambda t} \end{bmatrix} \quad \text{with } \lambda = -1 \pm i.$$

**12.** Consider the electric network shown in Figure 3.3. Find a system of differential equations that describes the currents $I_1$ and $I_2$.

**13.** Use Kirchhoff's voltage law to verify the equations in system (18).

**14.** Use Kirchhoff's voltage law to verify the equations in system (19).

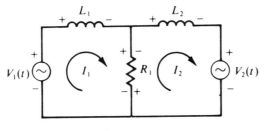

**Figure 3.3**

**15.** Professor Umbugio,[8] a mythical mathematics professor, has invented a remarkable scheme for reviewing books. He divides the time he allows himself for reviewing into three fractions, $\alpha$, $\beta$, $\gamma$. He devotes the fraction $\alpha$ of his time to a deep study of the title page and jacket. He devotes the fraction $\beta$ to a spirited search for his name and quotations from his own works. Finally, he spends the fraction $\gamma$ of his allotted time in a proportionally penetrating perusal of the remaining text. Knowing his characteristic taste for simple and direct methods, we cannot fail to be duly impressed by the differential equations on which he bases his scheme:

$$\frac{dx}{dt} = y - z, \qquad \frac{dy}{dt} = z - x, \qquad \frac{dz}{dt} = x - y. \tag{a}$$

He considers a system of solutions $x$, $y$, $z$ which is determined by initial conditions depending on a (small) parameter $\epsilon$ independent of $t$. Therefore, $x$, $y$, and $z$ depend on both $t$ and $\epsilon$, and we appropriately use the notation:

$$x = f(t, \epsilon), \qquad y = g(t, \epsilon), \qquad z = h(t, \epsilon). \tag{b}$$

The functions (b) satisfy the equations (a) and the initial conditions

$$f(0, \epsilon) = \tfrac{1}{3} - \epsilon, \qquad g(0, \epsilon) = \tfrac{1}{3}, \qquad h(0, \epsilon) = \tfrac{1}{3} + \epsilon. \tag{c}$$

Professor Umbugio defines his important fractions $\alpha$, $\beta$, and $\gamma$ by

$$\lim_{\epsilon \to 1} f(2, \epsilon) = \alpha, \qquad \lim_{\epsilon \to 0} g(5, \epsilon) = \beta, \qquad \lim_{\epsilon \to 0} h(279, \epsilon) = \gamma.$$

Deflate the professor! Find $\alpha$, $\beta$, and $\gamma$ without much numerical computation. [Hint: Observe that $\dot{x} + \dot{y} + \dot{z} = 0$ and $x\dot{x} + y\dot{y} + z\dot{z} = 0 \Rightarrow x + y + z = 1$ and $x^2 + y^2 + z^2 = \tfrac{1}{3} + 2\epsilon^2 \Rightarrow (x - \tfrac{1}{3})^2 + (y - \tfrac{1}{3})^2 + (z - \tfrac{1}{3})^2 = 2\epsilon^2$. Now take limits as $\epsilon \to 0$.]

**16. Reduction of Order** Assume that $(y_1(t), y_2(t))$ is a known solution of system (6) with $y_1(t) \neq 0$ for all $t$ in the interval $I$. Show that the transformation

$$x_1 = y_1 z_1$$

$$x_2 = y_2 z_1 + z_2$$

reduces the system (6) to the single differential equation

$$\dot{z}_2 = \left( a_{22} - \frac{y_2}{y_1} a_{12} \right) z_2.$$

[8]This problem is taken from *Amer. Math. Monthly* 54 (1974): 223.

Furthermore,[9] the function $z_1$ is given by

$$\dot{z}_1 = \frac{a_{12}}{y_1} z_2.$$

**17.** Given that $\begin{bmatrix} t \\ 1 \end{bmatrix}$ is a solution of the system

$$\dot{x}_1 = x_1 + (1 - t)x_2$$
$$\dot{x}_2 = \frac{1}{t}x_1 - x_2, \qquad t > 0$$

use reduction of order to compute the general solution of the system.

**18.** State Definitions 1 through 4 of Section 3.1 for system (2) and the corresponding homogeneous system

$$\dot{x}_i = \sum_{j=1}^{n} a_{ij}(t)x_j, \qquad i = 1, 2, \ldots, n. \tag{21}$$

**19.** State Theorem 2 of Section 3.1 for $n$ solutions of system (21).

**20.** State Theorem 3 of Section 3.1 for system (2).

**21.** State Theorem 4 of Section 3.1 for system (21).

**22.** State Theorem 5 of Section 3.1 for system (2).

In Exercises 23 through 30, answer true or false.

**23.** The column vectors

$$\begin{bmatrix} 1 \\ 0 \\ 0 \\ 0 \end{bmatrix}, \quad \begin{bmatrix} e^{-t} \\ -e^{-t} \\ 0 \\ 0 \end{bmatrix}, \quad \begin{bmatrix} e^t \\ 0 \\ -e^t \\ 0 \end{bmatrix}, \quad \begin{bmatrix} \frac{7}{2}e^{2t} \\ e^{2t} \\ -3e^{2t} \\ 3e^{2t} \end{bmatrix}$$

are linearly independent solutions of the system

$$\dot{x}_1 = x_2 - x_3 + x_4$$
$$\dot{x}_2 = -x_2 + x_4$$
$$\dot{x}_3 = x_3 - x_4$$
$$\dot{x}_4 = 2x_4.$$

**24.** The general solution of the system in Exercise 23 is

$$x_1 = c_1 + c_2 e^{-t} + c_3 e^t + \tfrac{7}{2}c_4 e^{2t}$$
$$x_2 = -c_2 e^{-t} + c_4 e^{2t}$$
$$x_3 = -c_3 e^t - 3c_4 e^{2t}$$
$$x_4 = 3c_4 e^{2t}.$$

[9]For a generalization, see J. W. Evans, *Amer. Math. Monthly* 75 (1968): 637.

**25.** The column vectors

$$\begin{bmatrix} -\frac{1}{2}e^{-3t} \\ e^{-3t} \\ e^{-3t} \end{bmatrix}, \quad \begin{bmatrix} 4e^{6t} \\ e^{6t} \\ e^{6t} \end{bmatrix}, \quad \text{and} \quad \begin{bmatrix} 0 \\ 0 \\ 0 \end{bmatrix}$$

are solutions of the system

$$\begin{aligned} \dot{x}_1 &= 5x_1 + 2x_2 + 2x_3 \\ \dot{x}_2 &= 2x_1 + 2x_2 - 4x_3 \\ \dot{x}_3 &= 2x_1 - 4x_2 + 2x_3. \end{aligned}$$

**26.** The general solution of the system in Exercise 25 is

$$\begin{aligned} x_1 &= -\tfrac{1}{2}c_1 e^{-3t} + 4c_2 e^{6t} \\ x_2 &= c_1 e^{-3t} + c_2 e^{6t} \\ x_3 &= c_1 e^{-3t} + c_2 e^{6t}. \end{aligned}$$

**27.** The column vector

$$\begin{bmatrix} 2e^{6t} \\ e^{6t} \\ 2 \end{bmatrix}$$

is the only solution of the IVP

$$\begin{aligned} \dot{x}_1 &= 5x_1 + 2x_2 + 2x_3 \\ \dot{x}_2 &= 2x_1 + 2x_2 - 4x_3 \\ \dot{x}_3 &= 2x_1 - 4x_2 + 2x_3 \end{aligned}$$

$$x_1(0) = 2, \quad x_2(0) = 1, \quad x_3(0) = 0.$$

**28.** The column vector

$$\begin{bmatrix} 2e^{6t} \\ e^{6t} \\ 0 \end{bmatrix}$$

is the only solution of the IVP in Exercise 27.

**29.** The column vectors

$$\begin{bmatrix} 2e^t \\ e^t \\ 3e^t \end{bmatrix}, \quad \begin{bmatrix} 3e^{2t} \\ e^{2t} \\ 5e^{2t} \end{bmatrix}, \quad \text{and} \quad \begin{bmatrix} e^{3t} \\ -e^{3t} \\ 2e^{3t} \end{bmatrix}$$

are linearly independent solutions of the system

$$\begin{aligned} \dot{x}_1 &= -10x_1 + x_2 + 7x_3 \\ \dot{x}_2 &= -9x_1 + 4x_2 + 5x_3 \\ \dot{x}_3 &= -17x_1 + x_2 + 12x_3. \end{aligned}$$

**30.** The unique solution of the IVP

$$\dot{x}_1 = -10x_1 + x_2 + 7x_3$$
$$\dot{x}_2 = -9x_1 + 4x_2 + 5x_3$$
$$\dot{x}_3 = -17x_1 + x_2 + 12x_3$$
$$x_1(0) = 5, \quad x_2(0) = 2, \quad x_3(0) = -2$$

is

$$x_1 = 2e^t + 3e^{2t}, \quad x_2 = e^t + e^{2t}, \quad x_3 = 3e^t - 5e^{2t}.$$

# 3.2 THE METHOD OF ELIMINATION

The most elementary method for solving a system of linear differential equations in two unknown functions and with constant coefficients is the *method of elimination.* In this method the aim is to reduce the given system to a single differential equation in one unknown function by eliminating the other dependent variable.

**EXAMPLE 1** Find the general solution of the homogeneous linear system

$$\dot{x}_1 = x_2$$
$$\dot{x}_2 = -2x_1 + 3x_2. \tag{1}$$

**Solution** Differentiating both sides of the first equation with respect to $t$ and using the second equation, we obtain

$$\ddot{x}_1 = \dot{x}_2 = -2x_1 + 3x_2.$$

The first equation in (1) states that $x_2 = \dot{x}_1$, hence

$$\ddot{x}_1 = -2x_1 + 3\dot{x}_1$$
$$\Rightarrow \quad \ddot{x}_1 - 3\dot{x}_1 + 2x_1 = 0. \tag{2}$$

Thus, we have succeeded in eliminating the function $x_2$ from system (1), obtaining in the process the differential equation (2), which contains one single function and which can be solved. The general solution of Eq. (2) is

$$x_1(t) = c_1 e^t + c_2 e^{2t}. \tag{3}$$

At this point we use the first equation in system (1) and the value of $x_1$ given in (3) to find the other unknown function $x_2 = \dot{x}_1$. Thus,

$$x_2(t) = c_1 e^t + 2c_2 e^{2t}. \tag{4}$$

Equations (3) and (4) give the general solution of system (1).

Other ways to eliminate one of the unknown functions and arrive at a single equation for the remaining unknown are the following: Solve the first equation of the system for $x_2$ and substitute the value of $x_2$ into the second equation of the system; or, solve the second equation of the system for $x_1$ and substitute the value of $x_1$ into the first equation of the system.

The method of elimination can also be applied efficiently to nonhomogeneous systems, as the following example illustrates.

**EXAMPLE 2**   Find the general solution of the nonhomogeneous system

$$\dot{x}_1 = x_2 + t$$

$$\dot{x}_2 = -2x_1 + 3x_2 + 1. \tag{5}$$

**Solution**   As in the previous example, differentiating both sides of the first equation with respect to $t$ and using the second equation, we obtain

$$\ddot{x}_1 = \dot{x}_2 + 1 = -2x_1 + 3x_2 + 2.$$

Using the first equation in (5) yields

$$\ddot{x}_1 = -2x_1 + 3(\dot{x}_1 - t) + 2$$

$$\Rightarrow \quad \ddot{x}_1 - 3\dot{x}_1 + 2x_1 = -3t + 2. \tag{6}$$

Thus, we have succeeded in eliminating the variable $x_2$ from system (5), obtaining the nonhomogeneous differential equation (6), which contains one single function and can be solved. The solution of the homogeneous equation associated with Eq. (6) is

$$x_{1h} = c_1 e^t + c_2 e^{2t}. \tag{7}$$

A particular solution of Eq. (6) is of the form $x_{1p} = At + B \Rightarrow \dot{x}_{1p} = A,$

$$\ddot{x}_{1p} = 0 \Rightarrow -3A + 2At + 2B = -3t + 2 \Rightarrow 2A = -3$$

and

$$-3A + 2B = 2 \Rightarrow A = -\tfrac{3}{2} \quad \text{and} \quad B = -\tfrac{5}{4} \Rightarrow x_{1p} = -\tfrac{3}{2}t - \tfrac{5}{4}. \tag{8}$$

Adding (7) and (8), we obtain the general solution of Eq. (6):

$$x_1(t) = c_1 e^t + c_2 e^{2t} - \tfrac{3}{2}t - \tfrac{5}{4}. \tag{9}$$

Now from the first equation in (5) we have $x_2 = \dot{x}_1 - t$, so

$$x_2(t) = c_1 e^t + 2c_2 e^{2t} - t - \tfrac{3}{2}. \tag{10}$$

Equations (9) and (10) give the general solution of (5).

**EXAMPLE 3**   Solve the IVP

$$\dot{x}_1 = -3x_1 + 4x_2 \tag{11}$$

$$\dot{x}_2 = -2x_1 + 3x_2 \tag{12}$$

$$x_1(0) = -1, \quad x_2(0) = 3. \tag{13}$$

**Solution**   Differentiating both sides of (11) and using (12) and then (11) again, we obtain

$$\ddot{x}_1 = -3\dot{x}_1 + 4\dot{x}_2 = -3\dot{x}_1 + 4(-2x_1 + 3x_2)$$
$$= -3\dot{x}_1 - 8x_1 + 12x_2 = -3\dot{x}_1 - 8x_1 + 3(\dot{x}_1 + 3x_1)$$

$$\Rightarrow \quad \ddot{x}_1 - x_1 = 0,$$

thus,

$$x_1(t) = c_1 e^t + c_2 e^{-t}. \tag{14}$$

Thus, from Eq. (11) we have

$$4x_2 = \dot{x}_1 + 3x_1 = c_1 e^t - c_2 e^{-t} + 3c_1 e^t + 3c_2 e^{-t} = 4c_1 e^t + 2c_2 e^{-t},$$

and so,

$$x_2(t) = c_1 e^t + \tfrac{1}{2} c_2 e^{-t}. \tag{15}$$

Using the initial conditions (13) in Eqs. (14) and (15), we obtain

$$c_1 + c_2 = -1$$

$$c_1 + \tfrac{1}{2} c_2 = 3.$$

Thus, $c_1 = 7$, $c_2 = -8$, and the solution of the IVP (11)–(13) is

$$x_1(t) = 7e^t - 8e^{-t}$$

$$x_2(t) = 7e^t - 4e^{-t}.$$

**REMARK 1**   It can be shown that the general solution of a system of $n$ first-order linear differential equations in $n$ unknown functions contains $n$ arbitrary constants. For $n = 2$, this follows from Theorem 5 of Section 3.1. Sometimes, however, the method of elimination introduces *extraneous* constants. The extraneous constants can be eliminated by substituting the solutions into the system and equating coefficients of similar terms. We illustrate this by the following example.

**EXAMPLE 4**   Find the general solution of the system

$$\dot{x}_1 = x_2 + 1$$

$$\dot{x}_2 = x_1.$$

**Solution**   Differentiating both sides of the first equation with respect to $t$ and using the second equation, we find $\ddot{x}_1 = \dot{x}_2 = x_1 \Rightarrow \ddot{x}_1 - x_1 = 0 \Rightarrow x_1(t) = c_1 e^t + c_2 e^{-t}$. Now that we have $x_1(t)$, we can find $x_2(t)$ either from the first equation of the system or from the second. If we use the first equation we find $x_2(t) = c_1 e^t - c_2 e^{-t} - 1$, and the general solution of the system contains, as expected, two arbitrary constants. However, if we find $x_2(t)$ from the second equation of the system, we have to integrate, which introduces a third arbitrary constant. In this case, $x_1(t) = c_1 e^t + c_2 e^{-t}$ and $x_2(t) = c_1 e^t - c_2 e^{-t} + c_3$. If we substitute these functions into the system and equate coefficients of similar terms, we find that $c_3$ should be $-1$. Thus the general solution is

$$x_1(t) = c_1 e^t + c_2 e^{-t}, \qquad x_2(t) = c_1 e^t - c_2 e^{-t} - 1.$$

## APPLICATIONS 3.2.1

■ Some investigations into the distribution, breakdown, and synthesis of albumin in animals has led to systems of ordinary differential equations.[10] Under normal conditions an animal is assumed to have a constant amount of albumin. Part of this albumin is assumed to be present in the animal's vascular system (plasma) and the remainder in the extravascular fluids (lymph and tissue fluids) (Figure 3.4). The albumin in the plasma (and in the extravascular fluids) is changing in the following manner. Some albumin is transferring to the extravascular fluids (or, respectively, to the plasma), and some albumin is being broken down, and this catabolized protein is replaced by newly synthesized protein. It is not completely known exactly where this breakdown takes place, and we therefore assume that some breakdown occurs in the plasma and some in the extravascular fluids. Thus, in its natural state the study of albumin can be viewed in terms of the compartment model given as Figure 3.4.

**Biokinetics**

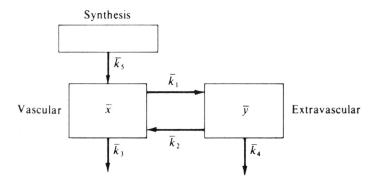

**Figure 3.4**

In this figure, $\bar{x}$ represents the number of grams of albumin in the plasma; $\bar{y}$ represents the number of grams of albumin in the extravascular fluids; and $\bar{k}_1, \bar{k}_2, \bar{k}_3, \bar{k}_4$, and $\bar{k}_5$ are constants associated with the rate at which the various processes take place. The newly synthesized protein is shown reentering the system via the plasma. $\bar{k}_3$ and $\bar{k}_4$ are associated with the breakdown process, and $\bar{k}_1$ and $\bar{k}_2$ are associated with the rates of interaction. We refer to the model above as being associated with unlabeled albumin.

In order to study the albumin process, some investigators proceed as follows. A certain amount of radioactive albumin, referred to as $I^{131}$-albumin, is injected

[10]The models and developments presented here have been extracted from E. B. Reeve and J. E. Roberts, "Kinetics of $I^{131}$-Albumin Distribution and Breakdown," *J. Gen. Physiol.* 43 (1959): 415.

into the vascular system, where it mixes with the vascular fluid. After a few minutes its behavior is assumed to follow that of the unlabeled albumin. At any time $t$ thereafter, there is a certain amount of radioactive substance $x$ in the plasma and a certain amount $y$ in the extravascular fluids.

$I^{131}$-albumin, being radioactive, is continuously being broken down as a result of the liberation of radioactive breakdown products, and certain of the breakdown products are being excreted by the animal. All these actions are continuous in nature, and therefore to speak of the rates of change, we must use derivatives with respect to time. The compartment model associated with $I^{131}$-albumin is shown in Figure 3.5. The quantities $x$, $y$, $z$, and $u$ in the figure represent fractions of the total amount originally injected. Since there is only one injection, no new protein is being synthesized in this model. The process described by this compartment model can be analyzed by studying the following system of differential equations:

$$\frac{dx}{dt} = k_2 y - (k_1 + k_3)x$$

$$\frac{dy}{dt} = k_1 x - (k_2 + k_4)y$$

$$\frac{dz}{dt} = k_3 x + k_4 y - k_5 z \tag{16}$$

$$\frac{du}{dt} = k_5 z.$$

The first equation is arrived at by noticing in the model that $x$ is changing due to gaining $k_2 y$ g/day and losing $k_1 x$ and $k_3 x$ g/day. The other equations are obtained by means of similar reasoning. This system also has initial and terminal conditions associated with it. Initially, all the radioactive substance is in the plasma; thus, $t = 0 \Rightarrow x = 1$, $y = z = u = 0$. After "infinite" time, all the substance has been excreted; thus, $t \to \infty \Rightarrow x = y = z = 0$, $u = 1$. The solution of this system is left to the exercises. We note, however, that (16), together with the initial conditions, determines a unique solution. This solution depends,

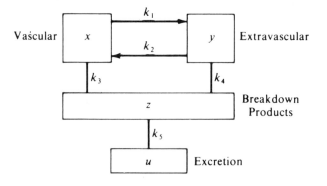

**Figure 3.5**

in general, on the rate constants $k_1, \ldots, k_5$. These constants are determined experimentally and by the terminal conditions.

■ In the electric circuit in Figure 3.6, assume that the currents are initially zero and that $R_1 = \frac{3}{2}$ ohms, $R_2 = 4$ ohms, $L_1 = L_2 = 1$ henry, and $V(t) = 2$ volts. Determine the currents $I_1(t)$ and $I_2(t)$ at any time $t$.

**Circuit Theory**

**Solution** From Eq. (18) of Section 3.1.1, the currents $I_1$ and $I_2$ satisfy the following linear system of differential equations:

$$\dot{I}_1 = -\frac{R_1}{L_1} I_1 + \frac{R_1}{L_1} I_2 + \frac{V(t)}{L_1} \tag{17}$$

$$\dot{I}_2 = \frac{R_1}{L_2} I_1 - \frac{R_1 + R_2}{L_2} I_2, \tag{18}$$

and the initial conditions

$$I_1(0) = 0, \qquad I_2(0) = 0. \tag{19}$$

Substituting the given values of $R_1$, $R_2$, $L_1$, $L_2$, and $V(t)$, we obtain

$$\dot{I}_1 = -\tfrac{3}{2} I_1 + \tfrac{3}{2} I_2 + 2 \tag{20}$$

$$\dot{I}_2 = \tfrac{3}{2} I_1 - \tfrac{11}{2} I_2. \tag{21}$$

Differentiating both sides of (20) and using (21) and then (20), we find that

$$\ddot{I}_1 = -\tfrac{3}{2}\dot{I}_1 + \tfrac{3}{2}\dot{I}_2$$

$$= -\tfrac{3}{2}\dot{I}_1 + \tfrac{3}{2}(\tfrac{3}{2}I_1 - \tfrac{11}{2}I_2) = -\tfrac{3}{2}\dot{I}_1 + \tfrac{9}{4}I_1 - \tfrac{33}{4}I_2$$

$$= -\tfrac{3}{2}\dot{I}_1 + \tfrac{9}{4}I_1 - \tfrac{11}{2}(\dot{I}_1 + \tfrac{3}{2}I_1 - 2)$$

$$\Rightarrow \quad \ddot{I}_1 + 7\dot{I}_1 + 6I_1 = 11$$

$$\Rightarrow \quad I_1(t) = c_1 e^{-t} + c_2 e^{-6t} + \tfrac{11}{6}.$$

**Figure 3.6**

Substituting the value of $I_1$ into (20), we find that

$$I_2 = \frac{2}{3}(-c_1 e^{-t} - 6c_2 e^{-6t} + \frac{3}{2}c_1 e^{-t} + \frac{3}{2}c_2 e^{-6t} + \frac{33}{12} - 2)$$

$$= \frac{1}{3}c_1 e^{-t} - 3c_2 e^{-6t} + \frac{1}{2}.$$

Using the initial conditions (19), we have

$$c_1 + c_2 + \frac{11}{6} = 0$$

$$\frac{1}{3}c_1 - 3c_2 + \frac{1}{2} = 0$$

$$\Rightarrow \quad c_1 = -\frac{9}{5}, \quad c_2 = -\frac{1}{30}.$$

Thus,

$$I_1(t) = -\frac{9}{5}e^{-t} - \frac{1}{30}e^{-6t} + \frac{11}{6}$$

$$I_2(t) = -\frac{3}{5}e^{-t} + \frac{1}{10}e^{-6t} + \frac{1}{2}.$$

### EXERCISES

In Exercises 1 through 8, find the general solution of the system.

**1.** $\dot{x}_1 = 4x_1 - x_2$
$\dot{x}_2 = 2x_1 + x_2$

**2.** $\dot{x}_1 = 3x_1 - 2x_2$
$\dot{x}_2 = x_1$

**3.** $\dot{x}_1 = 4x_1 - x_2 + 3e^{2t}$
$\dot{x}_2 = 2x_1 + x_2 + 2t$

**4.** $\dot{x}_1 = 3x_1 - 2x_2 + 1$
$\dot{x}_2 = x_1 + t$

**5.** $\dot{x} = 5x - 6y + 1$
$\dot{y} = 6x - 7y + 1$

**6.** $t\dot{x} = 3x - 2y$
$t\dot{y} = x + y - t^2$

**7.** $\dot{x} = 3x - 2y + 2t^2$
$\dot{y} = 5x + y - 1$

**8.** $\dot{x}_1 = x_1 + 2x_2$
$\dot{x}_2 = 4x_1 - x_2$

**9.** Determine the currents in the electric network of Exercise 12 of Section 3.1 if $R_1 = 4$ ohms, $L_1 = 2$ henries, $L_2 = 4$ henries, $V_1 = (2 \sin t)$ volts, $V_2(t) = (2 \sin t)$ volts, and the currents are initially zero.

**10.** Assume that the system

$$\dot{N}_1 = 4N_1 - 6N_2$$

$$\dot{N}_2 = 8N_1 - 10N_2$$

represents two competing populations, with $N_1$ being the desirable population and $N_2$ a parasite. Compute the populations $N_1(t)$ and $N_2(t)$ at any time $t$. Show that both populations are headed for extinction.

In Exercises 11 through 18, find the solution of the IVP.

**11.** $\dot{x}_1 = 4x_1 - x_2$
$\dot{x}_2 = 2x_1 + x_2$
$x_1(0) = 1, x_2(0) = 3.$

**12.** $\dot{x}_1 = 3x_1 - 2x_2$
$\dot{x}_2 = x_1$
$x_1(0) = -1, x_2(0) = 0.$

13. $\dot{x}_1 = 4x_1 - x_2 + 3e^{2t}$
    $\dot{x}_2 = 2x_1 + x_2 + 2t$
    $x_1(0) = -\frac{5}{18}, x_2(0) = \frac{47}{9}.$

14. $\dot{x}_1 = 3x_1 - 2x_2 + 1$
    $\dot{x}_2 = x_1 + t$
    $x_1(0) = -\frac{1}{2}, x_2(0) = -\frac{1}{4}.$

15. $\dot{x} = 5x - 6y + 1$
    $\dot{y} = 6x - 7y + 1$
    $x(0) = 0, y(0) = 0.$

16. $t\dot{x} = 3x - 2y$
    $t\dot{y} = x + y - t^2$
    $x(1) = 1, y(1) = \frac{1}{2}.$

17. $\dot{x} = 3x - 2y + 2t^2$
    $\dot{y} = 5x + y - 1$
    $x(0) = \frac{534}{2197}, y(0) = \frac{567}{2197}.$

18. $\dot{x}_1 = x_1 + 2x_2$
    $\dot{x}_2 = 4x_1 - x_2$
    $x_1(0) = 1, x_2(0) = 1.$

In Exercises 19 through 28, find the general solution of the system.

19. $\dot{x} = -x, \dot{y} = -y$

20. $\dot{x} = -2x + y, \dot{y} = x - 2y$

21. $\dot{x} = 3x - 2y, \dot{y} = 2x - 2y$

22. $\dot{x} = -x, \dot{y} = -x - y$

23. $\dot{x} = -x + y, \dot{y} = -x - y$

24. $\dot{x} = y, \dot{y} = -x$

25. $\dot{x} = y, \dot{y} = -x + 2y$

26. $\dot{x} = 2x - y, \dot{y} = 9x + 2y$

27. $\dot{x} = 2x + y, \dot{y} = -3x + 6y$

28. $\dot{x} = y, \dot{y} = z, \dot{z} = x$

29. In the chemistry application of Section 1.7.1 we arrived at the homogeneous system [See Eqs. (10), Section 1.7.1]

$$\dot{c}_1 = -\frac{k}{V_1}c_1 + \frac{k}{V_1}c_2$$

$$\dot{c}_2 = \frac{k}{V_2}c_1 - \frac{k}{V_2}c_2.$$

Use the method of elimination to obtain the general solution of this system.

30. (a) Using the method of elimination on the first two of Eqs. (16) show that $x$ satisfies the differential equation

$$\ddot{x} + (k_1 + k_2 + k_3 + k_4)\dot{x} + [(k_1 + k_3)(k_2 + k_4) - k_1k_2]x = 0.$$

Thus, if $r_1$ and $r_2$ represent the roots of the characteristic equation, then $x = c_1e^{r_1t} + c_2e^{r_2t}$. Find $r_1$ and $r_2$ algebraically.

(b) Use the terminal condition $(t \to \infty)$ to infer that $r_1 < 0$ and $r_2 < 0$.

(c) Since $r_1 < 0$ and $r_2 < 0$ write $r_1 = -a$ and $r_2 = -b$, where $a > 0$ and $b > 0$. Using the differential equation $dx/dt = k_2y - (k_1 + k_3)x$, show that

$$k_2y - (k_1 + k_3)x = -c_1ae^{-at} - c_2be^{-bt}.$$

(d) Use the initial conditions to find a relation between $k_1$ and $k_3$. [Hint: Use the result of part (c).]

(e) Solve for $y$.

(f) Solve for $z$.

(g) Solve for $u$.

**31. Learning Theory** In an article with applications to learning theory, Grossberg[11] obtained the following IVP:

$$\dot{x} = a(b - x) - cfy$$

$$\dot{y} = d(x - y) - cfy - ay$$

$$x(0) = b, \qquad y(0) = \frac{db}{a + d},$$

where $a$, $b$, $c$, $d$, and $f$ are positive constants. Show that for every fixed $f$ the unknown function $y$ decays monotonically to a positive minimum.

---

# 3.3   THE MATRIX METHOD

As we saw in Section 3.2, we have no difficulty solving systems with constant coefficients of *two* linear differential equations in *two* unknown functions. The method of elimination is a simple method to use for such systems. The method of this section, known as the *matrix method,* has the advantage that it can be easily extended to systems with constant coefficients of $n$ linear differential equations in $n$ unknown functions.

For the sake of simplicity, we present the main features of this method in the special case of linear homogeneous systems with constant coefficients of the form

$$\dot{x}_1 = a_{11}x_1 + a_{12}x_2$$
$$\dot{x}_2 = a_{21}x_1 + a_{22}x_2. \tag{1}$$

Illustrative examples will be given for systems of $n$ equations in $n$ unknown functions for $n = 2, 3$, and $4$.

Mimicking the method of solving linear homogeneous differential equations with constant coefficients, which was discussed in Sections 2.4 and 2.5, let us look for a solution of system (1) in the form

$$\begin{bmatrix} x_1 \\ x_2 \end{bmatrix} = \begin{bmatrix} A_1 e^{\lambda t} \\ A_2 e^{\lambda t} \end{bmatrix}. \tag{2}$$

The constants $A_1$, $A_2$, and $\lambda$ are to be determined by demanding that $x_1 = A_1 e^{\lambda t}$ and $x_2 = A_2 e^{\lambda t}$ satisfy system (1). Substituting (2) into (1), we find that $A_1$, $A_2$, and $\lambda$ must satisfy the system of equations

$$A_1 \lambda e^{\lambda t} = a_{11} A_1 e^{\lambda t} + a_{12} A_2 e^{\lambda t}$$

$$A_2 \lambda e^{\lambda t} = a_{21} A_1 e^{\lambda t} + a_{22} A_2 e^{\lambda t}.$$

---

[11]S. Grossberg, *J. Theoret. Biol.* (1969): 325–64.

Dividing through by $e^{\lambda t}$ and rearranging terms, we obtain the equivalent system

$$(a_{11} - \lambda)A_1 + a_{12}A_2 = 0$$

$$a_{21}A_1 + (a_{22} - \lambda)A_2 = 0. \tag{3}$$

System (3) consists of only two equations, and there are three unknown quantities involved: $\lambda$, $A_1$, and $A_2$. Nevertheless, one can solve this system by taking the following approach. We view system (3) as a linear homogeneous system in the unknowns $A_1$ and $A_2$. Naturally, we seek a nontrivial solution of this system, since the solution $(A_1, A_2) = (0, 0)$ gives rise to the trivial solution $x_1(t) = 0$, $x_2(t) = 0$ of system (1). Thus, $(A_1, A_2)$ should be different from (0, 0). We now recall (Appendix A) that a homogeneous system of algebraic equations has a nontrivial solution if and only if the determinant of its coefficients is equal to zero. In the case of system (3), this means that there is a solution $(A_1, A_2) \neq (0, 0)$ if and only if

$$\begin{vmatrix} a_{11} - \lambda & a_{12} \\ a_{21} & a_{22} - \lambda \end{vmatrix} = 0; \tag{4}$$

that is,

$$\lambda^2 - (a_{11} + a_{22})\lambda + a_{11}a_{22} - a_{12}a_{21} = 0. \tag{4'}$$

### DEFINITION 1

*Equation (4') is called the characteristic equation of system (1), and the roots of the characteristic equation are called the characteristic roots of the system.*

It will be useful to the reader to note at this point that the determinant in Eq. (4) is obtained from the determinant

$$\begin{vmatrix} a_{11} & a_{12} \\ a_{21} & a_{22} \end{vmatrix}$$

of the coefficients of system (1) by subtracting $\lambda$ from the main diagonal. [In matrix analysis, Eq. (4) is called the *characteristic equation* of the matrix

$$\begin{vmatrix} a_{11} & a_{12} \\ a_{21} & a_{22} \end{vmatrix}, \tag{5}$$

and the roots of the characteristic equation are called the *characteristic roots* or *eigenvalues* of the matrix (5).]

Let $\lambda_1$ and $\lambda_2$ be the characteristic roots of system (1). If the $\lambda$ in (3) is replaced by a characteristic root, the corresponding system (3) has a nontrivial solution $(A_1, A_2)$. Substituting these values of $A_1$ and $A_2$ in (2), we obtain a nontrivial solution of system (1). If Eq. (4') has two distinct roots (in other words, if $\lambda_1 \neq \lambda_2$), we can obtain two nontrivial solutions of system (1): one from $\lambda = \lambda_1$ and the other from $\lambda = \lambda_2$. Furthermore, it can be shown that these two solutions are linearly independent. On the other hand, if Eq. (4') does not have two distinct roots (in other words, if $\lambda_1 = \lambda_2$), the above procedure gives, in general, one nontrivial solution (and its multiples) of system (1). As we are interested in finding the general solution of system (1), we still need a second linearly independent solution of system (1). The trick in this case is to look for a second linearly independent solution of (1) in the form

$$\begin{bmatrix} x_1 \\ x_2 \end{bmatrix} = \begin{bmatrix} (a_1 t + b_1)e^{\lambda_1 t} \\ (a_2 t + b_2)e^{\lambda_1 t} \end{bmatrix}, \tag{6}$$

where the coefficients $a_1$, $b_1$, $a_2$, and $b_2$ are to be determined in such a way that (6) is a solution of (1) which is linearly independent to the first solution

$$\begin{bmatrix} x_1 \\ x_2 \end{bmatrix} = \begin{bmatrix} A_1 e^{\lambda_1 t} \\ A_2 e^{\lambda_1 t} \end{bmatrix}.$$

This is always possible. See Remark 4.

**EXAMPLE 1**  Find the general solution of the homogeneous system

$$\begin{aligned} \dot{x}_1 &= 2x_1 + x_2 \\ \dot{x}_2 &= -3x_1 + 6x_2. \end{aligned} \tag{7}$$

**Solution**  Let us look for a solution of system (7) of the form

$$\begin{bmatrix} x_1 \\ x_2 \end{bmatrix} = \begin{bmatrix} A_1 e^{\lambda t} \\ A_2 e^{\lambda t} \end{bmatrix}. \tag{8}$$

Then $\lambda$ should be a root of the characteristic equation (note that $a_{11} = 2$, $a_{12} = 1$, $a_{21} = -3$, and $a_{22} = 6$)

$$\begin{vmatrix} 2 - \lambda & 1 \\ -3 & 6 - \lambda \end{vmatrix} = 0, \quad \text{that is,} \quad \lambda^2 - 8\lambda + 15 = 0.$$

The characteristic roots are $\lambda_1 = 3$ and $\lambda_2 = 5$. When $\lambda = \lambda_1 = 3$, the constants $A_1$ and $A_2$ of the solution (8) should satisfy the homogeneous system (3); that is,

$$\begin{aligned} -A_1 + A_2 &= 0 \\ -3A_1 + 3A_2 &= 0 \end{aligned} \quad \Rightarrow A_1 = A_2.$$

Choosing $A_1 = A_2 = 1$, we have the solution

$$\begin{bmatrix} e^{3t} \\ e^{3t} \end{bmatrix}. \tag{9}$$

[Clearly, any other nontrivial solution of this system (for example, $A_1 = A_2 = 2$) would just give us another solution with which (9) is linearly dependent.] When $\lambda = \lambda_2 = 5$, system (3) becomes

$$\begin{aligned} -3A_1 + A_2 &= 0 \\ -3A_1 + A_2 &= 0 \end{aligned} \quad \Rightarrow A_2 = 3A_1.$$

The choice $A_1 = 1$ yields $A_2 = 3$, and we obtain another solution of (7):

$$\begin{bmatrix} e^{5t} \\ 3e^{5t} \end{bmatrix}. \tag{10}$$

It is easily seen (by applying Theorem 2 of Section 3.1) that the two solutions (9) and (10) of system (7) are linearly independent, and therefore the general solution of (7) is

$$\begin{bmatrix} x_1 \\ x_2 \end{bmatrix} = c_1 \begin{bmatrix} e^{3t} \\ e^{3t} \end{bmatrix} + c_2 \begin{bmatrix} e^{5t} \\ 3e^{5t} \end{bmatrix};$$

that is,

$$x_1 = c_1 e^{3t} + c_2 e^{5t}$$

$$x_2 = c_1 e^{3t} + 3c_2 e^{5t}.$$

**EXAMPLE 2** Find the general solution of the system

$$\dot{x}_1 = 2x_1 - x_2$$

$$\dot{x}_2 = 9x_1 + 2x_2. \tag{11}$$

**Solution** The characteristic equation of system (11) is

$$\begin{vmatrix} 2 - \lambda & -1 \\ 9 & 2 - \lambda \end{vmatrix} = 0, \quad \text{that is,} \quad \lambda^2 - 4\lambda + 13 = 0.$$

The characteristic roots are $\lambda_1 = 2 + 3i$ and $\lambda_2 = 2 - 3i$. When $\lambda_1 = 2 + 3i$, the constants $A_1$ and $A_2$ of the solution (8) satisfy the homogeneous system (3) with $a_{11} = 2$, $a_{12} = -1$, $a_{21} = 9$, $a_{22} = 2$, and $\lambda = 2 + 3i$. That is,

$$\begin{aligned} -3iA_1 - A_2 &= 0 \\ 9A_1 - 3iA_2 &= 0 \end{aligned} \Rightarrow A_2 = -3iA_1.$$

Choosing $A_1 = 1$, we find that $A_2 = -3i$, and therefore

$$\begin{bmatrix} e^{(2+3i)t} \\ -3ie^{(2+3i)t} \end{bmatrix} \tag{12}$$

is a solution of (11). When $\lambda_2 = 2 - 3i$, the constants $A_1$ and $A_2$ of the solution (8) satisfy the homogeneous system (3) with $a_{11} = 2$, $a_{12} = -1$, $a_{21} = 9$, $a_{22} = 2$, and $\lambda = 2 - 3i$. That is,

$$\begin{aligned} 3iA_1 - A_2 &= 0 \\ 9A_1 + 3iA_2 &= 0 \end{aligned} \Rightarrow A_2 = 3iA_1.$$

Choosing $A_1 = 1$, we find that $A_2 = 3i$; therefore

$$\begin{bmatrix} e^{(2-3i)t} \\ 3ie^{(2-3i)t} \end{bmatrix} \tag{13}$$

is another solution of (11). Applying Theorem 2 of Section 3.1, it is easily seen that the two solutions (12) and (13) of system (11) are linearly independent, and therefore the general solution of (11) is

$$\begin{bmatrix} x_1 \\ x_2 \end{bmatrix} = c_1 \begin{bmatrix} e^{(2+3i)t} \\ -3ie^{(2+3i)t} \end{bmatrix} + c_2 \begin{bmatrix} e^{(2-3i)t} \\ 3ie^{(2-3i)t} \end{bmatrix}, \tag{14}$$

where $c_1$ and $c_2$ are arbitrary constants.

**REMARK 1**  Using the Euler identity $e^{a+ib} = e^a(\cos b + i \sin b)$, the solution (14) can also be written as follows:

$$\begin{bmatrix} x_1 \\ x_2 \end{bmatrix} = c_1 \begin{bmatrix} e^{2t}\cos 3t + ie^{2t}\sin 3t \\ -3ie^{2t}\cos 3t + 3e^{2t}\sin 3t \end{bmatrix} + c_2 \begin{bmatrix} e^{2t}\cos 3t - ie^{2t}\sin 3t \\ 3ie^{2t}\cos 3t + 3e^{2t}\sin 3t \end{bmatrix}$$

$$= \begin{bmatrix} (c_1 + c_2)e^{2t}\cos 3t + i(c_1 - c_2)e^{2t}\sin 3t \\ 3(c_1 + c_2)e^{2t}\sin 3t - 3i(c_1 - c_2)e^{2t}\cos 3t \end{bmatrix}.$$

Setting $C_1 = c_1 + c_2$ and $C_2 = i(c_1 - c_2)$, we obtain

$$\begin{bmatrix} x_1 \\ x_2 \end{bmatrix} = \begin{bmatrix} C_1 e^{2t}\cos 3t + C_2 e^{2t}\sin 3t \\ 3C_1 e^{2t}\sin 3t - 3C_2 e^{2t}\cos 3t \end{bmatrix}.$$

where $C_1$ and $C_2$ are arbitrary constants.

**REMARK 2**  Comparing the solutions (12) and (13) of system (11), we observe that one is the complex conjugate of the other. This is always the case when the characteristic roots are complex conjugates and the coefficients of the system are real numbers.

**EXAMPLE 3**  Find the general solution of the system

$$\dot{x}_1 = x_2 \tag{15}$$
$$\dot{x}_2 = -x_1 + 2x_2.$$

**Solution**  The characteristic equation of (15) is

$$\begin{vmatrix} -\lambda & 1 \\ -1 & 2-\lambda \end{vmatrix} = 0, \quad \text{that is,} \quad \lambda^2 - 2\lambda + 1 = 0.$$

The characteristic roots are $\lambda_1 = \lambda_2 = 1$. Therefore, the constants $A_1$ and $A_2$ of the solution (8) satisfy the homogeneous system (3) with $a_{11} = 0$, $a_{12} = 1$, $a_{21} = -1$, $a_{22} = 2$, and $\lambda = 1$. That is,

$$\begin{aligned} -A_1 + A_2 &= 0 \\ -A_1 + A_2 &= 0 \end{aligned} \quad \Rightarrow A_1 = A_2.$$

Choosing $A_1 = A_2 = 1$, we obtain the solution

$$\begin{bmatrix} e^t \\ e^t \end{bmatrix}. \tag{16}$$

Since the roots of the characteristic equation are equal, we look for a second linearly independent solution of (15) of the form (6),

$$\begin{bmatrix} x_1 \\ x_2 \end{bmatrix} = \begin{bmatrix} (a_1 t + b_1)e^t \\ (a_2 t + b_2)e^t \end{bmatrix}, \tag{17}$$

where the constants $a_1$, $b_1$, $a_2$, and $b_2$ are to be determined in such a way that (17) is a solution of (15) that is linearly independent of (16). Substituting $x_1 = (a_1 t + b_1)e^t$ and $x_2 = (a_2 t + b_2)e^t$ into (15), we obtain

$$a_1 e^t + (a_1 t + b_1)e^t = (a_2 t + b_2)e^t$$

$$a_2 e^t + (a_2 t + b_2)e^t = -(a_1 t + b_1)e^t + 2(a_2 t + b_2)e^t.$$

Dividing by $e^t$ and equating coefficients of like powers of $t$, we have

$$a_1 + b_1 = b_2, \qquad a_1 = a_2$$
$$a_2 + b_2 = -b_1 + 2b_2, \qquad a_2 = -a_1 + 2a_2 \tag{18}$$

or [since the last two equations in (18) are equivalent to the first two]

$$a_1 + b_1 = b_2, \qquad a_1 = a_2. \tag{19}$$

Now any choice of $a_1$, $b_1$, $a_2$, and $b_2$ that satisfies (19) and gives a solution which is linearly independent to (16) is acceptable. For example, the choice $a_1 = a_2 = 1$, $b_1 = 0$, and $b_2 = 1$ yields the solution

$$\begin{bmatrix} te^t \\ (t + 1)e^t \end{bmatrix}, \tag{20}$$

which is linearly independent of (16). (Why?) Thus, the general solution of (15) is

$$\begin{bmatrix} x_1 \\ x_2 \end{bmatrix} = c_1 \begin{bmatrix} e^t \\ e^t \end{bmatrix} + c_2 \begin{bmatrix} te^t \\ (t + 1)e^t \end{bmatrix};$$

that is,

$$x_1 = c_1 e^t + c_2 te^t \qquad \text{and} \qquad x_2 = c_1 e^t + c_2(t + 1)e^t,$$

where $c_1$ and $c_2$ are arbitrary constants.

**REMARK 3** Notice that for a second linearly independent solution of (15), it is not enough to multiply (16) by $t$ as in the case of linear homogeneous differential equations with multiple characteristic roots. Indeed,

$$t \begin{bmatrix} e^t \\ e^t \end{bmatrix} = \begin{bmatrix} te^t \\ te^t \end{bmatrix}$$

is not a solution of system (15). In fact, system (15) does not even have a nontrivial solution of the form

$$\begin{bmatrix} x_1 \\ x_2 \end{bmatrix} = \begin{bmatrix} a_1 te^t \\ a_2 te^t \end{bmatrix}.$$

Also, unlike the case of linear homogeneous differential equations with multiple characteristic roots, a system of the form (1) with a double characteristic root may have two linearly independent solutions of the form (2), as the example below indicates. This is also true for more general linear homogeneous systems, but we will not go into the details in this book. See Example 5 and Remark 4.

**EXAMPLE 4**   Find the general solution of the system

$$\dot{x}_1 = x_1$$
$$\dot{x}_2 = x_2. \tag{21}$$

**Solution**   The characteristic equation of system (21) is

$$\begin{vmatrix} 1 - \lambda & 0 \\ 0 & 1 - \lambda \end{vmatrix} = 0, \quad \text{that is,} \quad \lambda^2 - 2\lambda + 1 = 0.$$

The characteristic roots are $\lambda_1 = \lambda_2 = 1$. Therefore, the constants $A_1$ and $A_2$ of the solution (8) satisfy the homogeneous system (3) with $a_{11} = 1$, $a_{12} = 0$, $a_{21} = 0$, $a_{22} = 1$, and $\lambda = 1$. That is,

$$0 \cdot A_1 + 0 \cdot A_2 = 0$$
$$\Rightarrow A_1 \text{ and } A_2 \text{ are arbitrary constants.}$$
$$0 \cdot A_1 + 0 \cdot A_2 = 0$$

Choosing first $A_1 = 1$, $A_2 = 0$ and then $A_1 = 0$, $A_2 = 1$, we obtain the two linearly independent solutions

$$\begin{bmatrix} e^t \\ 0 \end{bmatrix} \quad \text{and} \quad \begin{bmatrix} 0 \\ e^t \end{bmatrix}.$$

Thus, the general solution of system (21) is

$$\begin{bmatrix} x_1 \\ x_2 \end{bmatrix} = c_1 \begin{bmatrix} e^t \\ 0 \end{bmatrix} + c_2 \begin{bmatrix} 0 \\ e^t \end{bmatrix};$$

that is,

$$x_1 = c_1 e^t \quad \text{and} \quad x_2 = c_2 e^t.$$

As we explained in the introductory paragraph of this section, the matrix method has the distinct advantage that it can be easily extended to homogeneous systems of $n$ linear differential equations with constant coefficients in $n$ unknowns. The method of elimination is highly recommended in the case $n = 2$. We shall now present two examples with $n = 3$ and $n = 4$, respectively, and show how to apply the matrix method to compute their general solution.

**EXAMPLE 5**   Find the general solution of the system

$$\dot{x}_1 = 5x_1 + 2x_2 + 2x_3$$
$$\dot{x}_2 = 2x_1 + 2x_2 - 4x_3 \tag{22}$$
$$\dot{x}_3 = 2x_1 - 4x_2 + 2x_3.$$

**Solution**   Let us look for a solution of system (22) of the form

$$\begin{bmatrix} x_1 \\ x_2 \\ x_3 \end{bmatrix} = \begin{bmatrix} A_1 e^{\lambda t} \\ A_2 e^{\lambda t} \\ A_3 e^{\lambda t} \end{bmatrix}. \tag{23}$$

Proceeding exactly as we did for system (1), we find that in order to obtain nontrivial solutions, $\lambda$ should be a root of the characteristic equation

$$\begin{vmatrix} 5 - \lambda & 2 & 2 \\ 2 & 2 - \lambda & -4 \\ 2 & -4 & 2 - \lambda \end{vmatrix} = 0, \quad \text{that is,} \quad -\lambda^3 + 9\lambda^2 - 108 = 0.$$

The roots of the characteristic equation are $\lambda_1 = -3$ and $\lambda_2 = \lambda_3 = 6$. When $\lambda = -3$, the constants $A_1$, $A_2$, and $A_3$ should satisfy the homogeneous system

$$8A_1 + 2A_2 + 2A_3 = 0$$
$$2A_1 + 5A_2 - 4A_3 = 0 \Rightarrow \quad \begin{aligned} 4A_1 + A_2 + A_3 &= 0 \\ 2A_1 + 5A_2 - 4A_3 &= 0. \end{aligned}$$
$$2A_1 - 4A_2 + 5A_3 = 0$$

Choosing $A_3 = 1$, we then obtain $A_2 = 1$ and $A_1 = -\frac{1}{2}$. Thus one solution of (22) is

$$\begin{bmatrix} -\frac{1}{2} e^{-3t} \\ e^{-3t} \\ e^{-3t} \end{bmatrix}. \tag{24}$$

When $\lambda = 6$, the constants $A_1$, $A_2$, and $A_3$ should satisfy the homogeneous system

$$-A_1 + 2A_2 + 2A_3 = 0$$
$$2A_1 - 4A_2 - 4A_3 = 0 \Rightarrow -A_1 + 2A_2 + 2A_3 = 0. \tag{25}$$
$$2A_1 - 4A_2 - 4A_3 = 0$$

Choosing $A_3 = 1$ and $A_2 = 1$, we obtain $A_1 = 4$, and

$$\begin{bmatrix} 4e^{6t} \\ e^{6t} \\ e^{6t} \end{bmatrix} \tag{26}$$

is a second solution of (22). On the other hand, the choice $A_1 = 2$ and $A_2 = 1$ yields $A_3 = 0$, and therefore

$$\begin{bmatrix} 2e^{6t} \\ e^{6t} \\ 0 \end{bmatrix} \tag{27}$$

is a third solution of system (22). Let us check the solutions (24), (26), and (27) for linear independence. We have

$$\begin{vmatrix} -\frac{1}{2} e^{-3t} & 4e^{6t} & 2e^{6t} \\ e^{-3t} & e^{6t} & e^{6t} \\ e^{-3t} & e^{6t} & 0 \end{vmatrix} = e^{9t} \begin{vmatrix} -\frac{1}{2} & 4 & 2 \\ 1 & 1 & 1 \\ 1 & 1 & 0 \end{vmatrix} = \frac{9}{2} e^{9t} \neq 0,$$

which proves that the solutions are, in fact, linearly independent. The general solution of system (22) is, therefore,

$$
\begin{bmatrix} x_1 \\ x_2 \\ x_3 \end{bmatrix} = c_1 \begin{bmatrix} -\frac{1}{2}e^{-3t} \\ e^{-3t} \\ e^{-3t} \end{bmatrix} + c_2 \begin{bmatrix} 4e^{6t} \\ e^{6t} \\ e^{6t} \end{bmatrix} + c_3 \begin{bmatrix} 2e^{6t} \\ e^{6t} \\ 0 \end{bmatrix} ;
$$

that is,

$$
x_1 = -\tfrac{1}{2}c_1 e^{-3t} + 4c_2 e^{6t} + 2c_3 e^{6t}
$$

$$
x_2 = c_1 e^{-3t} + c_2 e^{6t} + c_3 e^{6t}
$$

$$
x_3 = c_1 e^{-3t} + c_2 e^{6t}.
$$

**REMARK 4**  In Example 5 the characteristic root $\lambda = -3$ is a simple root, and therefore system (22) has exactly one solution of the form (23) with $\lambda = -3$. However, the characteristic root $\lambda = 6$ is a double root; therefore system (22) should have one or two linearly independent solutions of the form (23) with $\lambda = 6$. Actually, in Example 5, for $\lambda = 6$, we found two linearly independent solutions, (26) and (27). Otherwise, we should have looked for another linearly independent solution of the form

$$
\begin{bmatrix} (a_1 t + b_1)e^{6t} \\ (a_2 t + b_2)e^{6t} \\ (a_3 t + b_3)e^{6t} \end{bmatrix}.
$$

In general, if $\lambda$ is a characteristic root of the linear homogeneous system

$$
\dot{x}_1 = a_{11}x_1 + a_{12}x_2 + \cdots + a_{1n}x_n
$$

$$
\dot{x}_2 = a_{21}x_1 + a_{22}x_2 + \cdots + a_{2n}x_n \tag{28}
$$

$$
\cdots \cdots \cdots \cdots \cdots \cdots \cdots \cdots \cdots
$$

$$
\dot{x}_n = a_{n1}x_1 + a_{n2}x_2 + \cdots + a_{nn}x_n,
$$

with multiplicity $k$ $(k \le n)$, then to this characteristic root there correspond $k$ linearly independent solutions of system (28). These $k$ solutions are found as follows. First, find all linearly independent solutions of (28) of the form

$$
\begin{bmatrix} A_1 e^{\lambda t} \\ A_2 e^{\lambda t} \\ \cdot \\ \cdot \\ \cdot \\ A_n e^{\lambda t} \end{bmatrix}. \tag{29}
$$

Then find all linearly independent solutions of (28) of the form

$$
\begin{bmatrix}
(A_1 t + B_1)e^{\lambda t} \\
(A_2 t + B_2)e^{\lambda t} \\
\cdot \\
\cdot \\
\cdot \\
(A_n t + B_n)e^{\lambda t}
\end{bmatrix}
\tag{30}
$$

that are linearly independent to solutions (29). Next, find all linearly independent solutions of (28) of the form

$$
\begin{bmatrix}
(A_1 t^2 + B_1 t + C_1)e^{\lambda t} \\
(A_2 t^2 + B_2 t + C_2)e^{\lambda t} \\
\cdot \\
\cdot \\
\cdot \\
(A_n t^2 + B_n t + C_n)e^{\lambda t}
\end{bmatrix}
\tag{31}
$$

that are linearly independent to solutions (29) and (30). Proceeding in this way, we will find exactly $k$ linearly independent solutions of (28) corresponding to this characteristic root. If we apply this procedure to each characteristic root of (28), we will find $n$ linearly independent solutions.[12]

**EXAMPLE 6** Find the general solution of the system

$$
\dot{x}_1 = x_2 - x_3 + x_4
$$

$$
\dot{x}_2 = -x_2 + x_4
\tag{32}
$$

$$
\dot{x}_3 = x_3 - x_4
$$

$$
\dot{x}_4 = 2x_4.
$$

**Solution** Let us look for a solution of system (32) of the form

$$
\begin{bmatrix}
x_1 \\
x_2 \\
x_3 \\
x_4
\end{bmatrix}
=
\begin{bmatrix}
A_1 e^{\lambda t} \\
A_2 e^{\lambda t} \\
A_3 e^{\lambda t} \\
A_4 e^{\lambda t}
\end{bmatrix}.
$$

In order that we obtain nontrivial solutions, $\lambda$ should be a root of the characteristic equation

$$
\begin{vmatrix}
-\lambda & 1 & -1 & 1 \\
0 & -1-\lambda & 0 & 1 \\
0 & 0 & 1-\lambda & -1 \\
0 & 0 & 0 & 2-\lambda
\end{vmatrix} = 0,
$$

[12]For a detailed presentation, see R. H. Cole, *Theory of Ordinary Differential Equations* (New York: Appleton-Century-Crofts, 1968), chapter 4.

that is, $-\lambda(-1 - \lambda)(1 - \lambda)(2 - \lambda) = 0$. The roots of the characteristic equation are

$$\lambda_1 = 0, \quad \lambda_2 = -1, \quad \lambda_3 = 1, \quad \lambda_4 = 2.$$

When $\lambda = 0$, the constants, $A_1, A_2, A_3$, and $A_4$ should satisfy the homogeneous system

$$A_2 - A_3 + A_4 = 0$$

$$-A_2 \qquad + A_4 = 0$$

$$A_3 - A_4 = 0 \qquad \Rightarrow A_4 = A_3 = A_2 = 0.$$

$$2A_4 = 0$$

Choosing $A_1 = 1$, we have the solution

$$\begin{bmatrix} 1 \\ 0 \\ 0 \\ 0 \end{bmatrix}. \tag{33}$$

When $\lambda = -1$, the constants $A_1, A_2, A_3$, and $A_4$ should satisfy the homogeneous system

$$A_1 + A_2 - A_3 + A_4 = 0$$

$$A_4 = 0$$

$$2A_3 - A_4 = 0 \qquad \Rightarrow A_4 = A_3 = 0 \quad \text{and} \quad A_1 + A_2 = 0.$$

$$3A_4 = 0$$

Choosing $A_1 = 1$, we have the solution

$$\begin{bmatrix} e^{-t} \\ -e^{-t} \\ 0 \\ 0 \end{bmatrix}. \tag{34}$$

When $\lambda = 1$, the constants $A_1, A_2, A_3$, and $A_4$ should satisfy the homogeneous system

$$-A_1 + A_2 - A_3 + A_4 = 0$$

$$-2A_2 \qquad + A_4 = 0$$

$$-A_4 = 0 \qquad \Rightarrow A_4 = A_2 = 0 \quad \text{and} \quad A_1 + A_3 = 0.$$

$$A_4 = 0$$

For this case, the choice $A_1 = 1$ yields the solution

$$
\begin{bmatrix} e^t \\ 0 \\ -e^t \\ 0 \end{bmatrix}.
\tag{35}
$$

Finally, when $\lambda = 2$, the constants $A_1$, $A_2$, $A_3$, and $A_4$ should satisfy the homogeneous system

$$
\begin{aligned}
-2A_1 + A_2 - A_3 + A_4 &= 0 \\
-3A_2 \qquad + A_4 &= 0 \quad \Rightarrow -2A_1 + 7A_2 = 0, \qquad A_4 = 3A_2 = -A_3. \\
-A_3 - A_4 &= 0
\end{aligned}
$$

Choosing $A_4 = 3$, we find that $A_2 = 1$, $A_1 = \frac{7}{2}$, and $A_3 = -3$. Thus, we have the solution

$$
\begin{bmatrix} \frac{7}{2}e^{2t} \\ e^{2t} \\ -3e^{2t} \\ 3e^{2t} \end{bmatrix}.
\tag{36}
$$

The four solutions (33), (34), (35), and (36) are linearly independent. In fact,

$$
\begin{vmatrix}
1 & e^{-t} & e^t & \frac{7}{2}e^{2t} \\
0 & -e^{-t} & 0 & e^{2t} \\
0 & 0 & -e^t & -3e^{2t} \\
0 & 0 & 0 & 3e^{2t}
\end{vmatrix} = 3e^{2t} \neq 0.
$$

Hence, the general solution of (32) is

$$
\begin{bmatrix} x_1 \\ x_2 \\ x_3 \\ x_4 \end{bmatrix} = c_1 \begin{bmatrix} 1 \\ 0 \\ 0 \\ 0 \end{bmatrix} + c_2 \begin{bmatrix} e^{-t} \\ -e^{-t} \\ 0 \\ 0 \end{bmatrix} + c_3 \begin{bmatrix} e^t \\ 0 \\ -e^t \\ 0 \end{bmatrix} + c_4 \begin{bmatrix} \frac{7}{2}e^{2t} \\ e^{2t} \\ -3e^{2t} \\ 3e^{2t} \end{bmatrix};
$$

that is,

$$
\begin{aligned}
x_1 &= c_1 + c_2 e^{-t} + c_3 e^t + \tfrac{7}{2}c_4 e^{2t} \\
x_2 &= \qquad -c_2 e^{-t} \qquad\qquad + c_4 e^{2t} \\
x_3 &= \qquad\qquad\qquad -c_3 e^t - 3c_4 e^{2t} \\
x_4 &= \qquad\qquad\qquad\qquad\qquad 3c_4 e^{2t}.
\end{aligned}
$$

## NONHOMOGENEOUS SYSTEMS—VARIATION OF PARAMETERS 3.3.1

Theorem 5 of Section 3.1 states that the general solution of a linear nonhomogeneous system is the sum of the general solution of the corresponding homogeneous system and a particular solution of the nonhomogeneous system. To find a particular solution of the nonhomogeneous system we can use a variation of parameters technique similar to the corresponding method for linear non-homogeneous differential equations. (See Section 2.12.) We illustrate the technique by means of an example.

Consider the nonhomogeneous linear system

$$\dot{x}_1 = 2x_1 + x_2 + f(t)$$

$$\dot{x}_2 = -3x_1 + 6x_2 + g(t). \tag{37}$$

In Example 1 of Section 3.3 we found that the general solution of the homogeneous system corresponding to (37) is

$$\begin{bmatrix} x_{1h} \\ x_{2h} \end{bmatrix} = c_1 \begin{bmatrix} e^{3t} \\ e^{3t} \end{bmatrix} + c_2 \begin{bmatrix} e^{5t} \\ 3e^{5t} \end{bmatrix}, \tag{38}$$

where $c_1$ and $c_2$ are arbitrary constants. The main idea in the method of variation of parameters is to seek a particular solution of (37) of the form

$$\begin{bmatrix} x_{1p} \\ x_{2p} \end{bmatrix} = u_1(t) \begin{bmatrix} e^{3t} \\ e^{3t} \end{bmatrix} + u_2(t) \begin{bmatrix} e^{5t} \\ 3e^{5t} \end{bmatrix}, \tag{39}$$

where $u_1$ and $u_2$ are functions to be determined. That is,

$$x_{1p} = u_1(t)e^{3t} + u_2(t)e^{5t}$$
$$x_{2p} = u_1(t)e^{3t} + 3u_2(t)e^{5t}.$$

Substituting into (37) and simplifying we obtain

$$\dot{u}_1(t)e^{3t} + \dot{u}_2(t)e^{5t} = f(t)$$

$$\dot{u}_1(t)e^{3t} + 3\dot{u}_2(t)e^{5t} = g(t).$$

Using Cramer's rule, we find

$$\dot{u}_1(t) = \frac{\begin{vmatrix} f(t) & e^{5t} \\ g(t) & 3e^{5t} \end{vmatrix}}{\begin{vmatrix} e^{3t} & e^{5t} \\ e^{3t} & 3e^{5t} \end{vmatrix}} = \frac{1}{2}[3f(t) - g(t)]e^{-3t} \tag{40}$$

and

$$\dot{u}_2(t) = \frac{\begin{vmatrix} e^{3t} & f(t) \\ e^{3t} & g(t) \end{vmatrix}}{\begin{vmatrix} e^{3t} & e^{5t} \\ e^{3t} & 3e^{5t} \end{vmatrix}} = \frac{1}{2}[g(t) - f(t)]e^{-5t}. \tag{41}$$

From (40) and (41) we find $u_1(t)$ and $u_2(t)$ by integration, and, substituting the results into (39), we obtain a particular solution of (37). The general solution of (37) is given by

$$\begin{bmatrix} x_1 \\ x_2 \end{bmatrix} = \begin{bmatrix} x_{1h} \\ x_{2h} \end{bmatrix} + \begin{bmatrix} x_{1p} \\ x_{2p} \end{bmatrix}.$$

In the special case where $f(t) = e^{3t}$ and $g(t) = 2e^{3t}$, we find (omitting the constants of integration which are unnecessary)

$$\dot{u}_1(t) = \tfrac{1}{2} \Rightarrow u_1(t) = \tfrac{1}{2}t$$

and

$$\dot{u}_2(t) = \tfrac{1}{2}e^{-2t} \Rightarrow u_2(t) = -\tfrac{1}{4}e^{-2t}.$$

Thus, from (39), we have

$$\begin{bmatrix} x_{1p} \\ x_{2p} \end{bmatrix} = \tfrac{1}{2}t \begin{bmatrix} e^{3t} \\ e^{3t} \end{bmatrix} - \tfrac{1}{4}e^{-2t} \begin{bmatrix} e^{5t} \\ 3e^{5t} \end{bmatrix}.$$

Hence, the general solution of (37), for $f(t) = e^{3t}$ and $g(t) = 2e^{3t}$, is given by

$$\begin{bmatrix} x_1 \\ x_2 \end{bmatrix} = c_1 \begin{bmatrix} e^{3t} \\ e^{3t} \end{bmatrix} + c_2 \begin{bmatrix} e^{5t} \\ 3e^{5t} \end{bmatrix} + \tfrac{1}{2}t \begin{bmatrix} e^{3t} \\ e^{3t} \end{bmatrix} - \tfrac{1}{4}e^{-2t} \begin{bmatrix} e^{5t} \\ 3e^{5t} \end{bmatrix}$$

or

$$x_1(t) = c_1 e^{3t} + c_2 e^{5t} + \tfrac{1}{2}te^{3t} - \tfrac{1}{4}e^{3t}$$

$$x_2(t) = c_1 e^{3t} + 3c_2 e^{5t} + \tfrac{1}{2}te^{3t} - \tfrac{3}{4}e^{3t}.$$

## APPLICATIONS 3.3.2

■ In the biokinetics application of Section 3.2.1, assume that the rate constants have the values

**Biokinetics**

$$k_1 = k_2 = k_3 = k_4 = 1 \quad \text{and} \quad k_5 = 2.$$

Then system (16) becomes

$$\begin{aligned} \dot{x} &= -2x + y \\ \dot{y} &= x - 2y \\ \dot{z} &= x + y - 2z \\ \dot{u} &= 2z. \end{aligned} \tag{42}$$

We also have the initial conditions

$$x(0) = 1, \quad y(0) = z(0) = u(0) = 0 \tag{43}$$

and the terminal conditions

$$x = y = z = 0 \quad \text{and} \quad u = 1 \quad \text{as } t \to \infty. \tag{44}$$

We want to solve the problem (42)–(44).

**Solution**  We use the matrix method. The characteristic equation of (42) is

$$\begin{vmatrix} -2 - \lambda & 1 & 0 & 0 \\ 1 & -2 - \lambda & 0 & 0 \\ 1 & 1 & -2 - \lambda & 0 \\ 0 & 0 & 2 & -\lambda \end{vmatrix} = \lambda(\lambda + 1)(\lambda + 2)(\lambda + 3) = 0,$$

with characteristic roots

$$\lambda_1 = 0, \qquad \lambda_2 = -1, \qquad \lambda_3 = -2, \qquad \lambda_4 = -3.$$

To each one of these four roots there corresponds a solution of (42) of the form

$$\begin{bmatrix} x \\ y \\ z \\ u \end{bmatrix} = \begin{bmatrix} A_1 e^{\lambda t} \\ A_2 e^{\lambda t} \\ A_3 e^{\lambda t} \\ A_4 e^{\lambda t} \end{bmatrix}, \tag{45}$$

where the constants $A_1$, $A_2$, $A_3$, and $A_4$ satisfy the system

$$(-2 - \lambda)A_1 + A_2 = 0$$

$$A_1 + (-2 - \lambda)A_2 = 0$$

$$A_1 + A_2 + (-2 - \lambda)A_3 = 0 \tag{46}$$

$$2A_3 - \lambda A_4 = 0.$$

For $\lambda = 0$, we find $A_1 = A_2 = A_3 = 0$, $A_4 = 1$, and therefore, from (45),

$$\begin{bmatrix} 0 \\ 0 \\ 0 \\ 1 \end{bmatrix}$$

is one solution of (42). For $\lambda = -1$, $-2$, and $-3$, solving the system (46) and using (45), we find, respectively, the solutions

$$\begin{bmatrix} e^{-t} \\ e^{-t} \\ 2e^{-t} \\ -4e^{-t} \end{bmatrix}, \qquad \begin{bmatrix} 0 \\ 0 \\ e^{-2t} \\ -e^{-2t} \end{bmatrix}, \qquad \begin{bmatrix} e^{-3t} \\ -e^{-3t} \\ 0 \\ 0 \end{bmatrix}.$$

The four solutions are linearly independent, and therefore the general solution of (42) is

$$\begin{bmatrix} x \\ y \\ z \\ u \end{bmatrix} = c_1 \begin{bmatrix} 0 \\ 0 \\ 0 \\ 1 \end{bmatrix} + c_2 \begin{bmatrix} e^{-t} \\ e^{-t} \\ 2e^{-t} \\ -4e^{-t} \end{bmatrix} + c_3 \begin{bmatrix} 0 \\ 0 \\ e^{-2t} \\ -e^{-2t} \end{bmatrix} + c_4 \begin{bmatrix} e^{-3t} \\ -e^{-3t} \\ 0 \\ 0 \end{bmatrix};$$

that is,

$$x = c_2 e^{-t} + c_4 e^{-3t}$$

$$y = c_2 e^{-t} - c_4 e^{-3t}$$

$$z = 2c_2 e^{-t} + c_3 e^{-2t} \tag{47}$$

$$u = c_1 - 4c_2 e^{-t} - c_3 e^{-2t}.$$

Using the initial conditions (43), we find that

$$c_2 + c_4 = 1$$

$$c_2 - c_4 = 0$$

$$2c_2 + c_3 = 0 \qquad (48)$$

$$c_1 - 4c_2 - c_3 = 0.$$

Because of the negative exponents, the terminal conditions $x = y = z = 0$ as $t \to \infty$ are satisfied without any restriction on the constants $c_1$, $c_2$, $c_3$, and $c_4$. The terminal condition $u = 1$ as $t \to \infty$ implies [from the last equation in (47)] that

$$c_1 = 1. \qquad (49)$$

From (48), we find that

$$c_1 = 1, \qquad c_2 = c_4 = \tfrac{1}{2}, \qquad c_3 = -1.$$

Thus, (49) is also satisfied and the only solution of problem (42)–(44) is

$$x = \tfrac{1}{2}e^{-t} + \tfrac{1}{2}e^{-3t}$$

$$y = \tfrac{1}{2}e^{-t} - \tfrac{1}{2}e^{-3t}$$

$$z = e^{-t} - e^{-2t}$$

$$u = 1 - 2e^{-t} + e^{-2t}.$$

## EXERCISES

In Exercises 1 through 6, find the general solution of the system by the matrix method.

**1.** $\dot{x}_1 = 4x_1 + 2x_2$
$\dot{x}_2 = -3x_1 - x_2$

**2.** $\dot{x}_1 = 4x_1 - 2x_2$
$\dot{x}_2 = x_1 + x_2$

**3.** $\dot{x}_1 = 3x_1 - 2x_2$
$\dot{x}_2 = 2x_1 - x_2$

**4.** $\dot{x}_1 = -4x_1 + x_2$
$\dot{x}_2 = -x_1 - 2x_2$

**5.** $\dot{x}_1 = 3x_1 - 2x_2$
$\dot{x}_2 = 17x_1 - 7x_2$

**6.** $\dot{x}_1 = 3x_1 - 5x_2$
$\dot{x}_2 = 4x_1 - 5x_2$

In Exercises 7 through 10, solve the initial value problem.

**7.** $\dot{x}_1 = 4x_1 - 2x_2$
$\dot{x}_2 = x_1 + x_2$
$x_1(0) = 1, x_2(0) = 0$

**8.** $\dot{x}_1 = 3x_1 - 2x_2$
$\dot{x}_2 = 2x_1 - x_2$
$x_1(0) = -1, x_2(0) = 1$

**9.** $\dot{x}_1 = 3x_1 - 2x_2$
$\dot{x}_2 = 17x_1 - 7x_2$
$x_1(0) = 0, x_2(0) = 3$

**10.** $\dot{x}_1 = 3x_1 - 5x_2$
$\dot{x}_2 = 4x_1 - 5x_2$
$x_1(\pi) = 1, x_2(\pi) = \tfrac{4}{5}$

In Exercises 11 through 16, find the general solution of the system by the matrix method.

**11.** $\dot{x}_1 = 2x_1$
$\quad\;\dot{x}_2 = -3x_2$

**12.** $\dot{x}_1 = 4x_2$
$\quad\;\dot{x}_2 = -x_1$

**13.** $\dot{x}_1 = -x_2$
$\quad\;\dot{x}_2 = 3x_1 + 4x_2$

**14.** $\dot{x}_1 = -x_2$
$\quad\;\dot{x}_2 = x_1 - 2x_2$

**15.** $\dot{x}_1 = 4x_1 + 3x_2$
$\quad\;\dot{x}_2 = -x_1$

**16.** $\dot{x}_1 = -2x_1 + x_2$
$\quad\;\dot{x}_2 = -x_1$

In Exercises 17 through 26, find the general solution of the system of differential equations.

**17.** $\dot{x}_1 = 3x_1 + 2x_2 + 2x_3$
$\quad\;\dot{x}_2 = x_1 + 4x_2 + x_3$
$\quad\;\dot{x}_3 = -2x_1 - 4x_2 - x_3$

**18.** $\dot{x}_1 = 3x_1 - x_2$
$\quad\;\dot{x}_2 = -x_1 + 2x_2 - x_3$
$\quad\;\dot{x}_3 = -x_2 + 3x_3$

**19.** $\dot{x}_1 = 2x_1 + x_2$
$\quad\;\dot{x}_2 = -x_1 + x_3$
$\quad\;\dot{x}_3 = x_1 + 3x_2 + x_3$

**20.** $\dot{x}_1 = 7x_1 + 4x_2 - 4x_3$
$\quad\;\dot{x}_2 = 4x_1 - 8x_2 - x_3$
$\quad\;\dot{x}_3 = -4x_1 - x_2 - 8x_3$

**21.** $\dot{x}_1 = x_1 + 2x_2 + x_3 - x_4$
$\quad\;\dot{x}_2 = -x_2 + 2x_3 + 2x_4$
$\quad\;\dot{x}_3 = 2x_2 + 2x_3 + 2x_4$
$\quad\;\dot{x}_4 = -3x_2 - 6x_3 - 6x_4$

**22.** $\dot{x}_1 = 3x_1 + x_2$
$\quad\;\dot{x}_2 = x_1 + 3x_2$
$\quad\;\dot{x}_3 = 2x_3 + x_4 + x_5$
$\quad\;\dot{x}_4 = x_3 + 2x_4 + x_5$
$\quad\;\dot{x}_5 = x_3 + x_4 + 2x_5$

**23.** $\dot{x}_1 = -10x_1 + x_2 + 7x_3$
$\quad\;\dot{x}_2 = -9x_1 + 4x_2 + 5x_3$
$\quad\;\dot{x}_3 = -17x_1 + x_2 + 12x_3$

**24.** $\dot{x}_1 = x_2$
$\quad\;\dot{x}_2 = x_3$
$\quad\;\dot{x}_3 = x_1$

In Exercises 25 through 30, solve the IVP.

**25.** $\dot{x}_1 = 3x_1 + 2x_2 + 2x_3$
$\quad\;\dot{x}_2 = x_1 + 4x_2 + x_3$
$\quad\;\dot{x}_3 = -2x_1 - 4x_2 - x_3$
$\quad\;x_1(0) = 1, x_2(0) = 0,$
$\quad\;x_3(0) = -1$

**26.** $\dot{x}_1 = 3x_1 - x_2$
$\quad\;\dot{x}_2 = -x_1 + 2x_2 - x_3$
$\quad\;\dot{x}_3 = -x_2 + 3x_3$
$\quad\;x_1(0) = 1, x_2(0) = 0,$
$\quad\;x_3(0) = -1$

**27.** $\dot{x}_1 = 2x_1 + x_2$
$\quad\;\dot{x}_2 = -x_1 + x_3$
$\quad\;\dot{x}_3 = x_1 + 3x_2 + x_3$
$\quad\;x_1(0) = 1, x_2(0) = 1,$
$\quad\;x_3(0) = 3$

**28.** $\dot{x}_1 = 7x_1 + 4x_2 - 4x_3$
$\quad\;\dot{x}_2 = 4x_1 - 8x_2 - x_3$
$\quad\;\dot{x}_3 = -4x_1 - x_2 - 8x_3$
$\quad\;x_1(0) = 3, x_2(0) = 5,$
$\quad\;x_3(0) = -1$

**29.** $\dot{x}_1 = -10x_1 + x_2 + 7x_3$
$\quad\;\dot{x}_2 = -9x_1 + 4x_2 + 5x_3$
$\quad\;\dot{x}_3 = -17x_1 + x_2 + 12x_3$
$\quad\;x_1(0) = 6, x_2(0) = 1,$
$\quad\;x_3(0) = 10$

**30.** $\dot{x}_1 = x_2$
$\quad\;\dot{x}_2 = x_3$
$\quad\;\dot{x}_3 = x_1$
$\quad\;x_1(0) = x_2(0) = x_3(0) = 1$

In Exercises 31 through 36, use the method of variation of parameters to find a particular solution. Then write down the general solution of the system.

**31.** $\dot{x}_1 = 2x_1 + x_2 + e^{5t}$
$\dot{x}_2 = -3x_1 + 6x_2 + e^{5t}$

**32.** $\dot{x}_1 = 4x_1 + 2x_2 + 4$
$\dot{x}_2 = -3x_1 - x_2 - 3$

**33.** $\dot{x}_1 = -x_2 + t$
$\dot{x}_2 = 3x_1 + 4x_2 - 2 - 4t$

**34.** $\dot{x}_1 = 4x_2 - 3\cos t$
$\dot{x}_2 = -x_1$

**35.** $\dot{x} = -4x + y$
$\dot{y} = -x - 2y + 9$

**36.** $\dot{x} = 4x + 3y - 2 - 4t$
$\dot{y} = -x + t$

**37.** In the chemistry application of Section 1.7.1, we arrived at the homogeneous system [see Eqs. (10), Section 1.7.1]

$$\dot{c}_1 = -\frac{k}{V_1} c_1 + \frac{k}{V_1} c_2$$

$$\dot{c}_2 = \frac{k}{V_2} c_1 - \frac{k}{V_2} c_2.$$

Use the matrix technique to obtain the general solution of this system.

**38.** In the biokinetics application of Section 3.2.1, assume that $k_1 = k_2 = k_3 = k_4 = 1$ and $k_5 = 5$. Then system (16) becomes

$$\dot{x} = -2x + y$$

$$\dot{y} = x - 2y$$

$$\dot{z} = x + y - 5z$$

$$\dot{u} = 5z.$$

Find the general solution of this system by the matrix method. Find the particular solution of the system above that satisfies the initial and terminal conditions mentioned in the biokinetics application,

$$x(0) = 1, \quad y(0) = z(0) = u(0) = 0$$

and

$$x = y = z = 0 \quad \text{and} \quad u = 1 \quad \text{as} \quad t \to \infty.$$

## REVIEW EXERCISES

In Exercises 1 through 10, find the general solution of the system.

**1.** $\dot{x} = y$
$\dot{y} = -2x + 3y$

**2.** $\dot{x} = -x + 4y$
$\dot{y} = 2x + y$

**3.** $\dot{x} = y + e^{3t}$
$\dot{y} = -2x + 3y$

**4.** $\dot{x} = -x + 4y + e^{3t}$
$\dot{y} = 2x + y$

**5.** $\dot{x}_1 = x_2 + t^2$
   $\dot{x}_2 = x_3$
   $\dot{x}_3 = x_4$
   $\dot{x}_4 = x_1$

**6.** $\dot{x}_1 = 2x_1$
   $\dot{x}_2 = x_1 + x_3 - x_4$
   $\dot{x}_3 = x_1 - x_3$
   $\dot{x}_4 = -x_1 + x_4$

**7.** $\dot{x} = 3x - 2y$
   $\dot{y} = -2x$
   $\dot{z} = x + 2z$

**8.** $\dot{x} = 3x + y$
   $\dot{y} = x + 2y + z$
   $\dot{z} = y + 3z$

**9.** $\dot{x}_1 = x_1 - x_2$
   $\dot{x}_2 = -2x_2 - x_3$
   $\dot{x}_3 = 3x_3$

**10.** $\dot{x}_1 = 2x_1 + x_2$
   $\dot{x}_2 = -x_1 - x_3$
   $\dot{x}_3 = -x_1 - 3x_2 + x_3$

In Exercises 11 through 15, solve the initial value problem.

**11.** $\dot{x} = y$
   $\dot{y} = -2x + 3y$
   $x(0) = 1$
   $y(0) = 3$

**12.** $\dot{x} = y + e^{3t}$
   $\dot{y} = -2x + 3y$
   $x(0) = 0$
   $y(0) = 0$

**13.** $\dot{x}_1 = x_2 + t^2$
   $\dot{x}_2 = x_3$
   $\dot{x}_3 = x_4$
   $\dot{x}_4 = x_1$
   $x_1(0) = 1, x_2(0) = 1, x_3(0) = 1, x_4(0) = -1$

**14.** $\dot{x} = 3x - 2y$
   $\dot{y} = -2x$
   $\dot{z} = x + 2z$
   $x(0) = 2, y(0) = -1, z(0) = 0$

**15.** $\dot{x}_1 = x_1 - x_2$
   $\dot{x}_2 = -2x_2 - x_3$
   $\dot{x}_3 = 3x_3$
   $x_1(0) = 5, x_2(0) = -3, x_3(0) = 1$

**16.** In the electric circuit in Figure 3.6, assume that the currents are initially zero and that $R_1 = 2$ ohms, $R_2 = 4$ ohms, $L_1 = 1$ henry, $L_2 = 2$ henries, and $V(t) = 4$ volts. Determine the currents $I_1(t)$ and $I_2(t)$ at any time $t$.

**17.** Assume that the system

$$\dot{N}_1 = 5N_1 - 2N_2$$
$$\dot{N}_2 = 6N_1 - 2N_2$$

represents two interracting populations. Are the populations headed for extinction? For explosion? Explain.

**18.** Two tanks, $A$ and $B$, are connected as shown in Figure 3.7. Tank $A$ contains 50 gallons of pure water, and tank $B$ contains 50 gallons of brine in which 20 pounds of salt is dissolved. Liquid circulates through the tanks at the rate of 3 gallons per minute, and the mixture in each tank is kept uniform by stirring.

(a) How much salt is in each tank after 20 minutes?

(b) How much salt is in each tank after a very long time?

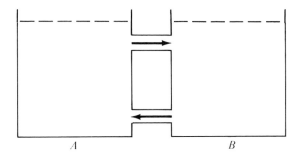

**Figure 3.7**

**19.** Solve[13] the system of differential equations

$$\frac{dx}{dt} = x + y - 3, \qquad \frac{dy}{dt} = -2x + 3y + 1,$$

subject to the condition $x = y = 0$ at $t = 0$.

**20.** In the network[14] shown in Figure 3.8, the switch is closed at $t = 0$ and the currents are initially zero. Find the current $i_2$ as a function of time.

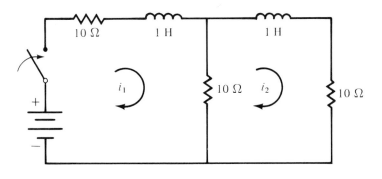

**Figure 3.8**

[13]This exercise is taken from the second William Lowell Putnam Mathematical Competition, *Amer. Math. Monthly* 46 (1939): 248.

[14]This is Example 14 in M. Van Valkenburg, *Network Analysis,* 2nd ed. (Englewood Cliffs, N.J.: Prentice-Hall, 1964), p. 179. © 1964. Reprinted by permission of Prentice-Hall, Inc.

# CHAPTER 4

# The Laplace Transform

## 4.1 INTRODUCTION

In this chapter we present another method for solving linear differential equations with constant coefficients and systems of such equations. It is called the method of Laplace transform. By this method an initial value problem is transformed into an algebraic equation or system of equations which we can solve by algebraic methods and a table of Laplace transforms. This method resembles in some ways the use of logarithms for solving exponential equations.

## 4.2 THE LAPLACE TRANSFORM AND ITS PROPERTIES

Assume that the function $g$ is defined for $0 \leq t < \infty$ and is bounded and integrable in every finite interval $0 \leq t \leq b$. Then by definition

$$\int_0^\infty g(t)dt = \lim_{b \to \infty} \int_0^b g(t)dt.$$

We say that the *improper integral* on the left converges or diverges according to whether the limit on the right does or does not exist. For example, the improper integral $\int_0^\infty e^{-3t}dt$ converges but $\int_0^\infty e^{3t}dt$ diverges. In fact,

$$\int_0^\infty e^{-3t}dt = \lim_{b \to \infty} \int_0^b e^{-3t}dt = \lim_{b \to \infty}\left( -\frac{1}{3}e^{-3t}\Big|_0^b \right)$$

$$= \lim_{b \to \infty}\left( -\frac{1}{3}e^{-3b} + \frac{1}{3} \right) = \frac{1}{3},$$

but

$$\int_0^\infty e^{3t}dt = \lim_{b \to \infty} \int_0^b e^{3t}dt = \lim_{b \to \infty}\left( \frac{1}{3}e^{3t}\Big|_0^b \right)$$

$$= \lim_{b \to \infty}\left( \frac{1}{3}e^{3b} - \frac{1}{3} \right) = \infty.$$

**DEFINITION 1**

*Assume that the function f is defined for $0 \le t < \infty$. The Laplace transform of f, which we denote by F or $\mathscr{L}[f]$, is defined by the improper integral*

$$F(s) = \mathscr{L}[f(t)] = \int_0^\infty e^{-st}f(t)dt, \tag{1}$$

*provided that the integral in (1) exists for all s larger than or equal to some value $s_0$.*

**EXAMPLE 1** Compute the Laplace transform of $f(t) = 1$.

**Solution**

$$\int_0^\infty e^{-st}f(t)dt = \int_0^\infty e^{-st}(1)dt = \int_0^\infty e^{-st}dt$$

$$= \lim_{b\to\infty}\int_0^b e^{-st}dt = \lim_{b\to\infty} -\left[\frac{e^{-st}}{s}\bigg|_0^b\right]$$

$$= \lim_{b\to\infty}\left[\frac{1}{s} - \frac{e^{-bs}}{s}\right].$$

If $s > 0$, the limit above exists and we obtain

$$\int_0^\infty e^{-st}f(t)dt = \frac{1}{s}, \qquad s > 0.$$

**EXAMPLE 2** Compute the Laplace transform of $f(t) = e^{2t}$.

**Solution**

$$\int_0^\infty e^{-st}f(t)dt = \int_0^\infty e^{-st}e^{2t}dt = \int_0^\infty e^{-(s-2)t}dt$$

$$= \lim_{b\to\infty}\int_0^b e^{-(s-2)t}dt = \lim_{b\to\infty} -\left[\frac{e^{-(s-2)t}}{s-2}\bigg|_0^b\right]$$

$$= \lim_{b\to\infty}\left[\frac{1}{s-2} - \frac{e^{-(s-2)b}}{s-2}\right].$$

This limit exists only when $s > 2$. Hence,

$$\int_0^\infty e^{-st}f(t)dt = \frac{1}{s-2}, \qquad s > 2.$$

In Examples 1 and 2 we observe that the Laplace transform is a function of $s$. We refer to $f$ as the *inverse Laplace transform* of $F$. For the purposes of notation for the inverse Laplace transform, we introduce the alternative symbolism $\mathcal{L}[f(t)] = F(s)$. We also write $f(t) = \mathcal{L}^{-1}[F(s)]$. From Examples 1 and 2 we see that

$$\mathcal{L}^{-1}\left[\frac{1}{s}\right] = 1 \quad \text{and} \quad \mathcal{L}^{-1}\left[\frac{1}{s-2}\right] = e^{2t}.$$

We note that for a given $f$ the corresponding $F$ is uniquely determined (when it exists). Concerning inverse Laplace transforms, we will state Theorem 1, but first we give the following definitions.

### DEFINITION 2

*A function f is said to be piecewise continuous on an interval I if I can be subdivided into a finite number of subintervals, in each of which f is continuous and has finite left- and right-hand limits.*

### DEFINITION 3

*A function f is of exponential order $\alpha$ as t tends to infinity if there exist numbers M, $\alpha$, and T such that*

$$|f(t)| \le Me^{\alpha t} \quad \text{when} \quad t \ge T.$$

Alternatively, $f$ is of exponential order $\alpha$ if there exists an $\alpha$ such that $\lim_{t \to \infty} |f(t)|/e^{\alpha t} = L$, where $L = 0$ or $L$ is a finite positive number.

We now quote without proof a theorem that guarantees the convergence of the integral (1).

### THEOREM 1

*If f is piecewise continuous on every finite interval in $[0, \infty)$, and if f is of exponential order $\alpha$ as t tends to infinity, the integral (1) converges for $s > \alpha$. Furthermore, if f and g are piecewise continuous functions whose Laplace transforms exist and satisfy $\mathcal{L}[f(t)] = \mathcal{L}[g(t)]$, then $f = g$ at their points of continuity. Thus, if F(s) has a continuous inverse f, then f is unique.*

The following theorems provide us with useful tools for manipulating Laplace transforms. The proof of the first theorem is a simple consequence of the definitions and will therefore be omitted (see Exercise 60). In what follows the symbols $c_1$, $c_2$, and $k$ denote arbitrary constants.

### THEOREM 2

$$\mathcal{L}[c_1 f(t) + c_2 g(t)] = c_1 F(s) + c_2 G(s).$$

### THEOREM 3

*If $F(s) = \mathcal{L}[f(t)]$, then $F(s + k) = \mathcal{L}[e^{-kt}f(t)]$.*

**Proof**

$$F(s + k) = \int_0^\infty e^{-(s+k)t}f(t)dt = \int_0^\infty e^{-st}e^{-kt}f(t)dt$$

$$= \int_0^\infty e^{-st}(e^{-kt}f(t))dt = \mathcal{L}[e^{-kt}f(t)].$$

**EXAMPLE 3**   Find (a) $\mathcal{L}[t^2]$.     (b) $\mathcal{L}[\cosh kt]$.

**Solution**

(a)
$$\mathcal{L}[t^2] = \int_0^\infty t^2 e^{-st}dt = \lim_{t\to\infty}\left[\frac{t^2 e^{-st}}{-s}\right] + \frac{2}{s}\int_0^\infty te^{-st}dt$$

$$= \lim_{t\to\infty}\left[\frac{t^2 e^{-st}}{-s}\right] + \frac{2}{s}\lim_{t\to\infty}\left[\frac{te^{-st}}{s}\right] + \frac{2}{s^2}\int_0^\infty e^{-st}dt$$

If $s > 0$, then both of the above limits are zero as can be verified by use of l'Hospital's rule. Thus

$$\mathcal{L}[t^2] = \frac{2}{s^3}, s > 0.$$

(b)     $\mathcal{L}[\cosh kt] = \mathcal{L}\left[\frac{1}{2}e^{kt} + \frac{1}{2}e^{-kt}\right] = \frac{1}{2}\mathcal{L}[e^{kt}] + \frac{1}{2}\mathcal{L}[e^{-kt}]$

$$= \frac{1}{2}\mathcal{L}[e^{kt}(1)] + \frac{1}{2}\mathcal{L}[e^{(-k)t}(1)] = \frac{1}{2}\frac{1}{s-k} + \frac{1}{2}\frac{1}{s-(-k)}$$

$$\mathcal{L}[\cosh kt] = \frac{1}{2}\frac{(s+k)+(s-k)}{(s-k)(s+k)} = \frac{1}{2}\frac{2s}{s^2-k^2} = \frac{s}{s^2-k^2}.$$

**EXAMPLE 4**   Find $\mathcal{L}[e^{-7t}t^2]$.

**Solution**   Since $\mathcal{L}[t^2] = 2/s^3$, we have, by Theorem 3,

$$\mathcal{L}[e^{-7t}t^2] = \frac{2}{(s+7)^3}.$$

In the next section we illustrate how Laplace transforms are used to solve initial value problems. Quite often the Laplace transform of the solution is expressed as the product of two known functions of $s$. To find the solution of the IVP, one needs to find the inverse Laplace transform of this product. To do so, one frequently uses either partial fractions (see Appendix B) or Theorem 4. The property expressed in Theorem 4 is known as the *convolution property of Laplace transforms*. We state the theorem without proof.

**THEOREM 4**

$$G(s)F(s) = \mathcal{L}\left[\int_0^t g(t-\tau)f(\tau)d\tau\right]. \qquad (2)$$

**EXAMPLE 5**   Find $\mathcal{L}[\int_0^t f(\tau)d\tau]$.

**Solution**   In order to find $\mathcal{L}[\int_0^t f(\tau)d\tau]$, we apply the definition. (Alternatively we could apply Theorem 4 with $g(t) = 1$.

$$\mathcal{L}\left[\int_0^t f(\tau)d\tau\right] = \int_0^\infty \left[\int_0^t f(\tau)d\tau\right]e^{-st}dt$$

$$= \lim_{t \to \infty}\left[\frac{\left(\int_0^t f(\tau)d\tau\right)e^{-st}}{-s}\right] + \frac{1}{s}\int_0^\infty f(t)e^{-st}dt.$$

Since $f$ is of exponential order, the integral of $f$ is also of exponential order and the above limit is zero. Therefore

$$\mathcal{L}\left[\int_0^t f(\tau)d\tau\right] = \frac{1}{s}F(s).$$

The result of Example 5 enables us to use Laplace transform methods to solve certain types of integral equations. (See the population growth application in Section 4.6.)

We summarize our results in Table 4.1. Verification of some of the entries is left to the exercises.

**Table 4.1**

|        | $F(s)$                                     | $f(t)$                            |
|--------|--------------------------------------------|-----------------------------------|
| (i)    | $\dfrac{1}{s}$, $s > 0$                      | $1$                               |
| (ii)   | $\dfrac{1}{s^2}$, $s > 0$                    | $t$                               |
| (iii)  | $\dfrac{n!}{s^{n+1}}$, $s > 0$              | $t^n$, $n = 1, 2, 3, \ldots$      |
| (iv)   | $\dfrac{1}{s - k}$, $s > k$                 | $e^{kt}$                          |
| (v)    | $\dfrac{n!}{(s - k)^{n+1}}$, $s > k$        | $e^{kt}t^n$, $n = 1, 2, 3, \ldots$ |
| (vi)   | $\dfrac{k}{s^2 + k^2}$, $s > 0$            | $\sin kt$                         |
| (vii)  | $\dfrac{s}{s^2 + k^2}$, $s > 0$            | $\cos kt$                         |
| (viii) | $\dfrac{m}{(s - k)^2 + m^2}$, $s > k$      | $e^{kt}\sin mt$                   |
| (ix)   | $\dfrac{s - k}{(s - k)^2 + m^2}$, $s > k$  | $e^{kt}\cos mt$                   |

| (x) | $\dfrac{s}{s^2 - k^2}, s > k$ | $\cosh kt$ |
| (xi) | $\dfrac{k}{s^2 - k^2}, s > k$ | $\sinh kt$ |
| (xii) | $\dfrac{k_1 - k_2}{(s - k_1)(s - k_2)}, s > k_1, k_2$ | $e^{k_1 t} - e^{k_2 t}$ |
| (xiii) | $\dfrac{2ks}{(s^2 + k^2)^2}, s > 0$ | $t \sin kt$ |
| (xiv) | $\dfrac{s^2 - k^2}{(s^2 + k^2)^2}, s > 0$ | $t \cos kt$ |
| (xv) | $c_1 F(s) + c_2 G(s)$ | $c_1 f(t) + c_2 g(t)$ |
| (xvi) | $F(s + k)$ | $e^{-kt} f(t)$ |
| (xvii) | $F(ks)$ | $\dfrac{1}{k} f\left(\dfrac{t}{k}\right), k > 0$ |
| (xviii) | $F^{(n)}(s)$ | $(-t)^n f(t)$ |
| (xix) | $G(s)F(s)$ | $\int_0^t g(t - \tau)f(\tau)d\tau = \int_0^t f(t - \tau)g(\tau)d\tau$ |
| (xx) | $\dfrac{1}{s} F(s)$ | $\int_0^t f(t)dt$ |
| (xxi) | $\dfrac{1}{s^2} F(s)$ | $\int_0^t \int_0^\tau f(u)du\, d\tau$ |
| (xxii) | $s^n F(s) - s^{n-1}f(0) - s^{n-2}f'(0) - \cdots - f^{(n-1)}(0)$ | $f^{(n)}(t)$ |

Our approach thus far has been to apply the definition of the Laplace transform to obtain the transform of certain specific functions or classes of functions. In much the same way as one uses a table of integrals or formulas for differentiation, one can also use a table of Laplace transforms. That is, we do not resort to the definition every time we wish to determine a particular Laplace transform, but rather we prefer to look up the transform in a table. Naturally, the reader realizes that not every function will appear in such a table. Therefore, if one is to be proficient at using the table, one must develop a capability for expressing the functions in the forms found in the tables. We take notice of the fact that all the specific functions that appear in Table 4.1, Laplace transforms, are continuous and of exponential order. Formula (xxii) of Table 4.1 is presented in Theorem 2 of Section 4.3.

**EXAMPLE 6**   Show that

$$\mathcal{L}[e^{k_1 t} - e^{k_2 t}] = \frac{k_1 - k_2}{(s - k_1)(s - k_2)}.$$

**Solution**   Using formula (xv) of Table 4.1, we have

$$\mathcal{L}[e^{k_1 t} - e^{k_2 t}] = \mathcal{L}[e^{k_1 t}] - \mathcal{L}[e^{k_2 t}].$$

Next we apply (iv) to obtain

$$\mathcal{L}[e^{k_1 t} - e^{k_2 t}] = \frac{1}{s - k_1} - \frac{1}{s - k_2} = \frac{s - k_2 - (s - k_1)}{(s - k_1)(s - k_2)} = \frac{k_1 - k_2}{(s - k_1)(s - k_2)}.$$

**EXAMPLE 7** Show that

$$\mathcal{L}[t \sin kt] = \frac{2ks}{(s^2 + k^2)^2}.$$

**Solution**   From formula (vi) we see that

$$\mathcal{L}[\sin kt] = \frac{k}{s^2 + k^2}. \tag{3}$$

Apply formula (xviii) with $n = 1$ to Eq. (3):

$$\frac{d}{ds}\left(\frac{k}{s^2 + k^2}\right) = \mathcal{L}[-t \sin kt] = -\mathcal{L}[t \sin kt].$$

Thus,

$$-\frac{k(2s)}{(s^2 + k^2)^2} = -\mathcal{L}[t \sin kt]$$

$$\Rightarrow \frac{2ks}{(s^2 + k^2)^2} = \mathcal{L}[t \sin kt].$$

**EXAMPLE 8** Find

$$\mathcal{L}^{-1}\left[\frac{5s^2 + s + 4}{(s+4)(s^2+4)}\right].$$

**Solution**   Using partial-fraction decomposition (see Appendix B), we have

$$\frac{5s^2 + s + 4}{(s+4)(s^2+4)} = \frac{4}{s+4} + \frac{s-3}{s^2+4}.$$

Thus

$$\mathcal{L}^{-1}\left[\frac{5s^2 + s + 4}{(s + 4)(s^2 + 4)}\right] = \mathcal{L}^{-1}\left[\frac{4}{s + 4} + \frac{s-3}{s^2 + 4}\right]$$

$$= 4\mathcal{L}^{-1}\left[\frac{1}{s + 4}\right] + \mathcal{L}^{-1}\left[\frac{s}{s^2 + 4}\right] - 3\mathcal{L}^{-1}\left[\frac{1}{s^2 + 4}\right]$$

$$= 4\mathcal{L}^{-1}\left[\frac{1}{s + 4}\right] + \mathcal{L}^{-1}\left[\frac{s}{s^2 + 4}\right] - \frac{3}{2}\mathcal{L}^{-1}\left[\frac{2}{s^2 + 4}\right]$$

$$= 4e^{-4t} + \cos 2t - \frac{3}{2}\sin 2t.$$

To obtain the last step we have used formulas iv, vi, and vii of Table 4.1.

We state the following theorem without proof.

**THEOREM 5**

*If $f(t)$ is a periodic function with period $T$, then*

$$\mathcal{L}[f(t)] = \frac{1}{1 - e^{-sT}} \int_0^T e^{-st} f(t)\,dt$$

**EXERCISES**

Using the definition, verify the Laplace transforms in Exercises 1 through 11. $c$, $c_1$, $c_2$, and $k$ denote arbitrary constants.

**1.** $\mathcal{L}[t] = \dfrac{1}{s^2}$     **2.** $\mathcal{L}[t^2] = \dfrac{2}{s^3}$

**3.** $\mathcal{L}[t^3] = \dfrac{3!}{s^4}$     **4.** $\mathcal{L}[ct] = \dfrac{c}{s^2}$

**5.** $\mathcal{L}[c_1 + c_2 t] = \dfrac{c_1}{s} + \dfrac{c_2}{s^2}$     **6.** $\mathcal{L}[\cos ct] = \dfrac{s}{s^2 + c^2}$

**7.** $\mathcal{L}[\sin ct] = \dfrac{c}{s^2 + c^2}$

**8.** $\mathcal{L}[\cosh kt] = \dfrac{s}{s^2 - k^2}$, $s > |k|$ $\left(\text{recall that } \cosh kt = \dfrac{e^{kt} + e^{-kt}}{2}\right)$

**9.** $\mathcal{L}[\sinh kt] = \dfrac{k}{s^2 - k^2}$, $s > |k|$     **10.** $\mathcal{L}[e^{kt}] = \dfrac{1}{s - k}$, $s > k$

**11.** $\mathcal{L}[0] = 0$

Using Exercises 1 through 11 as formulas, find the inverse Laplace transforms in Exercises 12 through 18.

**12.** $\mathcal{L}^{-1}\left[\dfrac{5}{s^2}\right]$     **13.** $\mathcal{L}^{-1}\left[\dfrac{s}{s^2 + 16}\right]$     **14.** $\mathcal{L}^{-1}\left[\dfrac{1}{s - 5}\right]$

**15.** $\mathcal{L}^{-1}\left[\dfrac{5}{s^2 - 25}\right]$     **16.** $\mathcal{L}^{-1}\left[\dfrac{2}{s} - \dfrac{5}{s^2}\right]$     **17.** $\mathcal{L}^{-1}\left[\dfrac{2}{s^2 + 4}\right]$

**18.** $\mathcal{L}^{-1}\left[\dfrac{6}{s^4}\right]$

In Exercises 19 through 38, find the Laplace transform of each function.

**19.** $5 - 8t^3$     **20.** $t \cos 3t$

**21.** $\frac{1}{8}\cos\frac{3}{8}t$     **22.** $e^{(5/2)t}\cosh\frac{7}{2}t$

**23.** $e^{3t}\cos 2t - e^t \sinh 5t$     **24.** $-te^{5t}\cosh t$

**25.** $\cos t - \sin t$     **26.** $e^t \cos t$

**27.** $t^7 - t^4 + 5t^2$

**28.** $e^{2t}(\cos t + \sin t)$

**29.** $t \sinh t$

**30.** $t^2 \sin 4t$

**31.** $\dfrac{d}{dt}[te^{5t}]$

**32.** $\dfrac{d}{dt}[t^2 e^{-t}]$

**33.** $\dfrac{d^2}{dt^2}[\cos t + te^t]$

**34.** $\dfrac{d^3}{dt^3}[t^3]$

**35.** $\displaystyle\int_0^t \cosh \tau \cos (t - \tau) \, d\tau$

**36.** $-t\displaystyle\int_0^t e^{-5(t-\tau)}\tau \, d\tau$

**37.** $\displaystyle\int_0^t e^\tau \cos 2\tau \, d\tau$

**38.** $\displaystyle\int_0^{t^{2/3}} \tau^{1/2} e^{\tau^{3/2}} \cos \tau^{3/2} \, d\tau$

In Exercises 39 through 58, find the inverse Laplace transform of each expression.

**39.** $\dfrac{2}{s^2 + k^2}$

**40.** $\dfrac{5}{s^2 - 5s + 4}$

**41.** $\dfrac{n!}{(s - k)^{n+1}}, n = 1, 2, 3, \ldots$

**42.** $\dfrac{5s^2 + 6s + 4}{(s + 4)(s^2 + 4)}$

**43.** $\dfrac{s}{s^2 - k^2}$

**44.** $\dfrac{3s + 2}{(s - 2)^2 + 4}$

**45.** $\dfrac{2}{(s^2 + 1)^2}$

**46.** $\dfrac{2s - 5}{s(s^2 - 6s + 34)}$

**47.** $\dfrac{s^2 + 3s + 36}{s(s^2 + 13s + 36)}$

**48.** $\dfrac{s^2 - 4s + 8}{s^2(s^2 - 4s + 4)}$

**49.** $\dfrac{s^2 + 4s + 36}{(s^2 - 4)^2}$

**50.** $\dfrac{2s^2 - s - 7}{(s + 1)^2(s^2 + 7s + 10)}$

**51.** $\dfrac{s^2 + 2s + 53}{(s + 2)(s^2 + 49)}$

**52.** $\dfrac{2}{(s - 1)(s^2 + 1)}$

**53.** $\dfrac{s^2 + 3s - 18}{s(s^2 - 6s + 9)}$

**54.** $\dfrac{s^2 + s + 4}{(s + 3)(s^2 - 10s + 25)}$

**55.** $\dfrac{s^2 - s + 1}{s^3(s + 1)}$

**56.** $\dfrac{s + 2}{s^2(s - 1)}$

**57.** $\dfrac{2s - 3}{(s + 1)^2 + 16}$

**58.** $\dfrac{5s + 7}{(s + 3)^2 - 16}$

**59.** Verify that $\displaystyle\int_0^t g(t - \tau)f(\tau)d\tau = \int_0^t f(t - \tau)g(\tau)d\tau$.

**60.** Prove Theorem 2.

**61.** If $F(s) = \mathcal{L}[f(t)]$, show that $F(ks) = \dfrac{1}{k}\mathcal{L}\left[f\left(\dfrac{t}{k}\right)\right]$, where $k > 0$ is a constant.

[Hint: set $t = ku$.]

**62.** If $F(s) = \mathscr{L}[f(t)]$, show that $\dfrac{d}{ds}F(s) = \mathscr{L}[-tf(t)]$.

**63.** If $n$ is a positive integer, show that $\dfrac{d^n}{ds^n}F(s) = \mathscr{L}[(-t)^n f(t)]$.

**64.** Show that $\mathscr{L}[t \cos kt] = \dfrac{s^2 - k^2}{(s^2 + k^2)^2}$.

**65.** Show that $\mathscr{L}\left[\displaystyle\int_0^t \int_0^\tau f(u)\, du\, d\tau\right] = \dfrac{1}{s^2}F(s)$.

**66.** Show that $\mathscr{L}[e^{kt} \sin mt] = \dfrac{m}{(s-k)^2 + m^2}$.

**67.** Show that $\mathscr{L}[e^{kt} \cos mt] = \dfrac{s-k}{(s-k)^2 + m^2}$.

**68.** Use Theorem 5 to compute $\mathscr{L}[\sin t]$.

**69.** Use Theorem 5 to compute $\mathscr{L}[\cos t]$.

**70.** Use Theorem 5 to compute $\mathscr{L}[\sin kt]$.

**71.** Use Theorem 5 to compute $\mathscr{L}[\cos kt]$.

---

## 4.3 THE LAPLACE TRANSFORM APPLIED TO DIFFERENTIAL EQUATIONS AND SYSTEMS

---

At the beginning of this chapter we emphasized that Laplace transforms provide us with a useful tool for solving certain types of differential equations. The theorems of Section 4.2 are helpful to us for the purpose of manipulating transforms. In order to apply Laplace transforms to differential equations, we need to know what the Laplace transform of a derivative (or second derivative, or higher) of a function is. The next two theorems provide us with this information. In both theorems we assume that all functions of $t$ that appear satisfy the hypotheses of Theorem 1 of Section 4.2, so that their Laplace transform exists.

**THEOREM 1**

$$\mathscr{L}[f'(t)] = sF(s) - f(0).$$

**Proof**

$$\mathscr{L}[f'(t)] = \int_0^\infty e^{-st} f'(t)\, dt.$$

Integrate by parts setting $u = e^{-st}$, $dv = f'(t)dt$. Then $du = -se^{-st} dt$ and $v = f(t)$. Hence,

$$\mathcal{L}[f'(t)] = e^{-st}f(t) \Big|_0^\infty - \int_0^\infty f(t)(-se^{-st}dt)$$

$$= -f(0) + s\int_0^\infty e^{-st}f(t)dt$$

$$= sF(s) - f(0).$$

In the boundary term, $e^{-st}f(t) \mid_0^\infty$, we know that $\lim_{t\to\infty} e^{-st}f(t) = 0$, since $f$ is of exponential order.

### THEOREM 2

$$\mathcal{L}[f^{(n)}(t)] = s^n F(s) - s^{n-1}f(0) - s^{n-2}f'(0) - \cdots - f^{(n-1)}(0).$$

**Proof**  We prove the result for $n = 2$ and leave the generalization to the reader.

$$\mathcal{L}[f''(t)] = \int_0^\infty e^{-st}f''(t)dt.$$

Integrate by parts, setting $u = e^{-st}$ and $dv = f''(t)dt$. Hence, $du = -se^{-st}dt$ and $v = f'(t)$. Therefore,

$$\mathcal{L}[f''(t)] = e^{-st}f'(t) \Big|_0^\infty - \int_0^\infty f'(t)(-se^{-st}dt)$$

$$= -f'(0) + s\int_0^\infty e^{-st}f'(t)dt$$

$$= -f'(0) + s\mathcal{L}[f'(t)] = -f'(0) + s(sF(s) - f(0))$$

$$= s^2 F(s) - sf(0) - f'(0).$$

**Note**  In the proof of Theorem 2 just given, it is assumed that $\lim_{t\to\infty} e^{-st}f'(t) = 0$. Similarly, in the general case it must be assumed that $\lim_{t\to\infty} e^{-st} f^{(k)}(t) = 0$, for $k = 1, 2, \ldots, n - 1$. Thus, we assume that $f$ and its derivatives are of exponential order as $t$ tends to infinity.

Basically, the method of solving differential equations by Laplace transforms is outlined as follows: Given a differential equation, we "take the Laplace transform of it" (in other words, we apply the Laplace transformation to each term in the differential equation). This results in an equation (usually a linear algebraic equation) for the Laplace transform of the unknown function. We then solve this equation for this transform, and, in the final step, we find the inverse of this transform. For systems of differential equations the procedure is similar and leads to a system of algebraic equations.

Formula (xxii) of Table 4.1 indicates that in order to obtain explicit results, one must know $f(0), f'(0), \ldots, f^{(n-1)}(0)$. That is, we must have an initial value problem. It is a fact that the Laplace transform has a rather wide range of applicability; however, we will restrict our attention here to linear differential equations and systems with constant coefficients.

The method outlined above will be clarified via examples.

**EXAMPLE 1** Solve the initial value problem

$$\ddot{y} - 3\dot{y} + 2y = 0, \tag{1}$$

$$y(0) = 1, \quad \dot{y}(0) = 0. \tag{2}$$

**Solution** Taking the Laplace transform of Eq. (1), we have

$$\mathcal{L}[\ddot{y} - 3\dot{y} + 2y] = \mathcal{L}[0]$$

$$\Rightarrow \mathcal{L}[\ddot{y}] - 3\mathcal{L}[\dot{y}] + 2\mathcal{L}[y] = 0$$

$$s^2 Y(s) - sy(0) - \dot{y}(0) - 3(sY(s) - y(0)) + 2Y(s) = 0.$$

Note that we use $Y(s)$ to denote $\mathcal{L}[y(t)]$. Substituting the initial values (2), we have

$$s^2 Y(s) - s(1) - 0 - 3(sY(s) - 1) + 2Y(s) = 0$$

$$s^2 Y(s) - s - 3sY(s) + 3 + 2Y(s) = 0$$

$$(s^2 - 3s + 2)Y(s) = s - 3$$

$$\Rightarrow \quad Y(s) = \frac{s - 3}{s^2 - 3s + 2}$$

$$\Rightarrow \quad y(t) = \mathcal{L}^{-1}\left[\frac{s - 3}{s^2 - 3s + 2}\right].$$

Thus, we have solved the IVP if we can determine the indicated inverse Laplace transform. There are perhaps a couple of approaches to this end, but the simplest is to apply partial-fraction decomposition. We proceed as follows (see Appendix B):

$$\frac{s - 3}{s^2 - 3s + 2} = \frac{s - 3}{(s - 2)(s - 1)} = \frac{A}{s - 2} + \frac{B}{s - 1}.$$

Now,

$$A = \frac{s - 3}{s - 1}\bigg|_{s=2} = -1 \quad \text{and} \quad B = \frac{s - 3}{s - 2}\bigg|_{s=1} = 2.$$

Hence,

$$\mathcal{L}^{-1}\left[\frac{s - 3}{(s - 2)(s - 1)}\right] = \mathcal{L}^{-1}\left[\frac{-1}{s - 2} + \frac{2}{s - 1}\right] = -\mathcal{L}^{-1}\left[\frac{1}{s - 2}\right] + 2\mathcal{L}^{-1}\left[\frac{1}{s - 1}\right]$$

$$= -e^{2t} + 2e^t.$$

Consequently, the solution to the IVP (1)–(2) is

$$y(t) = 2e^t - e^{2t}.$$

The reader no doubt recognizes that the IVP (1)–(2) can be solved easily by the methods of Chapter 2. So why should we take a seemingly difficult approach to a simple problem? Keep in mind two facts. First, we are at this point trying to illustrate the method. Second, the Laplace transform provides a simpler method of solution in many cases.

**EXAMPLE 2**   Solve the IVP

$$\ddot{y} - 5\dot{y} + 6y = te^{2t} + e^{3t}$$

$$y(0) = 0, \quad \dot{y}(0) = 1.$$

**Solution**

$$\mathcal{L}[\ddot{y} - 5\dot{y} + 6y] = \mathcal{L}[te^{2t} + e^{3t}]$$

$$\Rightarrow \qquad \mathcal{L}[\ddot{y}] - 5\mathcal{L}[\dot{y}] + 6\mathcal{L}[y] = \mathcal{L}[te^{2t}] + \mathcal{L}[e^{3t}]$$

$$\Rightarrow \qquad s^2Y(s) - sy(0) - \dot{y}(0) - 5[sY(s) - y(0)] + 6Y(s) = \frac{1}{(s-2)^2} + \frac{1}{s-3}$$

$$(s^2 - 5s + 6)Y(s) - 1 = \frac{1}{(s-2)^2} + \frac{1}{s-3}$$

$$\Rightarrow \qquad Y(s) = \frac{1}{(s^2 - 5s + 6)(s-2)^2} + \frac{1}{(s^2 - 5s + 6)(s-3)} + \frac{1}{s^2 - 5s + 6}$$

$$= \frac{1}{(s-2)^3(s-3)} + \frac{1}{(s-2)(s-3)^2} + \frac{1}{(s-2)(s-3)}$$

$$\Rightarrow \qquad y(t) = \mathcal{L}^{-1}\left[\frac{1}{(s-2)^3(s-3)} + \frac{1}{(s-2)(s-3)^2} + \frac{1}{(s-2)(s-3)}\right]$$

$$= \mathcal{L}^{-1}\left[\frac{1}{(s-2)^3(s-3)}\right] + \mathcal{L}^{-1}\left[\frac{1}{(s-2)(s-3)^2}\right]$$

$$+ \mathcal{L}^{-1}\left[\frac{1}{(s-2)(s-3)}\right].$$

Now

$$\mathcal{L}^{-1}\left[\frac{1}{(s-2)(s-3)}\right] = \mathcal{L}^{-1}\left[\frac{3-2}{(s-3)(s-2)}\right] = e^{3t} - e^{2t}.$$

Also, using partial fraction decomposition (see Appendix B),

$$\frac{1}{(s-2)^3(s-3)} = \frac{A}{(s-2)^3} + \frac{B}{(s-2)^2} + \frac{C}{s-2} + \frac{D}{s-3}$$

and

$$\frac{1}{(s - 2)(s - 3)^2} = \frac{E}{s - 2} + \frac{F}{s - 3} + \frac{G}{(s - 3)^2}.$$

We have,

$$A = \frac{1}{s - 3}\bigg|_{s=2} = -1, \qquad D = \frac{1}{(s - 2)^3}\bigg|_{s=3} = 1$$

$$E = \frac{1}{(s - 3)^2}\bigg|_{s=2} = 1, \qquad G = \frac{1}{s - 2}\bigg|_{s=3} = 1$$

and

$$C + D = 0 \Rightarrow C = -D = -1$$

$$-3A + 6B - 12C - 8D = 1 \Rightarrow B = \frac{1 + 12C + 8D + 3A}{6} = -1,$$

$$E + F = 0 \Rightarrow F = -E = -1.$$

Therefore,

$$y(t) = \mathcal{L}^{-1}\left[\frac{-1}{(s - 2)^3}\right] + \mathcal{L}^{-1}\left[\frac{-1}{(s - 2)^2}\right] + \mathcal{L}^{-1}\left[\frac{-1}{s - 2}\right] + \mathcal{L}^{-1}\left[\frac{1}{s - 3}\right]$$

$$+ \mathcal{L}^{-1}\left[\frac{-1}{s - 3}\right] + \mathcal{L}^{-1}\left[\frac{1}{(s - 3)^2}\right] + e^{3t} - e^{2t}$$

$$= \frac{-e^{2t}t^2}{2!} + (-1)e^{2t}t + (-1)e^{2t} + e^{3t} + e^{2t} - e^{3t} + e^{3t}t + e^{3t} - e^{2t}$$

$$= e^{2t}\left(-\frac{t^2}{2} - t - 1\right) + e^{3t}(t + 1).$$

   The following example illustrates that the Laplace-transform method can still be applied, even though the initial value is given at a point different than $t = 0$.


**EXAMPLE 3**   Solve the IVP

$$\dot{y} + 3y = 0 \tag{3}$$

$$y(3) = 1. \tag{4}$$

**Solution**   We note that the IVP can be attacked by first making the change of variables $\tau = t - 3$, thus obtaining an equivalent problem with the initial value given at $\tau = 0$. However, we can also proceed directly as follows.

$$\mathcal{L}[\dot{y} + 3y] = \mathcal{L}[0]$$

$$\Rightarrow \quad sY(s) - y(0) + 3Y(s) = 0$$

$$(s + 3) \, Y(s) = y(0)$$

$$Y(s) = \frac{y(0)}{s + 3}$$

$$\Rightarrow \quad y(t) = y(0)e^{-3t}$$

$$y(3) = 1 \Rightarrow 1 = y(0)e^{-3(3)} \Rightarrow y(0) = e^9$$

$$y(t) = e^9 e^{-3t} = e^{9-3t}.$$

**EXAMPLE 4**   Solve the IVP

$$\dot{x}_1 = -3x_1 + 4x_2 \tag{5}$$

$$\dot{x}_2 = -2x_1 + 3x_2 \tag{6}$$

$$x_1(0) = -1, \quad x_2(0) = 3. \tag{7}$$

**Solution**   Taking the Laplace transform of Eq. (5), we have

$$sX_1(s) - x_1(0) = -3X_1(s) + 4X_2(s). \tag{8}$$

Taking the Laplace transform of Eq. (6), we have

$$sX_2(s) - x_2(0) = -2X_1(s) + 3X_2(s). \tag{9}$$

Using the initial conditions (7), we can rewrite Eqs. (8) and (9) as follows:

$$(s + 3)X_1(s) - 4X_2(s) = -1 \tag{10}$$

$$2X_1(s) + (s - 3)X_2(s) = 3.$$

The system (10) is a linear system for $X_1(s)$, $X_2(s)$. Solving we have

$$X_1(s) = \frac{-s + 15}{s^2 - 1} = -\frac{s}{s^2 - 1} + \frac{15}{s^2 - 1}$$

$$X_2(s) = \frac{3s + 11}{s^2 - 1} = 3\frac{s}{s^2 - 1} + \frac{11}{s^2 - 1}.$$

Thus,

$$x_1(t) = -\cosh t + 15 \sinh t$$

$$x_2(t) = 3 \cosh t + 11 \sinh t,$$

or

$$x_1(t) = 7e^t - 8e^{-t},$$

$$x_2(t) = 7e^t - 4e^{-t}.$$

## EXERCISES

In Exercises 1 through 36, solve the IVP by the method of Laplace transforms.

**1.** $\dot{y} + 2y = 0$
$y(0) = 1$

**2.** $\ddot{y} + 4y = 4 \cos 2t$
$y(0) = 0, \dot{y}(0) = 6$

**3.** $\ddot{y} + 3\dot{y} - 4y = 0$
$y(0) = 0, \dot{y}(0) = -5$

**4.** $\dot{y} - 3y = 13 \cos 2t$
$y(0) = -1$

**5.** $\ddot{y} + 2\dot{y} + y = 2e^{-t}$
$y(0) = 3, \dot{y}(0) = -3$

**6.** $\ddot{y} + 3\dot{y} + 2y = 0$
$y(0) = 1, \dot{y}(0) = 1$

**7.** $\ddot{y} - y = 6e^{t}$
$y(0) = 2, \dot{y}(0) = 3$

**8.** $\ddot{y} - 4y = -3e^{t}$
$y(0) = -1, \dot{y}(0) = 1$

**9.** $\ddot{y} + 10\dot{y} + 25y = 2e^{-5t}$
$y(0) = 0, \dot{y}(0) = -1$

**10.** $\dddot{y} - 27y = 0$
$y(0) = -1, \dot{y}(0) = 6, \ddot{y}(0) = 18$

**11.** $\ddot{y} - 9\dot{y} + 18y = 54$
$y(0) = 0, \dot{y}(0) = -3$

**12.** $\ddot{y} - 9y = 20 \cos t$
$y(0) = 0, \dot{y}(0) = 18$

**13.** $\ddot{y} + 9y = e^{t}$
$y(0) = 0, \dot{y}(0) = 0$

**14.** $\ddot{y} - 3\dot{y} + 2y = 24 \cosh t$
$y(0) = 6, \dot{y}(0) = -3$

**15.** $\ddot{y} + 10\dot{y} + 26y = 37e^{t}$
$y(0) = 1, \dot{y}(0) = 2$

**16.** $\dddot{y} - y = -1$
$y(0) = 1, \dot{y}(0) = -2, \ddot{y}(0) = 2$

**17.** $\dddot{y} + y = 1$
$y(0) = 1, \dot{y}(0) = 3,$
$\ddot{y}(0) = -3$

**18.** $\dot{y} - 3y = 2e^{t}$
$y(1) = e^{3} - e$

**19.** $\ddot{y} + 6\dot{y} + 9y = 27t$
$y(0) = 1, \dot{y}(0) = 0$

**20.** $\ddot{y} - \dot{y} - 6y = \cos t + 57 \sin t$
$y(0) = 1, \dot{y}(0) = 7$

**21.** $\ddot{y} - 3\dot{y} - 4y = 25te^{-t}$
$y(0) = 0, \dot{y}(0) = 4$

**22.** $\ddot{y} + 13\dot{y} + 36y = 10 - 72t$
$y(0) = 2, \dot{y}(0) = -1$

**23.** $\ddot{y} + 2\dot{y} - 15y = 16te^{-t} - 15$
$y(0) = 1, \dot{y}(0) = -9$

**24.** $\ddot{y} - 10\dot{y} + 21y = 21t^{2} + t + 13$
$y(0) = 3, \dot{y}(0) = 11$

**25.** $\ddot{y} + 7\dot{y} + 10y = 3e^{-2t} - 6e^{-5t}$
$y(0) = 0, \dot{y}(0) = 0$

**26.** $4\ddot{y} - 3\dot{y} - y = 34 \sin t$
$y(0) = 1, \dot{y}(0) = -2$

**27.** $\ddot{y} - y = 12 \sinh t$
$y(0) = 6, \dot{y}(0) = -1,$
$\ddot{y}(0) = 7$

**28.** $\ddot{y} + 2\dot{y} - 3y = 3t^{3} - 9t^{2} - 5t + 1$
$y(0) = 0, \dot{y}(0) = 8$

**29.** $\ddot{y} + 4\dot{y} + 5y = 39e^{t} \sin t$
$y(0) = -1, \dot{y}(0) = -1$

**30.** $\ddot{y} + 2\dot{y} + 5y = 8e^{t} + 5t$
$y(0) = 3, \dot{y}(0) = 2$

**31.** $\ddot{y} - 4\dot{y} + 4y = 3te^{2t} - 4$
$y(0) = 0, \dot{y}(0) = 0$

**32.** $\dddot{y} + \dot{y} = 18e^{2t}$
$y(0) = -1, \dot{y}(0) = 13, \ddot{y}(0) = -1$

**33.** $\ddot{y} + 8y = -12e^{-2t}$
$y(0) = -8, \dot{y}(0) = 24,$
$\ddot{y}(0) = -46$

**34.** $\ddot{y} - y = 2t^{2} + 2e^{-t}$
$y(0) = 0, \dot{y}(0) = 1$

**35.** $\ddot{y} + 7\dot{y} + 6y = 250e^t \cos t$
$y(0) = 2, \dot{y}(0) = -7$

**36.** $\ddot{y} + 4\dot{y} + 13y = 13t + 17 + 40 \sin t$
$y(0) = 30, \dot{y}(0) = 4.$

Solve each of the initial value problems 37 through 46, using the Laplace transform method.

**37.** $\dot{x}_1 = 4x_1 - x_2$
$\dot{x}_2 = 2x_1 + x_2$
$x_1(0) = 1, x_2(0) = 3$

**38.** $\dot{x}_1 = 3x_1 - 2x_2$
$\dot{x}_2 = x_1$
$x_1(0) = -1, x_2(0) = 0$

**39.** $\dot{x}_1 = 4x_1 - 2x_2$
$\dot{x}_2 = x_1 + x_2$
$x_1(0) = 1, x_2(0) = 0$

**40.** $\dot{x}_1 = 3x_1 - 2x_2$
$\dot{x}_2 = 2x_1 - x_2$
$x_1(0) = -1, x_2(0) = 1$

**41.** $\dot{x}_1 = 3x_1 - 2x_2$
$\dot{x}_2 = 17x_1 - 7x_2$
$x_1(0) = 0, x_2(0) = 3$

**42.** $\dot{x}_1 = 3x_1 - 5x_2$
$\dot{x}_2 = 4x_1 - 5x_2$
$x_1(0) = 1, x_2(0) = \dfrac{4}{5}$

**43.** $\dot{x}_1 = 4x_1 - x_2 + 3e^{2t}$
$\dot{x}_2 = 2x_1 + x_2 + 2t$
$x_1(0) = -\dfrac{5}{18}, x_2(0) = \dfrac{47}{9}$

**44.** $\dot{x}_1 = 5x_1 - 6x_2 + 1$
$\dot{x}_2 = 6x_1 - 7x_2 + 1$
$x_1(0) = 0, x_2(0) = 0$

**45.** $\dot{x}_1 = 2x_1 + x_2$
$\dot{x}_2 = -x_1 + x_3$
$\dot{x}_3 = x_1 + 3x_2 + x_3$
$x_1(0) = 1, x_2(0) = 1$
$x_3(0) = 3$

**46.** $\dot{x}_1 = 3x_1 - x_2$
$\dot{x}_2 = -x_1 + 2x_2 - x_3$
$\dot{x}_3 = -x_2 + 3x_3$
$x_1(0) = 1, x_2(0) = 0$
$x_3(0) = -1$

# 4.4  THE UNIT STEP FUNCTION

The unit step function or the Heaviside function is defined as follows.

$$h_1(t-t_0) = \begin{cases} 0 & \text{if} \quad t < t_0 \\ 1 & \text{if} \quad t \geq t_0 \end{cases}. \tag{1}$$

We assume that $t_0 \geq 0$. The graph of $h_1(t-t_0)$ is given in Figure 4.1.

A generalization of the unit step function is the function $h_c(t-t_0)$, where

$$h_c(t-t_0) = \begin{cases} 0 & \text{if} \quad t < t_0 \\ c & \text{if} \quad t \geq t_0 \end{cases}.$$

It is easy to see that

$$h_c(t-t_0) = ch_1(t-t_0). \tag{2}$$

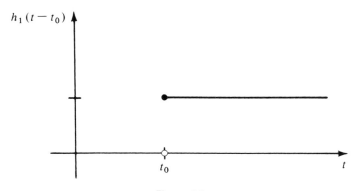

**Figure 4.1**

Applying the definition of the Laplace transform, we have

$$\mathcal{L}[h_1(t-t_0)] = \int_0^\infty h_1(t-t_0)e^{-st}dt$$

$$= \int_{t_0}^\infty e^{-st}dt = \left. \frac{-e^{-st}}{s} \right|_{t_0}^\infty$$

$$= \frac{e^{-st_0}}{s}. \tag{3}$$

For the special case $t_0 = 0$, we write $h_1(t) \equiv h_1(t-0)$, and from (3) we obtain

$$\mathcal{L}[h_1(t)] = \frac{1}{s}. \tag{4}$$

From (2) and (3) we obtain the Laplace transform of $h_c(t-t_0)$, namely,

$$\mathcal{L}[h_c(t-t_0)] = \frac{ce^{-st_0}}{s}.$$

One use of the Heaviside function is that many other functions can be expressed in terms of it. We illustrate with an example.

**EXAMPLE 1** Express the square-wave function (Figure 4.2) in terms of Heaviside functions.

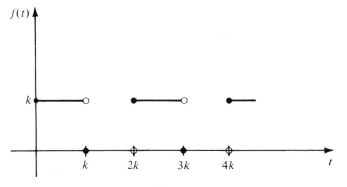

**Figure 4.2**

**Solution**  Beginning at $t = 0$, and considering in succession the intervals $(0,k)$, $(k,2k)$, $(2k,3k)$, $(3k,4k)$, . . . , we find that

$$f(t) = h_k(t) - h_k(t-k) + h_k(t-2k) - h_k(t-3k) + h_k(t-4k) - \cdots$$

$$= \sum_{i=0}^{\infty} (-1)^i h_k(t-ik) = k \sum_{i=0}^{\infty} (-1)^i h_1(t-ik).$$

**EXAMPLE 2**  If $f(t)$ is the square-wave function of Example 1, find $\mathcal{L}[f(t)]$.

**Solution**  From Example 1 we can write

$$\mathcal{L}[f(t)] = \mathcal{L}\left[k \sum_{i=0}^{\infty} (-1)^i h_1(t-ik)\right] = \int_0^\infty k \sum_{i=0}^{\infty} (-1)^i h_1(t-ik) e^{-st} dt.$$

If we assume that the integration and summation processes can be interchanged, we have

$$\mathcal{L}[f(t)] = k \sum_{i=0}^{\infty} (-1)^i \int_0^\infty h_1(t-ik) e^{-st} dt = k \sum_{i=0}^{\infty} (-1)^i \mathcal{L}[h_1(t-ik)].$$

Using (3) we obtain

$$\mathcal{L}[f(t)] = k \sum_{i=0}^{\infty} (-1)^i \frac{e^{-sik}}{s} = \frac{k}{s} \sum_{i=0}^{\infty} (-1)^i (e^{-sk})^i.$$

Since $sk > 0$, we know that $e^{-sk} < 1$; therefore, the infinite sum is a convergent geometric series with first term equal to 1 and ratio equal to $-e^{-sk}$. Consequently the sum is $1/[1-(-e^{-sk})] = 1/[1+e^{-sk}]$. Thus,

$$\mathcal{L}[f(t)] = \frac{k}{s(1+e^{-ks})}.$$

**EXAMPLE 3**  If $\mathcal{L}[f(t)] = F(s)$, show that

$$\mathcal{L}[f(t-t_0)h_1(t-t_0)] = e^{-st_0}F(s). \tag{5}$$

**Solution**

$$\mathcal{L}[f(t-t_0)h_1(t-t_0)] = \int_0^\infty f(t-t_0)h_1(t-t_0) e^{-st} dt$$

$$= \int_{t_0}^\infty f(t-t_0) e^{-st} dt.$$

In the latter integral make the substitution $z = t-t_0$ to obtain

$$\mathcal{L}[f(t-t_0)h_1(t-t_0)] = \int_0^\infty f(z) e^{-s(z+t_0)} dz$$

$$= e^{-st_0} \int_0^\infty f(z) e^{-sz} dz = e^{-st_0}F(s).$$

An important consequence of Eq. (5) is

$$\mathcal{L}^{-1}[e^{-st_0}F(s)] = f(t-t_0)h_1(t-t_0). \tag{6}$$

**EXAMPLE 4**   Solve the initial value problem

$$\ddot{y} - 3\dot{y} - 4y = h_1(t-1) + h_1(t-2)$$
$$y(0) = 0, \; \dot{y}(0) = 1. \tag{7}$$

**Solution**   Applying the Laplace transform to Eq. (7) we obtain

$$s^2 Y - sy(0) - \dot{y}(0) - 3sY - 3y(0) - 4Y = \frac{e^{-s}}{s} + \frac{e^{-2s}}{s}$$

$$(s^2 - 3s - 4)Y = \frac{e^{-s}}{s} + \frac{e^{-2s}}{s} + 1$$

$$Y = \frac{e^{-s}}{s(s-4)(s+1)} + \frac{e^{-2s}}{s(s-4)(s+1)} + \frac{1}{(s-4)(s+1)}$$

$$Y = e^{-s}\left[\frac{-1/4}{s} + \frac{1/20}{s-4} + \frac{1/5}{s+1}\right] + e^{-2s}\left[\frac{-1/4}{s} + \frac{1/20}{s-4} + \frac{1/5}{s+1}\right]$$

$$+ \frac{1/5}{s-4} + \frac{-1/5}{s+1}.$$

Employing (6) we can write

$$y(t) = -\frac{1}{4}h_1(t-1) + \frac{1}{20}e^{4(t-1)}h_1(t-1) + \frac{1}{5}e^{-(t-1)}h_1(t-1)$$

$$-\frac{1}{4}h_1(t-2) + \frac{1}{20}e^{4(t-2)}h_1(t-2) + \frac{1}{5}e^{-(t-2)}h_1(t-2)$$

$$+\frac{1}{5}e^{4t} - \frac{1}{5}e^{-t}.$$

**EXERCISES**

In Exercises 1 through 14, solve the given initial value problem.

**1.** $\ddot{y} + 2\dot{y} + y = h_1(t)$
$y(0) = 0, \; \dot{y}(0) = -3$

**2.** $\dot{y} - 3y = 13 \cos 2t + 3h_1(t)$
$y(0) = -1$

**3.** $\ddot{y} + 3\dot{y} - 4y = 20h_1(t-2)$
$y(0) = 0, \; \dot{y}(0) = 0$

**4.** $\ddot{y} - 9y = 20 \cos t + 18h_1(t-1)$
$y(0) = 2, \; \dot{y}(0) = 0$

**5.** $\ddot{y} - 9\dot{y} + 18y = 54t - 9h_1(t)$
$y(0) = 3, \; \dot{y}(0) = 0$

**6.** $\ddot{y} + 10\dot{y} + 25y = 2e^{-5t} +$
$25h_1(t-5)$
$y(0) = 0, \; \dot{y}(0) = -1$

**7.** $\ddot{y} - y = 4e^t + 2h_1(t-3)$
$y(0) = 0, \; \dot{y}(0) = 0$

**8.** $\ddot{y} + 3\dot{y} + 2y = 8h_1(t) + 2h_1(t-1)$
$y(0) = 1, \; \dot{y}(0) = 0$

**9.** $\ddot{y} + 4y = 8h_1(t-1) - 4h_1(t-2)$
$y(0) = -1, \; \dot{y}(0) = 0$

**10.** $\ddot{y} - 3\dot{y} + 2y = 24 \cosh t + h_2(t)$
$y(0) = 0, \; \dot{y}(0) = -18$

**11.** $\ddot{y} + 10\dot{y} + 26y = 52h_1(t) + 26h_1(t - 3)$
$y(0) = 1, \dot{y}(0) = -5$

**12.** $\ddot{y} + 6\dot{y} + 9y = 6h_3(t - 1) + 9h_1(t - 3)$
$y(0) = 0, \dot{y}(0) = 1$

**13.** $\ddot{y} - \dot{y} - 6y = 50 \cos t + 30h_1(t - 5)$
$y(0) = -7, \dot{y}(0) = 4$

**14.** $\ddot{y} - 3\dot{y} - 4y = 20h_1(t) + 40h_1(t - 1) + 10h_2(t - 2)$
$y(0) = 0, \dot{y}(0) = 0$

**15.** Use Theorem 5 of Section 4.2 to compute $\mathscr{L}[f(t)]$ where $f(t)$ is the square-wave function of Example 1.

**16.** Use Theorem 5 of Section 4.2 to compute $\mathscr{L}[f(t)]$ where $f(t)$ is the "sawtooth" function of Figure 4.3.

**Figure 4.3**

# 4.5  THE UNIT IMPULSE FUNCTION

The unit impulse function, or the Dirac delta function, happens not to be a function as the word is used in mathematics. For a first exposition, however, there are advantages to thinking of the unit impulse function as if it were a function. The idea behind the delta function is to provide an analytical model for certain types of physical phenomena that act for extremely short periods of time. For example, one might be interested in describing the force involved when a hammer strikes a nail. If the graph of this force were drawn, the ordinate would be zero up to the time the hammer struck the nail, say at $t = t_0$ ($t_0 \geq 0$); it then has some value for a very short period of time and is zero thereafter. Since the time interval of application of the force is small, it is reasonable to assume that during this period the force is a constant and that this constant is large in magnitude. It is conventional that the *impulse of the force* (that is, the integral of the force with respect to time) is unity. With this convention we can approximate the force graphically in Figure 4.4.

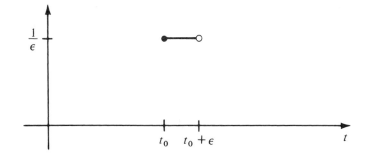

**Figure 4.4**

Utilizing the Heaviside function of Section 4.4, and denoting the force depicted in Figure 4.4 by $f_\epsilon(t - t_0)$, we can write

$$f_\epsilon(t - t_0) = \frac{1}{\epsilon}[h_1(t - t_0) - h_1(t - t_0 - \epsilon)]. \tag{1}$$

Note that

$$\int_{-\infty}^{\infty} f_\epsilon(t - t_0)dt = 1, \text{ for every } \epsilon > 0 \text{ and every } t_0 \geq 0. \tag{2}$$

The unit impulse function, or Dirac delta function $\delta(t - t_0)$ is defined by the conditions

$$\delta(t - t_0) = \lim_{\epsilon \to 0} f_\epsilon(t - t_0), \tag{3}$$

and

$$\int_{-\infty}^{\infty} \delta(t - t_0)dt = 1. \tag{4}$$

Thus we can think of $\delta(t - t_0)$ as representing a force of "infinite magnitude" that is applied instantaneously at $t = t_0$ and whose impulse is unity.

We use Eq. (3) to obtain the main properties of the unit impulse function.

**EXAMPLE 1**   Find the Laplace transform of $\delta(t - t_0)$.

**Solution**

$$\mathscr{L}[\delta(t - t_0)] = \int_0^{\infty} \delta(t - t_0)e^{-st}dt = \int_0^{\infty} \lim_{\epsilon \to 0} f_\epsilon(t - t_0)e^{-st}dt.$$

Assuming that the integration and limit processes can be interchanged, we have

$$\mathscr{L}[\delta(t - t_0)] = \lim_{\epsilon \to 0} \int_0^\infty f_\epsilon(t - t_0)e^{-st}dt$$

$$= \lim_{\epsilon \to 0} \int_0^\infty \frac{1}{\epsilon}[h_1(t - t_0) - h_1(t - t_0 - \epsilon)]e^{-st}dt$$

$$= \lim_{\epsilon \to 0} \frac{1}{\epsilon}\left\{ \mathscr{L}[h_1(t - t_0)] - \mathscr{L}[h_1(t - t_0 - \epsilon)]\right\}$$

$$= \lim_{\epsilon \to 0} \frac{1}{\epsilon}\left[\frac{e^{-st_0}}{s} - \frac{e^{-s(t_0 + \epsilon)}}{s}\right] = \frac{e^{-st_0}}{s}\lim_{\epsilon \to 0}\frac{1 - e^{-s\epsilon}}{\epsilon}.$$

Now $\lim\limits_{\epsilon \to 0} \dfrac{1 - e^{-s\epsilon}}{\epsilon}$ can be evaluated using l'Hospital's rule; thus

$$\lim_{\epsilon \to 0}\frac{1 - e^{-s\epsilon}}{\epsilon} = \lim_{\epsilon \to 0}\frac{se^{-s\epsilon}}{1} = s.$$

Consequently,

$$\mathscr{L}[\delta(t - t_0)] = e^{-st_0}. \tag{5}$$

In the special case $t_0 = 0$ we write $\delta(t) \equiv \delta(t - 0)$. From (5) we find that

$$\mathscr{L}[\delta(t)] = 1. \tag{6}$$

**EXAMPLE 2**   If $g$ is any continuous function, show that

$$\int_{-\infty}^\infty \delta(t - t_0)g(t)dt = g(t_0). \tag{7}$$

**Solution**

$$\int_{-\infty}^\infty \delta(t - t_0)g(t)dt = \int_{-\infty}^\infty \lim_{\epsilon \to 0} f_\epsilon(t - t_0)g(t)dt.$$

Assuming that the integration and limit processes can be interchanged, we have

$$\int_{-\infty}^\infty \delta(t - t_0)g(t)dt = \lim_{\epsilon \to 0}\int_{-\infty}^\infty f_\epsilon(t - t_0)g(t)dt = \lim_{\epsilon \to 0}\int_{t_0}^{t_0 + \epsilon}\left(\frac{1}{\epsilon}\right)g(t)dt$$

$$= \lim_{\epsilon \to 0}\frac{1}{\epsilon}\int_{t_0}^{t_0 + \epsilon}g(t)dt.$$

Since $g$ is a continuous function, we can apply the mean value theorem of integral calculus to obtain

$$\int_{-\infty}^\infty \delta(t - t_0)g(t)dt = \lim_{\epsilon \to 0}\frac{1}{\epsilon}g(\alpha)[(t_0 + \epsilon) - t_0],$$

where $t_0 \le \alpha \le t_0 + \epsilon$. Now $\dfrac{1}{\epsilon}[(t_0 + \epsilon) - t_0] = 1$, and since $g$ is continuous, $\lim\limits_{\epsilon \to 0} g(\alpha) = g(t_0)$; thus,

$$\int_{-\infty}^{\infty} \delta(t - t_0)g(t)dt = g(t_0).$$

**EXAMPLE 3**  Solve the initial value problem

$$\ddot{y} - 3\dot{y} + 2y = \delta(t-1) \tag{8}$$

$$y(0) = 0, \ \dot{y}(0) = 0.$$

**Solution**  Applying the Laplace transform to Eq. (8), we obtain

$$s^2 Y - sy(0) - \dot{y}(0) - 3sY + 3y(0) + 2Y = e^{-s}$$

$$(s^2 - 3s + 2)Y = e^{-s}$$

$$Y = \frac{e^{-s}}{(s - 1)(s - 2)} = e^{-s}\left[\frac{-1}{s - 1} + \frac{1}{s - 2}\right].$$

Thus,

$$y(t) = -e^{(t-1)}h_1(t - 1) + e^{2(t-1)}h_1(t - 1).$$

**EXERCISES**

In Exercises 1 through 14, solve the given initial value problem.

**1.** $\ddot{y} - 3\dot{y} - 4y = 5\delta(t)$
$y(0) = 0, \ \dot{y}(0) = 0$

**2.** $\ddot{y} - \dot{y} - 6y = 52 \cos 2t + \delta(t)$
$y(0) = -5, \ \dot{y}(0) = -8$

**3.** $\ddot{y} + 6\dot{y} + 9y = \delta(t - 2)$
$y(0) = 0, \ \dot{y}(0) = 1$

**4.** $\ddot{y} + 10\dot{y} + 26y = 145 \cos t +$
$\delta(t - 1)$
$y(0) = 1, \ \dot{y}(0) = -2$

**5.** $\ddot{y} - 3\dot{y} + 2y = 12 + \delta(t)$
$y(0) = 6, \ \dot{y}(0) = 0$

**6.** $\ddot{y} + 9y = 34e^{-5t} + 9\delta(t - 5)$
$y(0) = 0, \ \dot{y}(0) = -1$

**7.** $\ddot{y} + 3\dot{y} + 2y = 6e^t + \delta(t-3)$
$y(0) = 1, \ \dot{y}(0) = 0$

**8.** $\ddot{y} - y = \delta(t) + \delta(t - 1)$
$y(0) = 0, \ \dot{y}(0) = 3$

**9.** $\ddot{y} + 10\dot{y} + 25y = \delta(t - 1) - 3\delta(t - 2)$
$y(0) = 0, \ \dot{y}(0) = -1$

**10.** $\ddot{y} - 9\dot{y} + 18y = 280 \cosh t + \delta(t)$
$y(0) = 0, \ \dot{y}(0) = -1$

**11.** $\ddot{y} - 9y = 30\delta(t) + 6\delta(t - 3)$
$y(0) = 8, \ \dot{y}(0) = 0$

**12.** $\ddot{y} + 3\dot{y} - 4y = 5\delta(t - 1) + 10\delta(t - 3)$
$y(0) = 0, \ \dot{y}(0) = 0$

**13.** $\ddot{y} - 2\dot{y} - 8y = 85 \cos t + 6\delta(t - 5)$
     $y(0) = -9, \dot{y}(0) = 4$

**14.** $\ddot{y} + 2\dot{y} + y = \delta(t) + \delta(t - 1) + 3\delta(t - 2)$
     $y(0) = 0, \dot{y}(0) = -3$

# $4.6$   APPLICATIONS

In this section we illustrate a few practical applications of Laplace transforms. We do not restrict the treatment to differential equations exclusively. It is worth emphasizing that Laplace transforms have the nice feature that when they are applicable, they transform a given problem into a simpler problem.

**Electric Circuits** ■ In Section 2.11.1 we obtained, via Kirchhoff's voltage law, the equation

$$L\frac{di}{dt} + Ri + \frac{1}{C}\int_0^t i(\tau)d\tau = v(t). \tag{1}$$

Equation (1) is referred to as an *integrodifferential equation* for $i$. Here we have used lowercase $i$ and $v$ for current and voltage, respectively, to conform to the convention of labelling functions of $t$ in lower case and their Laplace transform in upper case. Note that Eq. (1) is Eq. (12) of Section 2.11.1, in which we have used the relation (11) of that section. In Section 2.11.1 we differentiated Eq. (1) to obtain a differential equation for $i$. Naturally, this requires that $v(t)$ be a differentiable function. Using Laplace transforms, we can solve Eq. (1) directly.

**EXAMPLE 1**   Find $i(t)$ for the electric circuit with $L = 20$ henries, $R = 40$ ohms, $C = 0.05$ farad, and $v(t) = \sin t$.

**Solution**   Applying the Laplace transform to Eq. (1), we obtain [recall that $i(0) = 0$]

$$LsI(s) + RI(s) + \frac{1}{Cs}I(s) = V(s)$$

$$\Rightarrow \quad I(s) = \frac{1}{Ls + R + 1/Cs}V(s)$$

$$= \frac{s/L}{s^2 + (R/L)s + 1/LC}V(s) \tag{2}$$

$$= G(s)V(s). \tag{3}$$

Substituting the given values for $L$, $R$, $C$, and $v(t)$ into Eq. (2), we obtain

$$I(s) = \frac{s/20}{s^2 + 2s + 1} \cdot \frac{1}{s^2 + 1} = \frac{s/20}{(s + 1)^2(s^2 + 1)}$$

$$= \frac{A}{s + 1} + \frac{B}{(s + 1)^2} + \frac{Cs + D}{s^2 + 1}.$$

Now (see Appendix B) $A = 0$, $B = -\frac{1}{40}$, $C = 0$, and $D = \frac{1}{40}$. Thus,

$$I(s) = \frac{-\frac{1}{40}}{(s + 1)^2} + \frac{\frac{1}{40}}{s^2 + 1}$$

$$i(t) = -\frac{1}{40}\mathscr{L}^{-1}\left[\frac{1}{(s + 1)^2}\right] + \frac{1}{40}\mathscr{L}^{-1}\left[\frac{1}{s^2 + 1}\right]$$

$$= -\frac{1}{40}te^{-t} + \frac{1}{40}\sin t.$$

**REMARK 1** The phenomenon illustrated by the electric circuit can be viewed in the following way. A voltage (*input*) is applied to the circuit (*system*) and a current (*output*) is produced; that is, a system acts upon an input to produce an output. In an abstract way many processes (biological, chemical, physical) can be viewed in this fashion. Following this model, it would be convenient if one could obtain an equation relating output to input. Then it would be simply a matter of substituting given values for the input and determining the values of the output via this relation. Equation (1) is such a relation, but unfortunately the relationship is not simple enough to easily obtain output from input. On the other hand, Eq. (2) is better suited for these purposes. Of course, Eq. (2) relates the Laplace transform of the output to the Laplace transform of the input. Nevertheless, it has the advantage that given an input $v(t)$ to obtain the output $i(t)$ requires some algebraic manipulation, use of tables of Laplace transforms, and in some cases performing integrations. Equation (3) is frequently called the *input–output relation*. The function $G(s)$ is called the *transfer function* or *system function*. Relation (3) is viewed diagrammatically as in Figure 4.5.

Engineers and biologists take the attitude that knowledge of the transfer function provides complete information of the system. For the most part, engineers know the transfer function and wish to analyze the output. Biologists observe input and output and try to determine the transfer function from this information (for more discussion in this direction, see Exercises 14 through 17).

**Figure 4.5**

It is in this context [Eq. (3)] that the Laplace transform is utilized in present-day investigations.[1]

**Mechanics**    ■ A particle is projected from a point on the surface of the earth with a given initial velocity. We wish to analyze the motion of this particle, taking into account the rotation of the earth. We choose the origin of a three-dimensional coordinate system to be at the point from which the particle was projected. The positive $x$ axis is to point south, the positive $y$ axis east, and the positive $z$ axis points opposite to the direction of the acceleration due to gravity. The origin is taken to have latitude $\beta$ and the angular velocity of the earth is denoted by $\omega$. Neglecting the mass of the particle, it can be shown[2] that the equations of motion are as follows:

$$\ddot{x} = 2\omega\dot{y}\sin\beta$$
$$\ddot{y} = -2\omega(\dot{x}\sin\beta + \dot{z}\cos\beta) \tag{4}$$
$$\ddot{z} = 2\omega\dot{y}\cos\beta - g,$$

with initial values $x(0) = y(0) = z(0) = 0$ and $\dot{x}(0) = u$, $\dot{y}(0) = v$, $\dot{z}(0) = w$.

Set $\cos\beta = a$, $\sin\beta = b$, and apply the Laplace transform to system (4) to obtain the following system (after rearrangement) (see Exercise 1):

$$s^2 X - 2\omega bs Y = u$$
$$2\omega bs X + s^2 Y + 2\omega as Z = v \tag{5}$$
$$-2\omega as Y + s^2 Z = w - \frac{g}{s}.$$

System (5) can be solved for $X$, $Y$, and $Z$ since the determinant of coefficients $\Delta$ has the value (see Exercise 2)

$$\Delta = s^4(s^2 + 4\omega^2). \tag{6}$$

Our restrictions on $s$ guarantee that $\Delta \neq 0$. Thus (see Exercises 3, 4, and 6),

$$X(s) = u\left(\frac{1}{s^2 + 4\omega^2} + \frac{4\omega^2 a^2}{s^2(s^2 + 4\omega^2)}\right) + v\frac{2\omega b}{s(s^2 + 4\omega^2)}$$
$$- w\frac{4\omega^2 ab}{s^2(s^2 + 4\omega^2)} + g\frac{4\omega^2 ab}{s^3(s^2 + 4\omega^2)} \tag{7}$$

$$Y(s) = -u\frac{2\omega b}{s(s^2 + 4\omega^2)} + v\frac{1}{s^2 + 4\omega^2} - w\frac{2\omega a}{s(s^2 + 4\omega^2)}$$
$$+ g\frac{2\omega a}{s^2(s^2 + 4\omega^2)} \tag{8}$$

[1]For a more complete discussion, see R. Rosen, *Optimality Principles in Biology* (New York: Plenum Press, 1967).
[2]See B.M. Planck, *Einführung in die allgemeine Mechanik*, 4th ed. (Leipzig: S. Hirzel, 1928), p. 81.

$$Z(s) = -u\frac{4\omega^2 ab}{s^2(s^2 + 4\omega^2)} + v\frac{2\omega a}{s(s^2 + 4\omega^2)}$$

$$+ w\left[\frac{1}{s^2 + 4\omega^2} + \frac{4\omega^2 b^2}{s^2(s^2 + 4\omega^2)}\right]$$

$$- g\left[\frac{1}{s(s^2 + 4\omega^2)} + \frac{4\omega^2 b^2}{s^3(s^2 + 4\omega^2)}\right]. \tag{9}$$

Consequently (see Exercises 7, 8, and 9),

$$x(t) = \frac{u}{2\omega}(2a^2\omega t + b^2\sin 2\omega t) + \frac{v}{\omega}b\sin^2\omega t$$

$$- \frac{w}{2\omega}ab(2\omega t - \sin 2\omega t) + \frac{g}{2\omega^2}ab(\omega^2 t^2 - \sin^2\omega t) \tag{10}$$

$$y(t) = -\frac{u}{\omega}b\sin^2\omega t + \frac{v}{2\omega}\sin 2\omega t - \frac{w}{\omega}a\sin^2\omega t$$

$$+ \frac{g}{4\omega^2}a(2\omega t - \sin 2\omega t) \tag{11}$$

$$z(t) = -\frac{u}{2\omega}ab(2\omega t - \sin 2\omega t) + \frac{v}{\omega}a\sin^2\omega t$$

$$+ \frac{w}{2\omega}(2b^2\omega t + a^2\sin 2\omega t) - \frac{g}{2\omega^2}(b^2\omega^2 t^2 + a^2\sin^2\omega t). \tag{12}$$

Note that if one considers the limit as $\omega$ tends to zero in Eqs. (10), (11), and (12), there results

$$x = ut, \quad y = vt, \quad z = wt - \tfrac{1}{2}gt^2. \tag{13}$$

Equations (13) would represent the motion of a particle relative to a still earth.[3]

■ In studying the population growth of some species, we assume that at time $t = 0$, there are a certain number of them, say $n(0)$, all of a certain age. For convenience, we refer to this age classification as zero age. At a future time $t$ there are $n_1(t)$ of these specimens still in the population. Then $n_1(t)$ and $n(0)$ are connected, with the use of a *survival function* $f(t)$ by the relation

**Population Growth**

$$n_1(t) = n(0)f(t).$$

Furthermore, we consider that individuals of age zero placed in this population at some time $t_1 > 0$ also survive according to this rule. That is, if $m(t_1)$ of these individuals are placed into the population at time $t_1$, then at a future time $t > t_1$ there will be $m(t)$ of these individuals left in the population, where

$$m(t) = m(t_1)f(t - t_1).$$

[3]For a more complete discussion of this illustration and other applications of the Laplace transform, see Gustav Doetsch, *Guide to the Applications of the Laplace and Z-Transforms* (London: Van Nostrand, 1967).

Consider now that age-zero individuals are placed in the population at a rate $r(t)$, commonly called the *replacement rate*. In the time interval from $\tau$ to $\tau + \Delta\tau$, $r(\tau_1)\Delta\tau$, individuals are placed in the population. $\tau_1$ must be such that $\tau \leq \tau_1 \leq \tau + \Delta\tau$. According to the survival law, $r(\tau_1)\Delta\tau\, f\,(t - \tau_1)$ of these individuals will still be present in the population at time $t$.

The consideration is as follows. At time $t = 0$ there are $n(0)$ zero-age individuals of a certain species given. As time goes on, zero-age individuals are placed into the population at the rate $r(t)$. We wish to determine the total population $n(t)$ at some later time $t$. Certainly one could approximate the answer for $n(t)$ by splitting the time interval $[0, t]$ into intervals of length $\Delta\tau$ and adding up the number of survivors for each interval. Naturally, the approximation is made more accurate by making the time interval $\Delta\tau$ smaller. In fact, we can say that the approximation tends to $n(t)$ as $\Delta\tau$ tends to zero. Thus,

$$n(t) = n(0)f(t) + \int_0^t r(\tau)\, f(t - \tau)d\tau. \tag{14}$$

Equation (14) is a model for studying the growth of populations governed by the considerations above.

**EXAMPLE 2**  Suppose that in 1976 there were 34,000 wild rabbits in Rhode Island, that the survival function for these rabbits is $e^{-t}$, and that wild rabbits are introduced into the population at a constant rate $r_0$. How many wild rabbits will be present at time $t$? What value should $r_0$ have if the population is to stay constant?

**Solution**

$$n(t) = 34{,}000e^{-t} + \int_0^t r_0 e^{-(t-\tau)}d\tau$$

$$= 34{,}000e^{-t} + r_0 e^{-t}\int_0^t e^{\tau}d\tau$$

$$= 34{,}000e^{-t} + r_0 e^{-t}\left(e^{\tau}\,\bigg|_0^t\right)$$

$$= 34{,}000e^{-t} + r_0(1 - e^{-t}).$$

It is easy to see that if $r_0 = 34{,}000$, then $n(t) = 34{,}000$ for all $t$.

**EXAMPLE 3**  Suppose that in Example 2 it is desired to determine a rate function $r(t)$ such that the wild rabbit population is a linear function of time; that is,

$$n(t) = n(0) + ct.$$

**Solution**   Here $n(0) = 34{,}000$ and $c$ is a constant. Then $r(t)$ must satisfy

$$n(0) + ct = n(0)e^{-t} + \int_0^t r(\tau)e^{-(t-\tau)}d\tau. \tag{15}$$

Although Eq. (15) is not a differential equation, Laplace transforms can be employed to determine the unknown function $r(t)$. Taking the Laplace transform of Eq. (15), we obtain

$$\frac{n(0)}{s} + \frac{c}{s^2} = \left[\frac{n(0)}{s+1} + R(s)\frac{1}{s+1}\right].$$

Solving for $R(s)$, we have

$$R(s) = (s+1)\left[\frac{n(0)}{s} + \frac{c}{s^2} - \frac{n(0)}{s+1}\right].$$

Simplifying algebraically, we obtain

$$R(s) = (s+1)\left[\frac{n(0)s(s+1) + c(s+1) - n(0)s^2}{s^2(s+1)}\right]$$

$$= \frac{[n(0) + c]s + c}{s^2}$$

$$= \frac{n(0) + c}{s} + \frac{c}{s^2}.$$

Taking the inverse Laplace transform, we obtain the rate function

$$r(t) = n(0) + c + ct.$$

**EXERCISES**

**1.** Verify Eq. (5).

**2.** Verify Eq. (6). [*Hint:* $a^2 + b^2 = 1$.]

**3.** Verify Eq. (7).

**4.** Verify Eq. (8).

**5.** Using the method of Laplace transforms, find the replacement rate function for a population whose survival function is $te^{-2t}$ and for which it is desired that the population numbers $n(0)t + 3t$ at any time $t > 0$.

**6.** Verify Eq. (9).

**7.** Verify Eq. (10).

**8.** Verify Eq. (11).

**9.** Verify Eq. (12). $\left[Hint:\ \dfrac{4\omega^2}{s^3(s^2 + 4\omega^2)} = \dfrac{1}{s^3} - \dfrac{1}{s(s^2 + 4\omega^2)}.\right]$

**10.** Verify the asserted limits in (13). Recall l'Hospital's rule.

**11.** Show that Eq. (2) can be written in the form

$$I(s) = \frac{s/L}{(s+a)^2 + (\omega_0^2 - a^2)}V(s) \tag{16}$$

with $a$ and $\omega_0$ appropriately defined.

**12.** Suppose that $v(t) = v_0$ (a constant). Determine $i(t)$ from Eq. (16) in each of the following cases:

(a) $\omega_0^2 = a^2$             (b) $\omega_0^2 > a^2$             (c) $\omega_0^2 < a^2$

**13.** Suppose that $v(t) = v_0 \sin \omega t$ ($v_0$, $\omega$ constants; $\omega \neq \omega_0$). Determine $i(t)$ from Eq. (16) in each of the following cases:

(a) $\omega_0^2 = a^2$             (b) $\omega_0^2 > a^2$             (c) $\omega_0^2 < a^2$

**Feedback Systems**  Feedback systems are comprised of two separate systems: a controlled system and a controlling system. A *controlled system* is one in which a specific activity is executed so as to maintain a certain fixed activity of the system. Such functions are performed by varying the input to the controlled system in amounts that depend on the discrepancy between the controlled system and its desired state at each time $t$. The amount of input to the controlled system is administered by another system whose function is to monitor the deviance of the controlled system and to supply the controlled system with inputs in such a way as to minimize the deviance. This other system is the *controlling system*. Such a feedback system can be represented diagrammatically as in Figure 4.6, where $Y_e(s) = Y_i(s) - Y_0(s)$ and $Y_c(s) = G_1(s)Y_e(s)$. From the transfer functions $G_1(s)$ and $G_2(s)$, a number of other transfer functions can be derived which characterize the behavior of the entire feedback system. Using the definition of transfer function, we can write

$$Y_0(s) = G_1(s)G_2(s)Y_e(s)$$

$$\Rightarrow \qquad \frac{Y_0(s)}{Y_e(s)} = G_1(s)G_2(s).$$

$G_1(s)G_2(s)$ is called the *open-loop* transfer function.

**Figure 4.6**

**14.** Show that

$$\frac{Y_0(s)}{Y_i(s)} = \frac{G_1(s)G_2(s)}{1 + G_1(s)G_2(s)}.$$

$$\left[ Hint: \frac{a}{c} = \frac{a}{b} \cdot \frac{b}{c}. \right]$$

**15.** Show that

$$\frac{Y_e(s)}{Y_i(s)} = \frac{1}{1 + G_1(s)G_2(s)}.$$

**Regulators**   For a "pure regulator" the input is a constant (thus it is generally chosen to be zero), and there is an additional signal which represents the effects of fluctuations in the environment of the system. The transform of this signal is denoted by $Y_f(s)$. Such feedback systems are represented diagrammatically as in Figure 4.7.

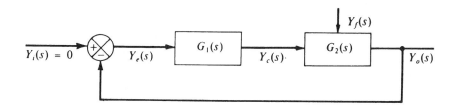

**Figure 4.7**

**16.** Show that

$$\frac{Y_0(s)}{Y_f(s)} = G_2(s)$$

is the open-loop transfer function.

[*Hint:* Follow an "initial signal" from input to output.]

**17.** Show that

$$\frac{Y_0(s)}{Y_f(s)} = \frac{G_2(s)}{1 + G_1(s)G_2(s)}$$

is the *closed-loop* transfer function.

**REMARK 2**   Many biological systems are modeled as feedback systems or as regulators. One such system is the eye.[4]

**REVIEW EXERCISES**

In Exercises 1 through 4, find the Laplace transform of the given function.

**1.** $4t^5 + 3t \sin t$ **2.** $h_1(t) \cos t$

[4]Many illustrations of this sort can be found in Rosen, *Optimality Principles*. The reader can see from the expressions developed in these exercises that researchers can attempt to determine the transfer function from the given input and the observed output. Efficient techniques for doing this are as yet not known.

**3.** $\delta(t) + h_1(t-2)$                     **4.** $\delta(t-1) - h_1(t)e^{3t} + \cos t$

In Exercises 5 through 8, find the inverse Laplace transform of the given function.

**5.** $\dfrac{1}{s^2 - 10s + 9}$                     **6.** $\dfrac{45s - 126}{s^3 - 5s^2 - 14s}$

**7.** $\dfrac{e^{-3(s-1)}}{s - 1}$                     **8.** $7 + \dfrac{2}{s^2 + 4s + 13}$

In Exercises 9 through 16, solve the given initial value problem by the method of Laplace transforms.

**9.** $\ddot{y} - 7\dot{y} - 8y = 384t$
   $y(0) = 0, \dot{y}(0) = 3$

**10.** $3\ddot{y} - 20\dot{y} - 7y = 925e^t \cos t$
   $y(0) = 10, \dot{y}(0) = 2$

**11.** $5\ddot{y} + 17\dot{y} + 6y = 676 \sin 2t - 36t$
   $y(0) = 0, \dot{y}(0) = -7$

**12.** $\ddot{y} - 8\dot{y} - 33y = 196e^{-3t}$
   $y(0) = 0, \dot{y}(0) = 0$

**13.** $\dddot{y} + y = 6e^t$
   $y(0) = 0, \dot{y}(0) = 0, \ddot{y}(0) = 0$

**14.** $\ddot{y} - 6\dot{y} + 9y = \delta(t) + 9$
   $y(0) = 0, \dot{y}(0) = 1$

**15.** $\ddot{y} - 11\dot{y} + 24y = 120h_1(t-2)$
   $y(0) = 5, \dot{y}(0) = 0$

**16.** $\ddot{y} + y = \delta(t-1)$
   $y(0) = 0, \dot{y}(0) = 0$

**17.** For a "pure regulator" show that the open-loop output is given by
$$\int_0^t g_2(t-\tau)y_f(\tau)d\tau.$$

**18.** Consider[5] the linear time invariant $RLC$-series circuit with input $e_s$ and response $i$ as shown in Figure 4.8. Calculate and sketch the impulse response $(q(0) = 0, \dot{q}(0) = 0)$.

**Figure 4.8**

[5]This is Exercise 22a in Charles A. Desoer and Ernest S. Kuh, *Basic Circuit Theory* (New York: McGraw-Hill, 1969), p. 268. Reprinted by permission of McGraw-Hill Book Company.

**19.** A force $f(t) = Qte^{-bt}h_1(t)$ is applied[6] to a simple (harmonic) oscillator of mass $m$, (spring) stiffness $K$, and (damping) resistance $R$, initially at rest $(x(0) = \dot{x}(0) = 0)$. Use the Laplace transform to calculate the displacement of the mass as a function of time for $t > 0$.

**20.** Show that

$$\mathscr{L}[f(t)] = \mathscr{L}\left[\sum_{n=0}^{\infty} \delta[t - \frac{\pi}{a}(2n+1)]\right] = \frac{1}{2\sinh\dfrac{\pi}{a}s},$$

where $a$ is a constant. [*Hint:* assume that $\mathscr{L}\left[\sum_{n=0}^{\infty}\cdots\right] = \sum_{n=0}^{\infty}\mathscr{L}[\cdots].$]

In Exercises 21 through 30, solve the initial value problem by the method of the Laplace transformation.

**21.** $\dot{x}_1 = -x_1$
$\dot{x}_2 = -x_2$
$x_1(0) = 1, x_2(0) = 5$

**22.** $\dot{x}_1 = -2x_1 + x_2$
$\dot{x}_2 = x_1 - 2x_2$
$x_1(0) = 3, x_2(0) = 1$

**23.** $\dot{x}_1 = 3x_1 - 2x_2$
$\dot{x}_2 = 2x_1 - 2x_2$
$x_1(0) = -1, x_2(0) = 1$

**24.** $\dot{x}_1 = -x_1$
$\dot{x}_2 = -x_1 - x_2$
$x_1(0) = -4, x_2(0) = 6$

**25.** $\dot{x}_1 = -x_1 + x_2$
$\dot{x}_2 = -x_1 - x_2$
$x_1(0) = 7, x_2(0) = 3$

**26.** $\dot{x}_1 = x_2$
$\dot{x}_2 = -x_1$
$x_1(0) = 9, x_2(0) = -8$

**27.** $\dot{x}_1 = x_2$
$\dot{x}_2 = -x_1 + 2x_2$
$x_1(0) = 1, x_2(0) = -5$

**28.** $\dot{x}_1 = 2x_1 - x_2$
$\dot{x}_2 = 9x_1 + 2x_2$
$x_1(0) = 2, x_2(0) = -3$

**29.** $\dot{x}_1 = 2x_1 + x_2$
$\dot{x}_2 = -3x_1 + 6x_2$
$x_1(0) = 0, x_2(0) = -2$

**30.** $\dot{x}_1 = x_2$
$\dot{x}_2 = x_3$
$\dot{x}_3 = x_1$
$x_1(0) = -1, x_2(0) = 5, x_3(0) = -1$

[6]This is Exercise 11 in Philip M. Morse and K. Uno Ingard, *Theoretical Acoustics* (New York: McGraw-Hill, 1968), p. 62. Copyright © 1968 by McGraw-Hill Book Company. Reprinted by permission of McGraw-Hill Book Company.

# Series Solutions of Second-Order Linear Equations

## 5.1   INTRODUCTION

In this chapter we present an effective method for solving many second-order linear differential equations with variable coefficients by means of infinite series. We shall refer to this as the method of *series solutions*. The method of series solutions can also be used to compute formal solutions and to approximate solutions of other linear and nonlinear differential equations with constant or variable coefficients. We concentrate here on series solutions of second-order linear differential equations with variable coefficients because of the importance of these equations in applications and because the method of series solutions has been well developed for such equations.

Second-order linear differential equations appear frequently in applied mathematics, especially in the process of solving some of the classical partial differential equations in mathematical physics. Following are some of the most important second-order linear differential equations with variable coefficients which occur in applications. It is common to refer to these equations by the name that appears to the left of the differential equation.

| | |
|---|---|
| Airy's equation: | $y'' - xy = 0$ |
| Bessel's equation: | $x^2 y'' + xy' + (x^2 - p^2)y = 0$ |
| Chebyshev's equation: | $(1 - x^2)y'' - xy' + p^2 y = 0$ |
| Gauss's hypergeometric equation: | $x(1 - x)y'' + [c - (a + b + 1)x]y' - aby = 0$ |
| Hermite's equation: | $y'' - 2xy' + 2py = 0$ |
| Laguerre's equation: | $xy'' + (1 - x)y' + py = 0$ |
| Legendre's equation: | $(1 - x^2)y'' - 2xy' + n(n + 1)y = 0$ |

With the exception of Airy's equation, each of these differential equations involves parameters whose description is associated with the problem that led to their formulation and the "separation constant" involved in the process of solving partial differential equations by separation of variables. All these equations can be solved by the method of series solutions. Before we explain how to obtain series solutions for the above as well as other differential equations with variable coefficients, we review some of the properties of power series that will be used in this chapter.

# 5.2 REVIEW OF POWER SERIES

A series of the form

$$a_0 + a_1(x - x_0) + a_2(x - x_0)^2 + \cdots + a_n(x - x_0)^n + \cdots$$

is called a *power series* in powers of $(x - x_0)$ and is denoted by

$$\sum_{n=0}^{\infty} a_n(x - x_0)^n. \tag{1}$$

The numbers $a_0, a_1, a_2, \ldots, a_n, \ldots$ are called the *coefficients* of the power series, and the point $x_0$ is called the *center* of the power series. We also say that (1) is a power series *about the point* $x_0$.

We say that a power series $\sum_{n=0}^{\infty} a_n(x - x_0)^n$ *converges* at a specific point $x_1$ if

$$\lim_{N \to \infty} \sum_{n=0}^{N} a_n(x_1 - x_0)^n$$

exists. In this case the value of the limit is called the *sum* of the series at the point $x_1$. If this limit does not exist, the series is said to *diverge* at the point $x_1$.

Given a power series (1), it is important to find all points $x$ for which the series converges. To do this we compute the *radius of convergence* of the power series. This term is denoted by $R$ and is given by the formula

$$R = \frac{1}{\lim_{n \to \infty} \sqrt[n]{|a_n|}} \tag{2}$$

or

$$R = \lim_{n \to \infty} \left| \frac{a_n}{a_{n+1}} \right|, \tag{3}$$

provided that the limit in (2) or (3) exists. (For series in a form other than (1), such as $\sum_{n=0}^{\infty} x^{2n+1}/2^n$, Eqs. (2) and (3) are not applicable in general. For such series the radius of convergence is determined by using the ratio test of calculus. See Exercises 3, 9, and 24.)

If $R = 0$, series (1) converges only at its center $x = x_0$. If $R = +\infty$, it converges for all $x$. Finally, if $0 < R < \infty$, the series converges in the interval $|x - x_0| < R$, that is, for

$$-R + x_0 < x < R + x_0, \tag{4}$$

and diverges for $|x - x_0| > R$. The interval (4), or the entire real line if $R = \infty$, is called the *interval of convergence* of series (1). It is within this interval that all operations that we will perform on series, in our attempt to find series solutions of differential equations, are legal. At the endpoints of the interval (4), that is, for $x = -R + x_0$ and $x = R + x_0$, the series may converge or diverge. To determine the behavior at these points, we set $x = -R + x_0$ in (1) and check the resulting series by one of the known tests, then repeat the procedure for $x = R + x_0$.

**EXAMPLE 1** Determine the radius of convergence of each of the following power series:

(a) $\sum_{n=1}^{\infty} n^n x^n$        (b) $\sum_{n=0}^{\infty} (-1)^n (x-1)^n$        (c) $\sum_{n=0}^{\infty} \frac{x^n}{n!}$ .

**Solution**    (a) Here $a_n = n^n$ and from formula (2),

$$R = \frac{1}{\lim\limits_{n\to\infty} \sqrt[n]{|a_n|}} = \frac{1}{\lim\limits_{n\to\infty} n} = 0.$$

Hence series (a) converges only for $x = 0$ and diverges for any other $x$.

(b) Here $a_n = (-1)^n$ and from formula (2),

$$R = \frac{1}{\lim\limits_{n\to\infty} \sqrt[n]{|(-1)^n|}} = \frac{1}{\lim\limits_{n\to\infty} 1} = 1.$$

Thus, series (b) converges for all points $x$ in the interval $|x-1| < 1$; that is, $-1 < x - 1 < 1$ or $0 < x < 2$. The series diverges for $|x-1| > 1$, that is, for $x < 0$ or $x > 2$. For $|x-1| = 1$, that is, for $x = 0$ or $x = 2$, one can see directly that the series becomes

$$\sum_{n=0}^{\infty} (-1)^n (\pm 1)^n = \sum_{n=0}^{\infty} (\mp 1)^n,$$

and both of these series diverge.

(c) Here $a_n = 1/n!$. It is more convenient to use formula (3) in this case. Then

$$R = \lim_{n\to\infty} \left| \frac{1/n!}{1/(n+1)!} \right| = \lim_{n\to\infty} (n+1) = \infty.$$

Thus, series (c) converges for all $x$.

If $R$ is the radius of convergence of a power series $\sum_{n=0}^{\infty} a_n(x-x_0)^n$, then for every $x$ in the interval of convergence $|x - x_0| < R$, the sum of the series exists and defines a function

$$f(x) = \sum_{n=0}^{\infty} a_n(x-x_0)^n \qquad \text{for } |x - x_0| < R. \tag{5}$$

The function $f(x)$ defined by power series (5) is continuous and has derivatives of all orders. Furthermore, the derivatives $f'(x), f''(x), \ldots$ of the function $f(x)$ can be found by differentiating series (5) term by term. That is,

$$f'(x) = \sum_{n=1}^{\infty} n a_n(x-x_0)^{n-1}$$

$$f''(x) = \sum_{n=2}^{\infty} n(n-1) a_n(x-x_0)^{n-2},$$

and so on. Finally, these series for $f'(x), f''(x), \ldots$ have the same radius of convergence $R$ as the initial series (5).

In the process of finding power-series solutions of differential equations, in addition to taking derivatives of power series, we have to add, subtract, multiply, and equate two or more power series. These operations are performed in a fashion that resembles very much the corresponding operations with polynomials. The additional restriction for power series is to perform the operation within the interval of convergence of all series involved. For example,

(a) $\displaystyle\sum_{n=0}^{\infty} a_n(x - x_0)^n + \sum_{n=0}^{\infty} b_n(x - x_0)^n = \sum_{n=0}^{\infty} (a_n + b_n)(x - x_0)^n.$

(b) $\displaystyle\sum_{n=0}^{\infty} a_n(x - x_0)^n - \sum_{n=0}^{\infty} b_n(x - x_0)^n = \sum_{n=0}^{\infty} (a_n - b_n)(x - x_0)^n.$

(c) $\displaystyle a(x - x_0)^k \sum_{n=0}^{\infty} a_n(x - x_0)^n = \sum_{n=0}^{\infty} a a_n(x - x_0)^{n+k}.$

(d) If $\displaystyle\sum_{n=0}^{\infty} a_n(x - x_0)^n = \sum_{n=0}^{\infty} b_n(x - x_0)^n$

for all $x$ in some interval $|x - x_0| < R$, then

$$a_n = b_n \quad \text{for } n = 0, 1, 2, \ldots .$$

In particular, if a series is identically equal to zero, all the coefficients of the series must be zero.

The operations in (a), (b), and (d) were performed in one step because the general terms of the series involved were of the same power. However, in practice, we have to combine series whose general terms are not of the same power. In such cases we make an appropriate change in the index of summation of the series which does not change the sum of the series but makes the general terms of the same power. The basic idea behind the change of index is incorporated in the following identity:

$$\sum_{n=0}^{\infty} a_n(x - x_0)^n = \sum_{n=k}^{\infty} a_{n-k}(x - x_0)^{n-k}, \tag{6}$$

which holds for every integer $k$. The simplest way to prove (6) is to write out the two series term by term. In words, (6) says that we can decrease by $k$ the $n$ in the general term $a_n(x - x_0)^n$ provided that we increase by $k$ the $n$ under the summation symbol, and vice versa. For example,

$$\sum_{n=1}^{\infty} n a_n(x - x_0)^{n-1} = \sum_{n=0}^{\infty} (n + 1)a_{n+1}(x - x_0)^n$$

and

$$\sum_{n=0}^{\infty} 3(n + 4)a_n x^{n+2} = \sum_{n=2}^{\infty} 3(n + 2)a_{n-2} x^n.$$

The reader should note that these changes in appearance of the summations are very much like making a simple substitution in a definite integral. Thus, in the first illustration the index $n$ in the summation on the left is replaced by the new index $j$, where $j + 1 = n$. The resulting summation is that on the right with the index being called $n$ again instead of $j$.

Finally, we define the concept of "analytic function," which will be widely used in this chapter.

A function $f$ is called *analytic* at a point $x_0$ if it can be written as a power series

$$f(x) = \sum_{n=0}^{\infty} a_n(x - x_0)^n \tag{7}$$

with a positive radius of convergence.

Within its interval of convergence the power series (7) can be differentiated term by term. Evaluating $f(x)$, $f'(x)$, $f''(x)$, ... at the point $x_0$ we obtain $f(x_0) = a_0$, $f'(x_0) = a_1$, $f''(x_0) = 2a_2$, ... and in general

$$f^{(n)}(x_0) = n!a_n \quad \text{for } n = 0, 1, 2, \ldots.$$

Hence, $a_n = f^{(n)}(x_0)/n!$ and the power series in (7) is the *Taylor-series expansion*

$$f(x) = \sum_{n=0}^{\infty} \frac{f^{(n)}(x_0)}{n!}(x - x_0)^n \tag{8}$$

of the function $f$ at the point $x_0$. Thus, a function $f$ is analytic at a point $x_0$ if its Taylor-series expansion (8) about $x_0$ exists and has a positive radius of convergence.

Let us give some examples of analytic functions. Every polynomial is an analytic function about any point $x_0$. Indeed, since the derivatives of order higher than $n$ of a polynomial of degree $n$ are equal to zero, the Taylor-series expansion of any polynomial function has only a finite number of nonzero terms, and so it converges everywhere. Also, a "rational function," that is, the quotient of two polynomials, is an analytic function at every point where the denominator is different from zero. For example, the function $3x^2 - 7x + 6$ is analytic everywhere, while the function

$$\frac{x^2 - 5x + 7}{x(x^2 - 9)}$$

is analytic about every point except $x = 0, 3,$ and $-3$. Also, the functions $e^x$, $\sin x$, and $\cos x$ are analytic everywhere, as we can see from their Taylor-series expansions.

### EXERCISES

In Exercises 1 through 6, determine the radius of convergence of the power series.

1. $\sum_{n=0}^{\infty} 3^n(x - 1)^n$    2. $\sum_{n=0}^{\infty} 3^n x^n$    3. $\sum_{n=0}^{\infty} \frac{(-1)^n}{(2n)!} x^{2n}$

4. $\sum_{n=0}^{\infty} \frac{x^n}{n!}$    5. $\sum_{n=0}^{\infty} \frac{(x + 1)^n}{n + 1}$    6. $\sum_{n=0}^{\infty} \frac{x^n}{5 \cdot 7 \cdots (2n + 3)}$

In Exercises 7 through 12, compute the first and second derivative of the series and find the radius of convergence of the resulting series.

**7.** $\displaystyle\sum_{n=0}^{\infty} 3^n(x-1)^n$    **8.** $\displaystyle\sum_{n=0}^{\infty} 3^n x^n$    **9.** $\displaystyle\sum_{n=0}^{\infty} \frac{(-1)^n}{(2n)!} x^{2n}$

**10.** $\displaystyle\sum_{n=0}^{\infty} \frac{x^n}{n!}$    **11.** $\displaystyle\sum_{n=0}^{\infty} \frac{(x+1)^n}{n+1}$    **12.** $\displaystyle\sum_{n=0}^{\infty} \frac{x^n}{5 \cdot 7 \cdots (2n+3)}$

**13.** If $y(x) = \displaystyle\sum_{n=0}^{\infty} a_n x^n$, show that

$$y'' - xy = 2a_2 + \sum_{n=3}^{\infty} [n(n-1)a_n - a_{n-3}]x^{n-2}.$$

**14.** If $y(x) = \displaystyle\sum_{n=0}^{\infty} a_n x^n$, show that

$$(1-x)y'' - y' + xy = (2a_2 - a_1) + \sum_{n=2}^{\infty} [(n+1)na_{n+1} - n^2 a_n + a_{n-2}]x^{n-1}.$$

**15.** If $y(x) = \displaystyle\sum_{n=0}^{\infty} a_n(x-1)^n$ is a solution of the differential equation $y' = 3y$, show that

$$a_{n+1} = \frac{3}{n+1} a_n, \quad n = 0, 1, 2, \ldots$$

and find the solution.

In Exercises 16 through 23, show that the function is analytic at the indicated point $x_0$ by computing its Taylor series about the point $x_0$ and finding its interval of convergence.

**16.** $x^2 - 1$, $x_0 = -2$    **17.** $\cos x$, $x_0 = 0$    **18.** $e^x$, $x_0 = 1$

**19.** $\sin x$, $x_0 = \pi$    **20.** $\dfrac{1}{x+3}$, $x_0 = 0$    **21.** $\dfrac{1}{x+3}$, $x_0 = 1$

**22.** $\dfrac{1}{1-x^2}$, $x_0 = 0$    **23.** $\dfrac{1}{x}$, $x_0 = -3$.

**24.** Find the radius of convergence of the series $\displaystyle\sum_{n=0}^{\infty} \frac{x^{2n+1}}{2^n}$.

---

# 5.3   ORDINARY POINTS AND SINGULAR POINTS

Consider a second-order linear differential equation with variable coefficients of the form

$$a_2(x)y'' + a_1(x)y' + a_0(x)y = 0. \tag{1}$$

In the subsequent sections we will seek series solutions of the differential equation (1) in powers of $(x - x_0)$, where $x_0$ is a real number. As we will see, the form of the solutions will depend very much on the kind of point that $x_0$ is with respect to the differential equation. A point $x_0$ can be either an ordinary point or a singular point, according to the following definition.

**DEFINITION 1**

*A point $x_0$ is called an ordinary point of the differential equation (1) if the two functions*

$$\frac{a_1(x)}{a_2(x)} \quad \text{and} \quad \frac{a_0(x)}{a_2(x)} \tag{2}$$

*are analytic at the point $x_0$. If at least one of the functions in (2) is not analytic at the point $x_0$, then $x_0$ is called a singular point of the differential equation (1).*

In most differential equations of the form (1) that occur in applications, the coefficients $a_2(x)$, $a_1(x)$, and $a_0(x)$ are polynomials. After cancelling common factors, the rational functions $a_1(x)/a_2(x)$ and $a_0(x)/a_2(x)$ are analytic at every point except where the denominator vanishes. The points at which the denominator vanishes are singular points of the differential equation, and all other real numbers are ordinary points.

With reference to the differential equations mentioned in Section 5.1 see Table 5.1, which gives their ordinary and singular points in the finite real line.

In connection with the theory of series solutions it is important to classify the singular points of a differential equation into two categories according to the following definition.

**DEFINITION 2**

*A point $x_0$ is called a regular singular point of the differential equation (1) if it is a singular point [if at least one of the functions in (2) is not analytic at $x_0$] and the two functions*

$$(x - x_0)\frac{a_1(x)}{a_2(x)} \quad \text{and} \quad (x - x_0)^2\frac{a_0(x)}{a_2(x)} \tag{3}$$

*are analytic at the point $x_0$. If at least one of the functions in (3) is not analytic at the point $x_0$, then $x_0$ is called an irregular singular point of the differential equation (1).*

In Exercises 9 through 15 the student is asked to verify that all singular points of the differential equations of Table 5.1 are regular singular points.

**EXAMPLE 1**  Locate the ordinary points, regular singular points, and irregular singular points of the differential equation

$$(x^4 - x^2)y'' + (2x + 1)y' + x^2(x + 1)y = 0. \tag{4}$$

| Differential equation | Ordinary points | Singular points |
|---|---|---|
| Airy | All points | None |
| Bessel | All points except $x_0 = 0$ | 0 |
| Chebyshev | All points except $x_0 = \pm 1$ | $\pm 1$ |
| Gauss | All points except $x_0 = 0, 1$ | 0, 1 |
| Hermite | All points | None |
| Laguerre | All points except $x_0 = 0$ | 0 |
| Legendre | All points except $x_0 = \pm 1$ | $\pm 1$ |

**Table 5.1**

**Solution**   Here

$$a_2(x) = x^4 - x^2, \qquad a_1(x) = 2x + 1, \qquad a_0(x) = x^2(x + 1),$$

and so

$$\frac{a_1(x)}{a_2(x)} = \frac{2x + 1}{x^4 - x^2} = \frac{2x + 1}{x^2(x - 1)(x + 1)}, \quad \frac{a_0(x)}{a_2(x)} = \frac{x^2(x + 1)}{x^4 - x^2} = \frac{1}{x - 1}. \qquad (5)$$

It follows from (5) that every real number except 0, 1, and $-1$ is an ordinary point of the differential equation (4). To see which of the singular points 0, 1, and $-1$ is a regular singular point and which is an irregular singular point for the differential equation (4), we need to examine the two functions in (3).
   For $x_0 = 0$, the two functions in (3) become

$$x \frac{2x + 1}{x^4 - x^2} = \frac{2x + 1}{x(x - 1)(x + 1)} \quad \text{and} \quad x^2 \frac{x^2(x + 1)}{x^4 - x^2} = \frac{x^2}{x - 1}.$$

The first of these expressions is not analytic at $x = 0$, hence we conclude that the point $x_0 = 0$ is an irregular singular point for the differential equation (4). For $x_0 = 1$, the two functions in (3) become

$$(x - 1) \frac{2x + 1}{x^4 - x^2} = \frac{2x + 1}{x^2(x + 1)} \quad \text{and} \quad (x - 1)^2 \frac{x^2(x + 1)}{x^4 - x^2} = x - 1.$$

Since both of these expressions are analytic at $x = 1$, we conclude that the point $x_0 = 1$ is a regular singular point for the differential equation (4). Finally, for $x_0 = -1$, the two functions in (3) become

$$(x + 1) \frac{2x + 1}{x^4 - x^2} = \frac{2x + 1}{x^2(x - 1)} \quad \text{and} \quad (x + 1)^2 \frac{x^2(x + 1)}{x^4 - x^2} = \frac{(x + 1)^2}{x - 1},$$

and since both of them are analytic at $x = -1$ (their denominators do not vanish at $x = -1$), we conclude that the point $x_0 = -1$ is a regular singular point for the differential equation (4).
   Our aim in the remaining sections of this chapter is to obtain series solutions about ordinary points and near regular singular points. The study of solutions near irregular singular points is difficult and beyond our scope.

**EXERCISES**

In Exercises 1 through 8, locate the ordinary points, regular singular points, and irregular singular points of the differential equation.

**1.** $xy'' - (2x + 1)y' + y = 0$          **2.** $y'' - 2(x - 1)y' + 2y = 0$

**3.** $(1 - x)y'' - y' + xy = 0$          **4.** $2x^2y'' + (x - x^2)y' - y = 0$

**5.** $(x - 1)^2y'' - (x^2 - x)y' + y = 0$     **6.** $x^2y'' - (x + 2)y = 0$

**7.** $x^3(1 - x^2)y'' + (2x - 3)y' + xy = 0$    **8.** $(x - 1)^4y'' - xy = 0$

In Exercises 9 through 15, verify that all singular points of the differential equation are regular singular points.

**9.** $y'' - xy = 0$    (Airy's equation)

**10.** $x^2y'' + xy' + (x^2 - p^2)y = 0$    (Bessel's equation)

**11.** $(1 - x^2)y'' - xy' + p^2y = 0$    (Chebyshev's equation)

**12.** $x(1 - x)y'' + [c - (a + b + 1)x]y' - aby = 0$    (Gauss's hypergeometric equation)

**13.** $y'' - 2xy' + 2py = 0$    (Hermite's equation)

**14.** $xy'' + (1 - x)y' + py = 0$    (Laguerre's equation)

**15.** $(1 - x^2)y'' - 2xy' + n(n + 1)y = 0$    (Legendre's equation)

In Exercises 16 through 21, answer true or false.

**16.** The point $x_0 = -1$ is a regular singular point for the differential equation $(1 - x^2)y'' - 2xy + 12y = 0$.

**17.** The point $x_0 = 0$ is an ordinary point for the differential equation $xy'' + (1 - x)y' + 2y = 0$.

**18.** The point $x_0 = 0$ is a singular point for the differential equation $(1 + x)y'' - 2y' + xy = 0$.

**19.** The point $x_0 = 0$ is an irregular singular point for the differential equation $x^3y'' - (x + 1)y = 0$.

**20.** The point $x_0 = 3$ is an ordinary point for the differential equation $(x + 3)y'' + xy' - y = 0$.

**21.** The point $x_0 = -3$ is a singular point for the differential equation $(x + 3)y'' + xy' - y = 0$.

---

# 5.4    POWER-SERIES SOLUTIONS ABOUT AN ORDINARY POINT

In this section we show how to solve any second-order linear differential equation with variable coefficients of the form

$$a_2(x)y'' + a_1(x)y' + a_0(x)y = 0 \qquad (1)$$

in some interval about any *ordinary point* $x_0$. The point $x_0$ is usually dictated by the specific problem at hand, which requires that we find the solution of the differential equation (1) that satisfies two given initial conditions of the form

$$y(x_0) = y_0 \tag{2}$$

and

$$y'(x_0) = y_1. \tag{3}$$

Let us recall that if the coefficients $a_2(x)$, $a_1(x)$, and $a_0(x)$ are polynomials in $x$, then a point $x_0$ is an ordinary point of the differential equation (1) when $a_2(x_0) \neq 0$. In general, $x_0$ is an ordinary point of the differential equation (1) when the functions $a_1(x)/a_2(x)$ and $a_0(x)/a_2(x)$ have power-series expansions of the form

$$\frac{a_1(x)}{a_2(x)} = \sum_{n=0}^{\infty} A_n(x - x_0)^n \qquad \text{for } |x - x_0| < R_1 \tag{4}$$

and

$$\frac{a_0(x)}{a_2(x)} = \sum_{n=0}^{\infty} B_n(x - x_0)^n \qquad \text{for } |x - x_0| < R_2 \tag{5}$$

with positive radii of convergence $R_1$ and $R_2$. The functions (4) and (5) are in particular continuous in the interval $|x - x_0| < R$, where $R$ is the smallest of $R_1$ and $R_2$, and therefore, by the existence theorem, Theorem 1 of Section 2.3, the IVP (1)–(3) has a unique solution throughout the interval $|x - x_0| < R$. Our task here is to compute (or approximate) this unique solution. The following theorem describes the form of any solution of the differential equation (1), in particular the form of the unique solution of the IVP (1)–(3).

**THEOREM 1**  (Solutions about an Ordinary Point)

*If $x_0$ is an ordinary point of the differential equation* (1), *then the general solution of the differential equation has a power-series expansion about $x_0$,*

$$y(x) = \sum_{n=0}^{\infty} a_n(x - x_0)^n, \tag{6}$$

*with positive radius of convergence. More precisely, if $R_1$ and $R_2$ are the radii of convergence of the series* (4) *and* (5), *then the radius of convergence of* (6) *is at least equal to the minimum of $R_1$ and $R_2$. The coefficients $a_n$ for $n = 2, 3, \ldots$ of the series* (6) *can be obtained in terms of $a_0$ and $a_1$ by direct substitution of* (6) *into the differential equation* (1) *and by equating coefficients of the same power. Finally, if* (6) *is the solution of the IVP* (1)–(3), *then $a_0 = y_0$ and $a_1 = y_1$.*

**EXAMPLE 1**  Find the general solution of the differential equation

$$y'' - 2(x - 1)y' + 2y = 0 \tag{7}$$

about the ordinary point $x_0 = 1$.

**Solution**   By Theorem 1 the general solution of Eq. (7) has a power-series expansion about $x_0 = 1$,

$$y(x) = \sum_{n=0}^{\infty} a_n(x - 1)^n, \qquad (8)$$

with a positive radius of convergence. To find a lower bound for the radius of convergence of the series (8), we need the radii of convergence $R_1$ and $R_2$ of the power-series expansions of the functions

$$\frac{a_1(x)}{a_2(x)} \quad \text{and} \quad \frac{a_0(x)}{a_2(x)}.$$

Here $a_2(x) = 1$, $a_1(x) = -2(x - 1)$, and $a_0(x) = 2$. Hence,

$$\frac{a_1(x)}{a_2(x)} = -2(x - 1) \quad \text{and} \quad \frac{a_0(x)}{a_2(x)} = 2,$$

and so $R_1 = R_2 = \infty$. Thus, the radius of convergence of the series (8) is also equal to $\infty$. That is, solution (8) will converge for all $x$. The coefficients of series (8) will be found by direct substitution of the series into the differential equation. Since (8) is the general solution of the second-order equation (7), it should contain two arbitrary constants. In fact, the coefficients $a_0$ and $a_1$ will remain unspecified, while every other coefficient $a_2, a_3, \ldots$ will be expressed in terms of $a_0$ and $a_1$. Differentiating series (8) term by term, we obtain

$$y' = \sum_{n=0}^{\infty} na_n(x - 1)^{n-1} = \sum_{n=1}^{\infty} na_n(x - 1)^{n-1}$$

and

$$y'' = \sum_{n=0}^{\infty} n(n - 1)a_n(x - 1)^{n-2} = \sum_{n=2}^{\infty} n(n - 1)a_n(x - 1)^{n-2}.$$

We are now ready to substitute $y$, $y'$, and $y''$ into the differential equation (7). As we can see from Eq. (7), $y'$ should be multiplied by $-2(x - 1)$ and $y$ by 2. For the sake of maintaining easy bookkeeping, we write the terms $y''$, $-2(x - 1)y'$, and $2y$ of the differential equation in a column as follows:

$$y'' = \sum_{n=2}^{\infty} n(n - 1)a_n(x - 1)^{n-2}$$

$$-2(x - 1)y' = -2(x - 1)\sum_{n=1}^{\infty} na_n(x - 1)^{n-1} = \sum_{n=1}^{\infty} -2na_n(x - 1)^n$$

$$2y = 2\sum_{n=0}^{\infty} a_n(x - 1)^n = \sum_{n=0}^{\infty} 2a_n(x - 1)^n.$$

The sum of the terms on the left-hand sides is zero, because $y$ is a solution of the differential equation (7). Hence, the sum of the three series on the right-hand side must be set equal to zero. Writing the expressions in columns is very handy in preparing the series in the summation process. It is easier to add the three series term by term if the general terms are of the same power in each series and that the lower index $n$ under the summation symbol is the same in the three series. With this in mind, we rewrite the three series above in the following suitable and equivalent form:

$$y'' = \sum_{n=0}^{\infty} (n + 2)(n + 1)a_{n+2}(x - 1)^n$$

$$= 2a_2 + \sum_{n=1}^{\infty} (n + 2)(n + 1)a_{n+2}(x - 1)^n$$

$$-2(x - 1)y' = \sum_{n=1}^{\infty} - 2na_n(x - 1)^n$$

$$2y = \sum_{n=0}^{\infty} 2a_n(x - 1)^n = 2a_0 + \sum_{n=1}^{\infty} 2a_n(x - 1)^n.$$

Adding the left-hand sides and the right-hand sides of these three equations, we obtain

$$0 = (2a_2 + 2a_0) + \sum_{n=1}^{\infty} [(n + 2)(n + 1)a_{n+2} - 2na_n + 2a_n](x - 1)^n.$$

The right-hand side of this equation is a power series that is identically equal to zero. Hence, all its coefficients must be zero. That is,

$$2a_2 + 2a_0 = 0 \tag{9}$$

and

$$(n + 2)(n + 1)a_{n+2} - 2na_n + 2a_n = 0 \quad \text{for} \quad n = 1, 2, \ldots. \tag{10}$$

The condition (10) is called a *recurrence formula* because it allows $a_{n+2}$ to be calculated once $a_n$ is known. Using Eq. (9) and the recurrence formula (10), we can express the coefficients $a_2, a_3, \ldots$ of the power series in terms of the coefficients $a_0$ and $a_1$. In fact, from (9) we have

$$a_2 = -a_0, \tag{11}$$

and from (10) we obtain

$$a_{n+2} = \frac{2(n - 1)}{(n + 2)(n + 1)} a_n \quad \text{for} \quad n = 1, 2, \ldots. \tag{12}$$

From Eq. (12) we find

$$a_3 = 0, \qquad a_4 = \frac{2}{4 \cdot 3} a_2 = -\frac{2}{4 \cdot 3} a_0 = -\frac{2^2}{4!} a_0$$

$$a_5 = 0, \qquad a_6 = \frac{2 \cdot 3}{6 \cdot 5} a_4 = -\frac{2^2 \cdot 3}{6 \cdot 5 \cdot 4 \cdot 3} a_0 = -\frac{2^3 \cdot 3}{6!} a_0$$

$$a_7 = 0, \qquad a_8 = \frac{2 \cdot 5}{8 \cdot 7} a_6 = -\frac{2^3 \cdot 5 \cdot 3}{8 \cdot 7 \cdot 6 \cdot 5 \cdot 4 \cdot 3} a_0 = -\frac{2^4 \cdot 5 \cdot 3 \cdot 1}{8!} a_0$$

.........................................................................

Thus,

$$a_{2n+1} = 0, \qquad n = 1, 2, \ldots$$

and

$$a_{2n} = -\frac{2^n \cdot 1 \cdot 3 \cdot 5 \cdots (2n-3)}{(2n)!} a_0, \qquad n = 2, 3, \ldots.$$

Hence, the general solution of the differential equation (7) is

$$y(x) = a_0 + a_1(x-1) + a_2(x-1)^2 + a_4(x-1)^4 + a_6(x-1)^6 + \cdots$$
$$= a_1(x-1) + a_0[1 - (x-1)^2 - \frac{2^2}{4!}(x-1)^4 - \frac{2^3 \cdot 3}{6!}(x-1)^6 - \cdots].$$

**REMARK 1**   As we expected, the general solution involves the two arbitrary constants $a_1$ and $a_0$. Therefore, the functions $x - 1$ and $1 - (x-1)^2 - (2^2/4!)(x-1)^4 - \cdots$ are two linearly independent solutions of Eq. (7).

**EXAMPLE 2**   Solve the IVP

$$(1-x)y'' - y' + xy = 0 \tag{13}$$

$$y(0) = 1 \tag{14}$$

$$y'(0) = 1. \tag{15}$$

**Solution**   Since the initial conditions are given at the point 0, we are interested in a solution of the IVP (13)–(15) about the point $x_0 = 0$. The only singular point of the differential equation (13) is $x = 1$, and so the point $x_0 = 0$ is an ordinary point. Hence, the IVP (13)–(15) has a unique solution of the form:

$$y(x) = \sum_{n=0}^{\infty} a_n x^n. \tag{16}$$

If we want at this time to find a lower estimate of the radius of convergence of the power series (16), we must compute the radii of convergence of the power-series expansions of the functions $a_1(x)/a_2(x)$ and $a_0(x)/a_2(x)$. Observe that

$$\frac{a_1(x)}{a_2(x)} = -\frac{1}{1-x} = -\sum_{n=0}^{\infty} x^n, \qquad |x| < 1$$

and

$$\frac{a_0(x)}{a_2(x)} = \frac{x}{1-x} = x\sum_{n=0}^{\infty} x^n = \sum_{n=0}^{\infty} x^{n+1}, \qquad |x| < 1.$$

Thus, series (16) converges at least for $|x| < 1$.

By direct substitution of (16) into (13) and equating coefficients, the reader can verify that

$$2a_2 - a_1 = 0$$

and

$$a_{n+1} = \frac{n^2 a_n - a_{n-2}}{(n+1)n} \qquad n = 2, 3, \ldots .$$

From the initial condition (14) we obtain $a_0 = 1$, and from (15) we find that $a_1 = 1$.

$$\Rightarrow \qquad a_2 = \frac{1}{2!}, a_3 = \frac{1}{3!}, \ldots, a_n = \frac{1}{n!}, \ldots .$$

Thus, the solution of the IVP (13)–(15) is

$$y(x) = \sum_{n=0}^{\infty} a_n x^n = \sum_{n=0}^{\infty} \frac{1}{n!} x^n = e^x. \tag{17}$$

By Theorem 1 the radius of convergence of the power-series solution (17) is at least equal to 1. However, it can be larger. In fact, the radius of convergence of the solution (17) is equal to $\infty$.

**REMARK 2**   In Examples 1 and 2 we were able to compute all the coefficients $a_n$ of the power-series solution. However, this is a luxury that is not always possible. Of course, we always have a recurrence formula that we can use to compute as many coefficients of the power-series solution as we please. In general, we compute enough coefficients $a_n$ of the power-series solution to obtain a "good approximation" to the solution.

**EXAMPLE 3**   Compute the first five coefficients of the power-series solution $y(x) = \sum_{n=0}^{\infty} a_n x^n$ of the IVP

$$y'' - 2x^2 y' + 8y = 0 \tag{18}$$
$$y(0) = 0 \tag{19}$$
$$y'(0) = 1. \tag{20}$$

**Solution**   We have

$$y'(x) = \sum_{n=0}^{\infty} n a_n x^{n-1} = \sum_{n=1}^{\infty} n a_n x^{n-1}$$

and

$$y''(x) = \sum_{n=0}^{\infty} n(n-1) a_n x^{n-2} = \sum_{n=2}^{\infty} n(n-1) a_n x^{n-2}.$$

Thus,

$$y'' = \sum_{n=2}^{\infty} n(n-1) a_n x^{n-2} = \sum_{n=0}^{\infty} (n+2)(n+1) a_{n+2} x^n \tag{21}$$

$$-2x^2 y' = \sum_{n=1}^{\infty} -2n a_n x^{n+1} = \sum_{n=2}^{\infty} -2(n-1) a_{n-1} x^n \tag{22}$$

$$8y = \sum_{n=0}^{\infty} 8a_n x^n \tag{23}$$

---

$$0 = (2a_2 + 8a_0) + (6a_3 + 8a_1)x$$

$$+ \sum_{n=2}^{\infty} [(n + 2)(n + 1)a_{n+2} - 2(n - 1)a_{n-1} + 8a_n]x^n,$$

where the terms $(2a_2 + 8a_0)$ and $(6a_3 + 8a_1)x$ are the contributions of the series (21) and (23) for $n = 0$ and $n = 1$. Hence, $a_2 = -4a_0$, $a_3 = -\frac{4}{3}a_1$, and

$$(n + 2)(n + 1)a_{n+2} - 2(n - 1)a_{n-1} + 8a_n = 0 \quad \text{for} \quad n = 2, 3, \ldots . \tag{24}$$

From the initial conditions we obtain $a_0 = 0$ and $a_1 = 1$. Then $a_2 = 0$ and $a_3 = -\frac{4}{3}$. Finally, from the recurrence formula (24) for $n = 2$, we find $12a_4 - 2a_1 + 8a_2 = 0$ and so $a_4 = \frac{1}{6}$. Thus,

$$y(x) = a_0 + a_1 x + a_2 x^2 + a_3 x^3 + a_4 x^4 + \cdots$$
$$= x - \tfrac{4}{3}x^3 + \tfrac{1}{6}x^4 + \cdots .$$

In this example, using the recurrence formula, we can compute as many coefficients as time permits. However, there is no known closed-form expression for the general coefficients $a_n$ for all $n$.

### APPLICATIONS 5.4.1

The method of power-series solutions about an ordinary point provides an effective tool for obtaining the solutions of some differential equations that occur in applications.

**Legendre's Equation**    ■ The differential equation

$$(1 - x^2)y'' - 2xy' + p(p + 1)y = 0, \tag{25}$$

where $p$ is a constant, is *Legendre's equation*. The solutions of Eq. (25) are very important in many branches of applied mathematics. For example, Legendre's equation appears in the study of the potential equation in spherical coordinates. In fact, the potential equation

$$\frac{\partial^2 V}{\partial x^2} + \frac{\partial^2 V}{\partial y^2} + \frac{\partial^2 V}{\partial z^2} = 0,$$

transformed to spherical polar coordinates

$$x = r \sin \theta \cos \phi, \qquad y = r \sin \theta \sin \phi, \qquad z = r \cos \theta,$$

becomes

$$\frac{\partial^2 V}{\partial r^2} + \frac{2}{r} \frac{\partial V}{\partial r} + \frac{1}{r^2} \frac{\partial^2 V}{\partial \theta^2} + \frac{\cot \theta}{r^2} \frac{\partial V}{\partial \theta} + \frac{1}{r^2 \sin^2 \theta} \frac{\partial^2 V}{\partial \phi^2} = 0.$$

If we are interested in a solution that is independent of $\phi$ of the form $V = r^p \Theta$, where $\Theta$ is a function of $\theta$ only, we find

$$\frac{d^2 \Theta}{d\theta^2} + \cot \theta \frac{d\Theta}{d\theta} + p(p + 1)\Theta = 0.$$

Using the change of variables $x = \cos \theta$ and replacing $\Theta$ by $y$, we obtain Legendre's equation (25).

When $p$ is a nonnegative integer, one of the solutions of Eq. (25) about the ordinary point $x_0 = 0$ is a polynomial. When suitably normalized (as we shall explain below), these polynomial solutions are called *Legendre's polynomials*. Legendre's polynomials are widely used in applications. They appear, for example, in quantum mechanics in the study of the hydrogen atom.

Now we proceed to obtain two linearly independent solutions of Legendre's equation about the point $x_0 = 0$. Here $a_2(x) = 1 - x^2$, $a_1(x) = -2x$, and $a_0(x) = p(p + 1)$. Since $a_2(0) = 1 \neq 0$, the point $x_0 = 0$ is an ordinary point for the differential equation (25). The form of any solution of the differential equation (25) about the point $x_0 = 0$ is

$$y(x) = \sum_{n=0}^{\infty} a_n x^n. \tag{26}$$

To find a lower bound for the radius of convergence of the solution (26), we need to compute the radii of convergence of the Taylor-series expansions about zero of the functions $a_1(x)/a_2(x)$ and $a_0(x)/a_2(x)$. We have

$$\frac{a_1(x)}{a_2(x)} = -\frac{2x}{1 - x^2} = -2x(1 + x^2 + x^4 + \cdots)$$

$$= \sum_{n=0}^{\infty} -2x^{2n+1}, \qquad |x| < 1$$

and

$$\frac{a_0(x)}{a_2(x)} = \frac{p(p + 1)}{1 - x^2} = p(p + 1)(1 + x^2 + x^4 + \cdots)$$

$$= \sum_{n=0}^{\infty} p(p + 1)x^{2n}, \qquad |x| < 1.$$

Hence, the radius of convergence of the solution (26) is at least equal to 1; that is, the series (26) converges at least for $|x| < 1$. From Eq. (26) we have

$$y'(x) = \sum_{n=1}^{\infty} na_n x^{n-1} \quad \text{and} \quad y''(x) = \sum_{n=2}^{\infty} n(n-1)a_n x^{n-2}$$

$$\Rightarrow \quad y'' = \sum_{n=2}^{\infty} n(n-1)a_n x^{n-2} = \sum_{n=0}^{\infty} (n+2)(n+1)a_{n+2} x^n$$

$$-x^2 y'' = \sum_{n=2}^{\infty} -n(n-1)a_n x^n$$

$$-2xy' = \sum_{n=1}^{\infty} -2na_n x^n$$

$$p(p+1)y = \sum_{n=0}^{\infty} p(p+1)a_n x^n$$

---

$$0 = [2a_2 + p(p+1)a_0] + [6a_3 - 2a_1 + p(p+1)a_1]x$$

$$+ \sum_{n=2}^{\infty} [(n+2)(n+1)a_{n+2} - n(n-1)a_n - 2na_n + p(p+1)a_n]x^n$$

$$\Rightarrow \quad 2a_2 + p(p+1)a_0 = 0, \qquad 6a_3 - 2a_1 + p(p+1)a_1 = 0$$

and

$$(n+2)(n+1)a_{n+2} - n(n-1)a_n - 2na_n + p(p+1)a_n = 0,$$

$$n = 2, 3, \ldots$$

or

$$a_2 = -\frac{p(p+1)}{2}a_0, \qquad a_3 = \frac{2 - p(p+1)}{6}a_1 = -\frac{(p-1)(p+2)}{3!}a_1$$

and

$$a_{n+2} = \frac{n(n-1) + 2n - p(p+1)}{(n+2)(n+1)}a_n = -\frac{(p-n)(p+n+1)}{(n+2)(n+1)}a_n,$$

$$n = 2, 3, \ldots \qquad (27)$$

$$\Rightarrow \quad a_4 = -\frac{(p-2)(p+3)}{4 \cdot 3}a_2 = \frac{p(p-2)(p+1)(p+3)}{4!}a_0,$$

and, in general,

$$a_{2n} = (-1)^n \frac{p(p-2) \cdots (p-2n+2)(p+1)(p+3) \cdots (p+2n-1)}{(2n)!}a_0,$$

$$n = 1, 2, \ldots.$$

Also,

$$a_5 = -\frac{(p-3)(p+4)}{5 \cdot 4}a_3 = \frac{(p-1)(p-3)(p+2)(p+4)}{5!}a_1$$

and, in general,

$$a_{2n+1} = (-1)^n \frac{(p-1)(p-3) \cdots (p-2n+1)(p+2)(p+4) \cdots (p+2n)}{(2n+1)!} a_1,$$

$$n = 1, 2, \ldots . \qquad (28)$$

Hence, two linearly independent solutions of Legendre's equation about the point 0 are

$$y_1(x) = 1 +$$

$$\sum_{n=1}^{\infty} (-1)^n \cdot \frac{p(p-2) \cdots (p-2n+2)(p+1)(p+3) \cdots (p+2n-1)}{(2n)!} x^{2n}$$

$$(29)$$

and

$$y_2(x) = x +$$

$$\sum_{n=1}^{\infty} (-1)^n \cdot \frac{(p-1)(p-3) \cdots (p-2n+1)(p+2)(p+4) \cdots (p+2n)}{(2n+1)!} x^{2n+1}$$

$$(30)$$

and they converge for $x < 1$.

As we see from Eqs. (27) and (28), when $p$ is equal to a nonnegative integer $n$, one of the solutions above is a polynomial of degree $n$. A multiple of this polynomial solution that takes the value 1 at $x = 1$ is called a *Legendre polynomial* and is denoted by $P_n(x)$. For example,

$$P_0(x) = 1, \qquad P_1(x) = x, \qquad P_2(x) = \tfrac{3}{2} x^2 - \tfrac{1}{2}, \qquad P_3(x) = \tfrac{5}{2} x^3 - \tfrac{3}{2} x$$

are Legendre polynomials.

■ The differential equation

$$y'' - xy = 0 \qquad (31)$$

**Airy's Equation**

is *Airy's equation*. The solutions of Airy's equation about the ordinary point $x_0 = 0$ are called *Airy functions* and have applications in the theory of diffraction. Airy functions were originally studied by Airy in connection with his calculations of light intensity in the neighborhood of a caustic surface.

From Theorem 1 it follows that any solution of the differential equation (31) about the point $x_0 = 0$ is of the form

$$y(x) = \sum_{n=0}^{\infty} a_n x^n \qquad (32)$$

and converges for all $x$. Substituting (32) into the differential equation (31) and equating coefficients, the reader can verify that $a_2 = 0$ and for $n = 1, 2, \ldots ,$

$$a_{3n} = \frac{1 \cdot 4 \cdots (3n-2)}{(3n)!} a_0, \quad a_{3n+1} = \frac{2 \cdot 5 \cdots (3n-1)}{(3n+1)!} a_1, \quad a_{3n+2} = 0. \quad (33)$$

Hence, the general solution of Airy's equation is

$$y(x) = \sum_{n=0}^{\infty} a_n x^n = \sum_{n=0}^{\infty} a_{3n} x^{3n} + \sum_{n=0}^{\infty} a_{3n+1} x^{3n+1} + \sum_{n=0}^{\infty} a_{3n+2} x^{3n+2}$$

$$= a_0 y_1(x) + a_1 y_2(x),$$

where

$$y_1(x) = 1 + \sum_{n=1}^{\infty} \frac{1 \cdot 4 \cdots (3n-2)}{(3n)!} x^{3n} \text{ and } y_2(x) = x + \sum_{n=1}^{\infty} \frac{2 \cdot 5 \cdots (3n-1)}{(3n+1)!} x^{3n+1}$$

are two linearly independent solutions of (31).

**Chebyshev's Equation**

■ The differential equation

$$(1 - x^2)y'' - xy' + p^2 y = 0, \tag{34}$$

where $p$ is a constant, is *Chebyshev's equation* (the spelling Tschebyscheff is also used). As we will see, if the constant $p$ is a nonnegative integer, Eq. (34) has a polynomial solution about the point $x_0 = 0$. When suitably normalized (when the leading coefficient is chosen as we will explain), these polynomial solutions are called *Chebyshev's polynomials*. The Chebyshev polynomials are very useful in numerical analysis. Chebyshev obtained the polynomials that bear his name in 1857 while seeking the polynomial of degree $n$ and leading coefficient 1 that deviates least from zero on the interval $-1 \le x \le 1$.

From Theorem 1 it follows that any solution of the differential equation (34) about the point $x_0 = 0$ is of the form

$$y(x) = \sum_{n=0}^{\infty} a_n x^n \tag{35}$$

and converges for $x < 1$. By direct substitution of (35) into (34) and equating of coefficients, the reader can verify that for $n = 1, 2, \ldots,$

$$a_{2n} = (-1)^n \frac{p^2(p^2 - 2^2)(p^2 - 4^2) \cdots [p^2 - (2n-2)^2]}{(2n)!} a_0, \tag{36}$$

and

$$a_{2n+1} = (-1)^n \frac{(p^2 - 1^2)(p^2 - 3^2) \cdots [p^2 - (2n-1)^2]}{(2n+1)!} a_1. \tag{37}$$

Thus,

$$y(x) = \sum_{n=0}^{\infty} a_n x^n = \sum_{n=0}^{\infty} a_{2n} x^{2n} + \sum_{n=0}^{\infty} a_{2n+1} x^{2n+1}$$

$$= a_0 \left[ 1 + \sum_{n=1}^{\infty} (-1)^n \frac{p^2(p^2 - 2^2) \cdots [p^2 - (2n-2)^2]}{(2n)!} x^{2n} \right]$$

$$+ a_1 \left[ x + \sum_{n=1}^{\infty} (-1)^n \frac{(p^2 - 1^2)(p^2 - 3^2) \cdots [p^2 - (2n-1)^2]}{(2n+1)!} x^{2n+1} \right].$$

Hence, two linearly independent solutions of Chebyshev's equation are

$$y_1(x) = 1 + \sum_{n=1}^{\infty} (-1)^n \frac{p^2(p^2 - 2^2) \cdots [p^2 - (2n - 2)^2]}{(2n)!} x^{2n} \qquad (38)$$

and

$$y_2(x) = x + \sum_{n=1}^{\infty} (-1)^n \frac{(p^2 - 1^2)(p^2 - 3^2) \cdots [p^2 - (2n - 1)^2]}{(2n + 1)!} x^{2n+1}. \qquad (39)$$

It is clear from Eqs. (36) and (37) that if $p$ is a nonnegative integer, one of the solutions is a polynomial of degree $n$. If we multiply this polynomial by $2^{n-1}$, we obtain a polynomial solution called a *Chebyshev polynomial* and denoted by $T_n(x)$. For example, the polynomials 1, $x$, $2x^2 - 1$, and $4x^3 - 3x$ are the Chebyshev polynomials $T_0(x)$, $T_1(x)$, $T_2(x)$, and $T_3(x)$, respectively.

■ The differential equation                                          **Hermite's**

$$y'' - 2xy' + 2py = 0, \qquad (40) \qquad \text{**Equation**}$$

where $p$ is a constant, is *Hermite's equation*. As we shall see, if the constant $p$ is a nonnegative integer, Eq. (40) has a polynomial solution about the point $x_0 = 0$. When suitably normalized, these polynomial solutions are called *Hermite's polynomials*. The Hermite polynomials are very important in quantum mechanics in the investigation of acceptable solutions of the Schrödinger equation for a harmonic oscillator. Hermite's polynomials are also useful in probability and statistics in obtaining the Gram–Charlier series expansions, that is, expansions in terms of Hermite's polynomials.

Clearly, the point $x_0 = 0$ is an ordinary point of the differential equation (40) and any solution is of the form

$$y(x) = \sum_{n=0}^{\infty} a_n x^n \qquad (41)$$

and converges for all $x$. By direct substitution of (41) into (40) and equating coefficients, the reader can verify that for $n = 1, 2, \ldots$,

$$a_{2n} = (-1)^n \frac{2^n p(p - 2) \cdots (p - 2n + 2)}{(2n)!} a_0 \qquad (42)$$

and

$$a_{2n+1} = (-1)^n \frac{2^n (p - 1)(p - 3) \cdots (p - 2n + 1)}{(2n + 1)!} a_1. \qquad (43)$$

The general solution of the differential equation (40) is

$$y(x) = \sum_{n=0}^{\infty} a_n x^n = \sum_{n=0}^{\infty} a_{2n} x^{2n} + \sum_{n=0}^{\infty} a_{2n+1} x^{2n+1}$$

or

$$y(x) = a_0 \left[ 1 + \sum_{n=1}^{\infty} (-1)^n \frac{2^n p(p - 2) \cdots (p - 2n + 2)}{(2n)!} x^{2n} \right]$$

$$+ a_1 \left[ x + \sum_{n=1}^{\infty} (-1)^n \frac{2^n (p - 1)(p - 3) \cdots (p - 2n + 1)}{(2n + 1)!} x^{2n+1} \right].$$

Hence, two linearly independent solutions of Hermite's differential equation are

$$y_1(x) = 1 + \sum_{n=1}^{\infty} (-1)^n \frac{2^n p(p-2) \cdots (p-2n+2)}{(2n)!} x^{2n} \tag{44}$$

and

$$y_2(x) = x + \sum_{n=1}^{\infty} (-1)^n \frac{2^n(p-1)(p-3) \cdots (p-2n+1)}{(2n+1)!} x^{2n+1}. \tag{45}$$

The series converge for all $x$.

As we see from Eq. (42), if $p$ is zero or a positive even integer, say $p = 2k$, the coefficients $a_{2n}$ vanish for $n \ge k + 1$, and so the solution (44) is a polynomial of degree $2k$. Similarly, from Eq. (43), we see that if $p$ is a positive odd integer, say $p = 2k + 1$, the coefficients $a_{2n+1}$ vanish for $n \ge k + 1$, and so the solution (45) is a polynomial of degree $2k + 1$.

Thus, when $p$ is a nonnegative integer $n$, Hermite's differential equation has a polynomial solution of degree $n$. These polynomials, normalized so that the leading coefficient (the coefficient of $x^n$) is $2^n$, are called the *Hermite polynomials* and are denoted by $H_n(x)$. For example,

$$H_0(x) = 1, \quad H_1(x) = 2x, \quad H_2(x) = 4x^2 - 2, \quad H_3(x) = 8x^3 - 12x, \text{ and so on.}$$

## EXERCISES

In Exercises 1 through 10, solve the IVP by the method of power series about the initial point $x_0$.

**1.** $y'' - 2xy' + 4y = 0$
$\qquad y(0) = 1$
$\qquad y'(0) = 0$

**2.** $(1 - x^2)y'' - 2xy' + 6y = 0$
$\qquad y(0) = 1$
$\qquad y'(0) = 0$

**3.** $y'' - 2(x + 2)y' + 4y = 0$
$\qquad y(-2) = 1$
$\qquad y'(-2) = 0$

**4.** $(-x^2 + 4x - 3)y'' - 2(x - 2)y' + 6y = 0$
$\qquad y(2) = 1$
$\qquad y'(2) = 0$

**5.** $(1 - x^2)y'' - xy' + y = 0$
$\qquad y(0) = 0$
$\qquad y'(0) = 1$

**6.** $(1 - x^2)y'' - xy' + 4y = 0$
$\qquad y(0) = 1$
$\qquad y'(0) = 0$

**7.** $y'' - 2xy' + 2y = 0$
$\qquad y(0) = 0$
$\qquad y'(0) = 1$

**8.** $y'' - 2(x - 1)y' + 2y = 0$
$\qquad y(1) = 0$
$\qquad y'(1) = 1$

**9.** $(1 - x^2)y'' - 2xy' + 2y = 0$
$\qquad y(0) = 0$
$\qquad y'(0) = -1$

**10.** $(x^2 + 4x + 3)y'' + 2(x + 2)y' - 2y = 0$ .
$$y(-2) = 0$$
$$y'(-2) = -1$$

In Exercises 11 through 19, compute the first four coefficients of the power-series solution about the initial point.

**11.** $y'' - xy = 0$
$$y(0) = 0$$
$$y'(0) = 1$$

**12.** $y'' - xy = 0$
$$y(0) = 1$$
$$y'(0) = 0$$

**13.** $(x^2 + 2)y'' - 3y' + (x - 1)y = 0$
$$y(1) = -20$$
$$y'(1) = -2$$

**14.** $xy'' - 2(x + 1)y' + 2y = 0$
$$y(3) = 2$$
$$y'(3) = 0$$

**15.** $(x - 1)y'' - xy' + y = 0$
$$y(0) = 0$$
$$y'(0) = 1$$

**16.** $y'' - 2xy' + 4y = 0$
$$y(0) = 0$$
$$y'(0) = 1$$

**17.** $(1 - x^2)y'' - 2xy' + 6y = 0$
$$y(0) = 0$$
$$y'(0) = 1$$

**18.** $(1 - x^2)y'' - xy' + y = 0$
$$y(0) = 1$$
$$y'(0) = 0$$

**19.** $y'' - 2(x - 1)y' + 2y = 0$
$$y(1) = 1$$
$$y'(1) = 0$$

In Exercises 20 through 24, prove the statement without finding explicit solutions.

**20.** The power-series solution $\sum_{n=0}^{\infty} a_n x^n$ of the IVP

$$y'' + (x + 1)y' - y = 0$$
$$y(0) = 1$$
$$y'(0) = -1$$

converges for all $x$.

**21.** Any power-series solution of the differential equation
$$y'' + (x + 1)y' - y = 0$$

converges for all $x$.

**22.** The power-series solution $\sum_{n=0}^{\infty} a_n(x - 3)^n$ of the IVP

$$xy'' - 2y' + xy = 0$$
$$y(3) = 1$$
$$y'(3) = -2$$

converges for all $x$ in the interval $0 < x < 6$.

**23.** The radius of convergence of the power-series solution of Exercise 18 is at least equal to 1.

**24.** The power-series solution of Exercise 15 converges for all $x$ in the interval $-1 < x < 1$.

In Exercises 25 through 27, compute the Legendre polynomial corresponding to Legendre's equation.

**25.** $(1 - x^2)y'' - 2xy' + 2y = 0$.     **26.** $(1 - x^2)y'' - 2xy' + 6y = 0$.

**27.** $(1 - x^2)y'' - 2xy' + 12y = 0$.

**28.** Verify the recurrence formulas (33).

**29.** Verify the recurrence formulas (36) and (37).

**30.** Verify the recurrence formulas (42) and (43).

In Exercises 31 through 33, compute the Chebyshev polynomial corresponding to Chebyshev's equation.

**31.** $(1 - x^2)y'' - xy' + y = 0$     **32.** $(1 - x^2)y'' - xy' + 4y = 0$

**33.** $(1 - x^2)y'' - xy' + 9y = 0$

In Exercises 34 through 36, compute the Hermite polynomial corresponding to Hermite's equation.

**34.** $y'' - 2xy' + 2y = 0$     **35.** $y'' - 2xy' + 4y = 0$

**36.** $y'' - 2xy' + 6y = 0$

**37.** In an $RLC$-series circuit (see the electric circuit application in Section 2.11.1), assume that $L = 20$ henries, $R = (60 + 20t)$ ohms, $C = 0.05$ farad, and $V(t) = 0$. Compute a recurrence formula that can be used to evaluate approximately the current $I(t)$ in the circuit.

---

## 5.5   SERIES SOLUTIONS ABOUT A REGULAR SINGULAR POINT

In this section we show how to solve any second-order linear differential equation with variable coefficients of the form

$$a_2(x)y'' + a_1(x)y' + a_0(x)y = 0 \tag{1}$$

in a small deleted interval about a regular singular point $x_0$. A deleted interval about $x_0$ is a set of the form $0 < |x - x_0| < R$ for some positive number $R$. This set consists of the interval $|x - x_0| < R$, from which we delete its center $x_0$ (see Figure 5.1).

**Figure 5.1**

Let us recall that when the point $x_0$ is a regular singular point of the differential equation (1), then the functions

$$(x - x_0)\frac{a_1(x)}{a_2(x)} \quad \text{and} \quad (x - x_0)^2 \frac{a_0(x)}{a_2(x)}$$

have power-series expansions of the form

$$(x - x_0)\frac{a_1(x)}{a_2(x)} = \sum_{n=0}^{\infty} A_n(x - x_0)^n \qquad \text{for } |x - x_0| < R_1 \qquad (2)$$

and

$$(x - x_0)^2 \frac{a_0(x)}{a_2(x)} = \sum_{n=0}^{\infty} B_n(x - x_0)^n \qquad \text{for } |x - x_0| < R_2, \qquad (3)$$

with positive radii of convergence $R_1$ and $R_2$. Since the point $x_0$ is a singular point of the differential equation (1), its solutions, in general, are not defined at $x_0$. However, the differential equation (1) has two linearly independent solutions in the deleted interval $0 < |x - x_0| < R$, where $R$ is the smallest of $R_1$ and $R_2$. Our problem in this section is to compute (or approximate) these two solutions near every regular singular point. Before we state a theorem that describes the form of the two linearly independent solutions of the differential equation (1) near a regular singular point, we need the following definition.

**DEFINITION 1**

*Assume that $x_0$ is a regular singular point of the differential equation (1), and assume that the expansions (2) and (3) hold. Then the quadratic equation*

$$\lambda^2 + (A_0 - 1)\lambda + B_0 = 0$$

*is called the indicial equation of (1) at $x_0$.*

**THEOREM 1**  (Solutions near a Regular Singular Point)

*Assume that $x_0$ is a regular singular point of the differential equation (1) and assume that the expansions (2) and (3) hold. Let $\lambda_1$ and $\lambda_2$ be the two roots of the indicial equation*

$$\lambda^2 + (A_0 - 1)\lambda + B_0 = 0, \qquad (4)$$

*indexed in such a way that $\lambda_1 \geq \lambda_2$ in case that both roots are real numbers. Then, one of the solutions of Eq. (1) is of the form*

$$y_1(x) = |x - x_0|^{\lambda_1} \sum_{n=0}^{\infty} a_n(x - x_0)^n, \qquad (5)$$

with $a_0 = 1$, and is valid in the deleted interval $0 < |x - x_0| < R$, where $R$ = min $(R_1, R_2)$. A second linearly independent solution $y_2(x)$ of Eq. (1) in the deleted interval $0 < |x - x_0| < R$ is found as follows.

**CASE 1**   If $\lambda_1 - \lambda_2 \neq integer$, then

$$y_2(x) = |x - x_0|^{\lambda_2} \sum_{n=0}^{\infty} b_n(x - x_0)^n, \tag{6}$$

with $b_0 = 1$.

**CASE 2**   If $\lambda_1 = \lambda_2$, then

$$y_2(x) = y_1(x) \ln|x - x_0| + |x - x_0|^{\lambda_2} \sum_{n=0}^{\infty} b_n(x - x_0)^n, \tag{7}$$

with $b_0 = 1$.

**CASE 3**   If $\lambda_1 = \lambda_2 + (positive\ integer)$, then

$$y_2(x) = Cy_1(x) \ln|x - x_0| + |x - x_0|^{\lambda_2} \sum_{n=0}^{\infty} b_n(x - x_0)^n, \tag{8}$$

with $b_0 = 1$. The constant $C$ is sometimes equal to zero. (See Exercise 29.)

As in the case of ordinary points, the coefficients of the series solutions above can be obtained by direct substitution of the solution into the differential equation and equating coefficients. First we compute the solution (5). Series of the form of Eq. (5) are called *Frobenius series,* and the method of finding such solutions of differential equations is customarily called the *method of Frobenius.* A second solution can be computed from (6), (7), or (8), as the case may be. A second solution can also be computed by using the method of reduction of order described in Section 2.8. The following three examples correspond to Cases 1, 2, and 3 of Theorem 1.

**EXAMPLE 1**   Compute the general solution of the differential equation

$$2x^2y'' + (x - x^2)y' - y = 0 \tag{9}$$

near the point $x_0 = 0$.

**Solution**   Here $a_2(x) = 2x^2$, $a_1(x) = x - x^2$, and $a_0(x) = -1$. Since $a_2(0) = 0$, the point $x_0 = 0$ is a singular point of the differential equation (9). Since

$$(x - x_0)\frac{a_1(x)}{a_2(x)} = x\frac{x - x^2}{2x^2} = \frac{1}{2} - \frac{1}{2}x$$

and

$$(x - x_0)^2\frac{a_0(x)}{a_2(x)} = x^2\frac{-1}{2x^2} = -\frac{1}{2}$$

are analytic functions (with radius of convergence equal to $\infty$), the point $x_0 = 0$ is a regular singular point of the differential equation (9). Here $A_0 = \frac{1}{2}$ and $B_0 = -\frac{1}{2}$, and therefore the indicial equation of the differential equation (9) at the regular singular point 0 is $\lambda^2 + (\frac{1}{2} - 1)\lambda - \frac{1}{2} = 0$, that is,

$$2\lambda^2 - \lambda - 1 = 0.$$

The roots of the indicial equation are $-\frac{1}{2}$ and 1, and we must index them so that $\lambda_1 \geq \lambda_2$, that is,

$$\lambda_1 = 1 \quad \text{and} \quad \lambda_2 = -\frac{1}{2}.$$

By Theorem 1, one solution of the differential equation (9) is of the form

$$y_1(x) = x \sum_{n=0}^{\infty} a_n x^n, \tag{10}$$

with $a_0 = 1$. Since $\lambda_1 - \lambda_2 = \frac{3}{2}$, the two roots of the indicial equation do not differ by an integer, and so a second linearly independent solution $y_2(x)$ is of the form (Case 1)

$$y_2(x) = |x|^{-1/2} \sum_{n=0}^{\infty} b_n x^n, \tag{11}$$

with $b_0 = 1$. Since $R_1 = R_2 = \infty$, it follows that the power series in (10) converges for all $x$. However, the solution (11) is not defined at $x = 0$. It is defined (by Theorem 1) in the deleted interval $0 < |x| < \infty$, that is, for $x < 0$ or $x > 0$. We shall now compute the coefficients of the solutions (10) and (11) by direct substitution into the differential equation (9) and by equating coefficients of the same power of $x$.

We first compute the coefficients $a_n$ of the solution (10). We have

$$y(x) = x \sum_{n=0}^{\infty} a_n x^n = \sum_{n=0}^{\infty} a_n x^{n-1}$$

$$y'(x) = \sum_{n=0}^{\infty} (n+1)a_n x^n \quad \text{and} \quad y''(x) = \sum_{n=0}^{\infty} n(n+1)a_n x^{n-1}$$

$$\Rightarrow 2x^2 y'' = \sum_{n=0}^{\infty} 2n(n+1)a_n x^{n-1}$$

$$xy' = \sum_{n=0}^{\infty} (n+1)a_n x^{n-1}$$

$$-x^2 y' = \sum_{n=0}^{\infty} -(n+1)a_n x^{n-2} = \sum_{n=1}^{\infty} -na_{n-1}x^{n-1}$$

$$-y = \sum_{n=0}^{\infty} -a_n x^{n-1}$$

$$0 = (a_0 - a_0)x + \sum_{n=1}^{\infty} [2n(n+1)a_n + (n+1)a_n - na_{n-1} - a_n]x^{n+1}.$$

The term $(a_0 - a_0)x$ corresponds to the sum of terms for $n = 0$. Equating coefficients to zero, we obtain the recurrence formula

$$2n(n + 1)a_n + (n + 1)a_n - na_{n-1} - a_n = 0, \qquad n = 1, 2, \ldots$$

or

$$a_n = \frac{na_{n-1}}{2n(n + 1) + (n + 1) - 1} = \frac{1}{2n + 3} a_{n-1}, \qquad n = 1, 2, \ldots.$$

We take $a_0 = 1$. Then,

$$\text{for } n = 1: \quad a_1 = \frac{1}{5}$$

$$\text{for } n = 2: \quad a_2 = \frac{1}{7} a_1 = \frac{1}{5 \cdot 7}$$

$$\text{for } n = 3: \quad a_3 = \frac{1}{9} a_2 = \frac{1}{5 \cdot 7 \cdot 9}$$

and, in general,

$$a_n = \frac{1}{5 \cdot 7 \cdots (2n + 3)} \qquad \text{for } n = 1, 2, \ldots.$$

Thus, one of the solutions of the differential equation (9) near $x_0 = 0$ is

$$y_1(x) = x \left( 1 + \frac{x}{5} + \frac{x^2}{5 \cdot 7} + \cdots + \frac{x^n}{5 \cdot 7 \cdots (2n + 3)} + \cdots \right)$$

or

$$y_1(x) = x \left[ 1 + \sum_{n=1}^{\infty} \frac{x^n}{5 \cdot 7 \cdots (2n + 3)} \right]. \tag{12}$$

Next, we compute the coefficients $b_n$ of the solution (11). This solution is defined in the deleted interval $0 < |x|$, that is, for $x > 0$ or $x < 0$. We first assume that $x > 0$. Then

$$y_2(x) = x^{-1/2} \sum_{n=0}^{\infty} b_n x^n = \sum_{n=0}^{\infty} b_n x^{n-(1/2)},$$

$$y_2'(x) = \sum_{n=0}^{\infty} (n - \tfrac{1}{2}) b_n x^{n-(3/2)}$$

and

$$y_2''(x) = \sum_{n=0}^{\infty} (n - \tfrac{1}{2})(n - \tfrac{3}{2})b_n x^{n-(5/2)}.$$

$$\Rightarrow 2x^2 y'' = \sum_{n=0}^{\infty} 2(n - \tfrac{1}{2})(n - \tfrac{3}{2})b_n x^{n-(1/2)}$$

$$xy' = \sum_{n=0}^{\infty} (n - \tfrac{1}{2})b_n x^{n-(1/2)}$$

$$-x^2 y' = \sum_{n=0}^{\infty} -(n - \tfrac{1}{2})b_n x^{n+(1/2)} = \sum_{n=1}^{\infty} -(n - \tfrac{3}{2})b_{n-1} x^{n-(1/2)}$$

$$-y = \sum_{n=0}^{\infty} -b_n x^{n-(1/2)}$$

$$0 = (2 \cdot \tfrac{3}{4}b_0 - \tfrac{1}{2}b_0 - b_0)x^{-1/2}$$

$$+ \sum_{n=1}^{\infty} [2(n - \tfrac{1}{2})(n - \tfrac{3}{2})b_n + (n - \tfrac{1}{2})\, b_n$$

$$- (n - \tfrac{3}{2})b_{n-1} - b_n]x^{n-(1/2)}.$$

As before, the first term on the right-hand side is identically zero. Equating each coefficient of the series to zero, we obtain the recurrence formula

$$2(n - \tfrac{1}{2})(n - \tfrac{3}{2})b_n + (n - \tfrac{1}{2})b_n - (n - \tfrac{3}{2})b_{n-1} - b_n = 0, \qquad n = 1, 2, \ldots$$

or

$$b_n = \frac{(n - \tfrac{3}{2})b_{n-1}}{2(n - \tfrac{1}{2})(n - \tfrac{3}{2}) + (n - \tfrac{1}{2}) - 1} = \frac{1}{2n}b_{n-1}, \qquad n = 1, 2, \ldots.$$

We take $b_0 = 1$. Then

for $n = 1$: $\qquad b_1 = \dfrac{1}{2}$

for $n = 2$: $\qquad b_2 = \dfrac{1}{2 \cdot 2}b_1 = \dfrac{1}{2^2} \cdot \dfrac{1}{2}$

for $n = 3$: $\qquad b_3 = \dfrac{1}{2 \cdot 3}b_2 = \dfrac{1}{2 \cdot 3}\dfrac{1}{2^2 \cdot 2} = \dfrac{1}{2^3 \cdot 3!}$

and, in general,

$$b_n = \frac{1}{2^n \cdot n!} \qquad n = 1, 2, \ldots.$$

Thus,

$$y_2(x) = x^{-1/2} \left( 1 + \frac{x}{2} + \frac{x^2}{2^2 \cdot 2!} + \cdots + \frac{x^n}{2^n \cdot n!} + \cdots \right)$$

$$= x^{-1/2} \sum_{n=0}^{\infty} \frac{x^n}{2^n n!} = x^{-1/2} \sum_{n=0}^{\infty} \frac{(x/2)^n}{n!}.$$

or

$$y_2(x) = x^{-1/2} e^{x/2}. \tag{13}$$

Now we must compute the solution (11) when $x < 0$. To do this, we use the transformation $x = -t$ in the differential equation (9). By the chain rule and using dots for derivatives with respect to $t$, we obtain

$$y' = \frac{dy}{dx} = \frac{dy}{dt} \cdot \frac{dt}{dx} = -\dot{y}$$

and

$$y'' = \frac{d}{dt}(-\dot{y})\frac{dt}{dx} = \ddot{y}.$$

Thus, Eq. (9) becomes

$$2t^2\ddot{y} - (-t - t^2)\,\dot{y} - y = 0. \tag{9'}$$

Since $x < 0$ in Eq. (9), we have $t > 0$ in Eq. (9'). The indicial equation of Eq. (9') is the same as for Eq. (9). Its roots are again

$$\lambda_1 = 1 \quad \text{and} \quad \lambda_2 = -\tfrac{1}{2}.$$

We only have to find the solution corresponding to $\lambda_2 = -\frac{1}{2}$ because for $\lambda_1 = 1$ we have already found a solution. As before, we look for a solution in the form

$$y_2(t) = |t|^{-1/2} \sum_{n=0}^{\infty} b_n t^n.$$

Because $t > 0$ here, we have

$$y_2(t) = t^{-1/2} \sum_{n=0}^{\infty} b_n t^n.$$

Computing the coefficients by direct substitution of $y_2(t)$ into the differential equation (9'), we obtain (the calculations are omitted)

$$y_2(t) = t^{-1/2} e^{-t/2}, \qquad t > 0.$$

But $t = -x$, and so

$$y_2(x) = (-x)^{-1/2} e^{x/2}, \qquad x < 0. \tag{13'}$$

Combining (13) and (13'), we see that for $x > 0$ or $x < 0$, we have

$$y_2(x) = |x|^{-1/2} e^{x/2}. \tag{14}$$

Thus, the general solution of Eq. (9) is

$$y(x) = c_1 x \left[ 1 + \sum_{n=1}^{\infty} \frac{x^n}{5 \cdot 7 \cdots (2n+3)} \right] + c_2 |x|^{-1/2} e^{x/2},$$

where $c_1$ and $c_2$ are arbitrary constants.

It will be very convenient to use the following remark in the sequel.

**REMARK 1**  It can be shown that if

$$y(x) = x^\lambda \sum_{n=0}^{\infty} a_n x^n \qquad (15)$$

is a solution of Eq. (1) for $x > 0$, then

$$y(x) = (-x)^\lambda \sum_{n=0}^{\infty} a_n x^n \qquad (16)$$

is also a solution for $x < 0$. This was observed in the previous example in Eqs. (13) and (13′). Combining (15) and (16), we obtain for $x > 0$ or $x < 0$,

$$y(x) = |x|^\lambda \sum_{n=0}^{\infty} a_n x^n.$$

**EXAMPLE 2**  Compute the general solution of the differential equation

$$(x-1)^2 y'' - (x^2 - x)y' + y = 0 \qquad (17)$$

near the point $x_0 = 1$.

**Solution**  Since computations about the point zero are simpler, we set $t = x - 1$ and find the general solution of the resulting equation near 0. Since $t = x - 1$, we obtain $x = t + 1$, $y' = \dot{y}$, and $y'' = \ddot{y}$. Thus, Eq. (17) becomes

$$t^2 \ddot{y} - (t^2 + t)\dot{y} + y = 0. \qquad (18)$$

The point $t_0 = 0$ is a singular point of Eq. (18), and because

$$t \frac{a_1(t)}{a_2(t)} = t \frac{-(t^2 + t)}{t^2} = -1 - t$$

and

$$t^2 \frac{a_0(t)}{a_2(t)} = t^2 \frac{1}{t^2} = 1,$$

it follows that $t_0 = 0$ is a regular singular point. Here $A_0 = -1$ and $B_0 = 1$. Therefore, the indicial equation is

$$\lambda^2 - 2\lambda + 1 = 0 \qquad \text{and} \qquad \lambda_1 = \lambda_2 = 1.$$

By Theorem 1, one solution of the differential equation (17) is of the form

$$y_1(t) = t \sum_{n=0}^{\infty} a_n t^n, \tag{19}$$

with $a_0 = 1$. Since $\lambda_1 - \lambda_2 = 0$, the two roots of the indicial equation are equal, and so a second linearly independent solution $y_2(t)$ is of the form (Case 2)

$$y_2(t) = y_1(t) \ln |t| + |t| \sum_{n=0}^{\infty} b_n t^n. \tag{20}$$

We first compute the coefficients $a_n$ of the solution (19). By direct substitution of (19) into (18) and equating coefficients, the reader can verify that

$$a_n = \frac{1}{n!} a_0, \quad n = 1, 2, \ldots .$$

Thus, taking $a_0 = 1$, we find that

$$y_1(t) = t \left( 1 + \frac{t}{1!} + \frac{t^2}{2!} + \cdots + \frac{t^n}{n!} + \cdots \right) = t \sum_{n=0}^{\infty} \frac{t^n}{n!} \tag{21}$$

or

$$y_1(t) = te^t. \tag{22}$$

Next, we compute the coefficients $b_n$ of the solution (20). In Eq. (20) we substitute $y_1(t)$ from Eq. (21), and for $t > 0$, we obtain

$$y_2(t) = t \left( \sum_{n=0}^{\infty} \frac{t^n}{n!} \right) \ln t + t \sum_{n=0}^{\infty} b_n t^n = \left( \sum_{n=0}^{\infty} \frac{t^{n+1}}{n!} \right) \ln t + \sum_{n=0}^{\infty} b_n t^{n+1}$$

$$\dot{y}_2(t) = \left( \sum_{n=0}^{\infty} \frac{n+1}{n!} t^n \right) \ln t + \sum_{n=0}^{\infty} \frac{t^n}{n!} + \sum_{n=0}^{\infty} (n+1) b_n t^n$$

$$\ddot{y}_2(t) = \left( \sum_{n=0}^{\infty} \frac{n(n+1)}{n!} t^{n-1} \right) \ln t + \sum_{n=0}^{\infty} \frac{n+1}{n!} t^{n-1} + \sum_{n=0}^{\infty} \frac{n}{n!} t^{n-1}$$

$$+ \sum_{n=0}^{\infty} (n+1) n b_n t^{n-1}$$

$$\Rightarrow \quad t^2 \ddot{y}_2 = \left( \sum_{n=0}^{\infty} \frac{n(n+1)}{n!} t^{n+1} \right) \ln t + \sum_{n=0}^{\infty} \frac{n+1}{n!} t^{n+1} + \sum_{n=0}^{\infty} \frac{n}{n!} t^{n+1}$$

$$+ \sum_{n=0}^{\infty} (n+1) n b_n t^{n+1}$$

$$-t^2 \dot{y}_2 = \left( \sum_{n=0}^{\infty} -\frac{n+1}{n!} t^{n+2} \right) \ln t + \sum_{n=0}^{\infty} -\frac{1}{n!} t^{n+1}$$

$$+ \sum_{n=0}^{\infty} -(n+1) b_n t^{n+2}$$

$$-t\dot{y}_2 = \left( \sum_{n=0}^{\infty} -\frac{n+1}{n!} t^{n+1} \right) \ln t + \sum_{n=0}^{\infty} -\frac{1}{n!} t^{n+1}$$

$$+ \sum_{n=0}^{\infty} -(n+1)b_n t^{n+1}$$

$$y_2 = \left( \sum_{n=0}^{\infty} \frac{1}{n!} t^{n+1} \right) \ln t + \sum_{n=0}^{\infty} b_n t^{n+1}.$$

The sum of the coefficients of $\ln t$ is easily seen to be equal to zero. (This is the case in all examples when we are in Case 2.) Also, as always, the sum of all terms corresponding to $n = 0$ is zero. Thus, equating coefficients to zero, we obtain

$$\frac{n+1}{n!} + \frac{n}{n!} - \frac{1}{(n-1)!} - \frac{1}{n!} + (n+1)nb_n - nb_{n-1} - (n+1)b_n + b_n = 0,$$

$$n = 1, 2, \ldots .$$

Simplifying, we obtain

$$\frac{n}{n!} + n^2 b_n - nb_{n-1} = 0$$

or

$$b_n = \frac{1}{n} b_{n-1} - \frac{1}{n \cdot n!}, \qquad n = 1, 2, \ldots .$$

Thus, with $b_0 = 1$, we obtain

$$b_1 = \frac{1}{1} - \frac{1}{1 \cdot 1!} = 0, \qquad b_2 = -\frac{1}{2 \cdot 2!} = -\frac{1}{4},$$

$$b_3 = \frac{1}{3} b_2 - \frac{1}{3 \cdot 3!} = -\frac{5}{36}, \qquad b_4 = \frac{1}{4} b_3 - \frac{1}{4 \cdot 4!} = -\frac{13}{288},$$

and so on. Hence,

$$y_2(t) = \left( t \sum_{n=0}^{\infty} \frac{t^n}{n!} \right) \ln t + t \left( 1 - \frac{t^2}{4} - \frac{5t^3}{36} - \frac{13t^4}{288} + \cdots \right)$$

$$= te^t \ln t + t - \frac{t^3}{4} - \frac{5t^4}{36} - \frac{13t^5}{288} - \cdots . \qquad (23)$$

Equations (22) and (23) gives two linearly independent solutions of Eq. (18) for $t > 0$. Setting $t = x - 1$ in these equations, we get two linearly independent solutions of Eq. (17) near $x_0 = 1$. The general solution of Eq. (17) is (using Remark 1)

$$y(x) = c_1(x - 1)e^{x-1} + c_2 \left[ (x - 1)e^{x-1} \ln |x - 1| + (x - 1) - \frac{(x-1)^3}{4} - \cdots \right].$$

**EXAMPLE 3**  Compute two linearly independent solutions of the differential equation

$$x^2y'' - (x + 2)y = 0 \tag{24}$$

near the point $x_0 = 0$.

**Solution**  Here $a_2(x) = x^2$, $a_1(x) = 0$, and $a_0(x) = -(x + 2)$. The point $x_0 = 0$ is a singular point of the differential equation (24), because $a_2(0) = 0$. Since

$$(x - x_0)\frac{a_1(x)}{a_2(x)} = 0$$

and

$$(x - x_0)^2\frac{a_0(x)}{a_2(x)} = x^2\frac{-(x + 2)}{x^2} = -2 - x,$$

the point $x_0 = 0$ is a regular singular point. The indicial equation is (here $A_0 = 0$ and $B_0 = -2$) $\lambda^2 - \lambda - 2 = 0$. Its roots are

$$\lambda_1 = 2 \quad \text{and} \quad \lambda_2 = -1.$$

By Theorem 1, one solution of the differential equation (24) is of the form

$$y_1(x) = x^2 \sum_{n=0}^{\infty} a_n x^n, \tag{25}$$

with $a_0 = 1$. Since $\lambda_1 - \lambda_2 = +3$, the two roots of the indicial equation differ by a positive integer, and so a second linearly independent solution $y_2(x)$ is of the form (Case 3)

$$y_2(x) = Cy_1(x) \ln |x| + x^{-1} \sum_{n=0}^{\infty} b_n x^n, \tag{26}$$

with $b_0 = 1$ and $C$ possibly equal to zero.

We first compute the coefficients $a_n$ of the solution (25). We have

$$y_1(x) = x^2 \sum_{n=0}^{\infty} a_n x^n = \sum_{n=0}^{\infty} a_n x^{n+2}$$

$$y_1'(x) = \sum_{n=0}^{\infty} (n + 2)a_n x^{n+1} \quad \text{and} \quad y_1''(x) = \sum_{n=0}^{\infty} (n + 2)(n + 1)a_n x^n$$

$$\Rightarrow x^2 y_1'' = \sum_{n=0}^{\infty} (n + 2)(n + 1)a_n x^{n+2}$$

$$-xy_1 = \sum_{n=0}^{\infty} -a_n x^{n+3} = \sum_{n=1}^{\infty} -a_{n-1} x^{n+2}$$

$$-2y_1 = \sum_{n=0}^{\infty} -2a_n x^{n+2}$$

$$0 = (2a_0 - 2a_0)x^2 + \sum_{n=1}^{\infty} [(n + 2)(n + 1)a_n - a_{n-1} - 2a_n]x^{n+2}.$$

Equating coefficients to zero, we obtain the recurrence formula

$$(n + 2)(n + 1)a_n - a_{n-1} - 2a_n = 0, \qquad\qquad n = 1, 2, \ldots$$

or

$$a_n = \frac{a_{n-1}}{(n + 2)(n + 1) - 2} = \frac{1}{n(n + 3)} a_{n-1}, \qquad n = 1, 2, \ldots.$$

We take $a_0 = 1$. Then $a_1 = \frac{1}{4}, a_2 = \frac{1}{40}, \ldots$. Therefore,

$$y_1(x) = x^2 + \frac{x^3}{4} + \frac{x^4}{40} + \cdots.$$

Next, we compute the coefficients $b_n$ of the solution (26). We do it for $x > 0$. Substituting $y_1$ in (26), we obtain

$$y_2(x) = C\left(x^2 + \frac{x^3}{4} + \frac{x^4}{40} + \cdots\right)\ln x + \sum_{n=0}^{\infty} b_n x^{n-1}$$

$$y_2'(x) = C\left(2x + \frac{3}{4}x^2 + \frac{1}{10}x^3 + \cdots\right)\ln x$$

$$+ C\left(x + \frac{1}{4}x^2 + \frac{1}{40}x^3 + \cdots\right) + \sum_{n=0}^{\infty} (n - 1)b_n x^{n-2}$$

$$y_2''(x) = C\left(2 + \frac{3}{2}x + \frac{3}{10}x^2 + \cdots\right)\ln x + C\left(2 + \frac{3}{4}x + \frac{1}{10}x^2 + \cdots\right)$$

$$+ C\left(1 + \frac{1}{2}x + \frac{3}{40}x^2 + \cdots\right) + \sum_{n=0}^{\infty} (n - 1)(n - 2)b_n x^{n-3}$$

$$\Rightarrow \quad x^2 y_2'' = C\left(2x^2 + \frac{3}{2}x^3 + \cdots\right)\ln x + C\left(3x^2 + \frac{5}{4}x^3 + \cdots\right)$$

$$+ \left(2b_0\frac{1}{x} + 2b_3 x^2 + \cdots\right)$$

$$-xy_2 = C\left(-x^3 - \frac{1}{4}x^4 - \cdots\right)\ln x + (-b_0 - b_1 x - b_2 x^2 - \cdots)$$

$$-2y_2 = C\left(-2x^2 - \frac{1}{2}x^3 - \cdots\right)\ln x$$

$$+ \left(-2b_0\frac{1}{x} - 2b_1 - 2b_2 x - 2b_3 x^2 - \cdots\right)$$

---

$$0 = C\left(3x^2 + \frac{5}{4}x^3 + \cdots\right)$$

$$+ [(-b_0 - 2b_1) + (-b_1 - 2b_2)x - b_2 x^2 - \cdots]$$

or

$$(-b_0 - 2b_1) + (-b_1 - 2b_2)x + (3C - b_2)x^2 + \cdots = 0.$$

Taking $b_0 = 1$ and equating coefficients to zero, we obtain

$$b_1 = -\tfrac{1}{2}, \qquad b_2 = \tfrac{1}{4}, \qquad C = \tfrac{1}{12}, \cdots .$$

Therefore,

$$y_2(x) = \frac{1}{12}\left(x^2 + \frac{x^3}{4} + \frac{x^4}{40} + \cdots\right)\ln|x| + \left(\frac{1}{x} - \frac{1}{2} + \frac{1}{4}x + \cdots\right).$$

### APPLICATIONS 5.5.1

The method of Frobenius is a powerful technique for obtaining solutions of certain differential equations which occur in applications.

**Bessel's Equation**

■ The differential equation

$$x^2y'' + xy' + (x^2 - p^2)y = 0, \tag{27}$$

where $p$ is a constant, is *Bessel's equation* of *order p*. The solutions of Bessel's equation, called *Bessel functions,* are very important in applied mathematics and especially in mathematical physics. Before we study the solutions of Eq. (27) we show briefly how Bessel's equation appears in a specific application. The important method of separation of variables in solving partial differential equations is also mentioned in the process.

**Temperature Distribution in a Cylinder**

■ If we know the temperature distribution in a cylinder at time $t = 0$, then it is proved in physics that the temperature $u = u(r, \theta, t)$ at the point $(r, \theta)$ at any time $t$ satisfies the following partial differential equation (in polar coordinates):

$$u_{rr} + \frac{1}{r}u_r + \frac{1}{r^2}u_{\theta\theta} = \frac{1}{k}u_t. \tag{28}$$

Here we assume that the temperature is independent of the height of the cylinder. The quantity $k$ in Eq. (28) is a constant that depends on the thermal conductivity and, in general, the materials used in making the cylinder. We now employ the method of *separation of variables* to solve the partial differential equation (28). According to this method, we assume that Eq. (28) has a solution $u(r, \theta, t)$ which is a product of a function of $r$, a function of $\theta$, and a function of $t$. That is,

$$u(r, \theta, t) = R(r)\Theta(\theta)T(t), \tag{29}$$

where the functions $R$, $\Theta$, and $T$ are to be determined. As we will see below, the function $R$ will satisfy Bessel's equation (27). Substituting (29) in (28) and suppressing the arguments $r$, $\theta$, and $t$, we obtain

$$R''\Theta T + \frac{1}{r}R'\Theta T + \frac{1}{r^2}R\Theta''T = \frac{1}{k}R\Theta T'. \tag{30}$$

The crucial step in the method of separation of variables is to be able to rewrite Eq. (30) in such a way that the variables separate. In fact, we can write Eq. (30) as follows:

$$r^2 \frac{R''}{R} + r \frac{R'}{R} - \frac{r^2}{k} \frac{T'}{T} = -\frac{\Theta''}{\Theta}. \tag{31}$$

The left side of Eq. (31) is a function independent of $\theta$, while the right side is a function of $\theta$ alone. The only way that this can happen is if both sides of Eq. (31) are equal to a constant, say $p^2$. Then we obtain the equations

$$\Theta'' + p^2 \Theta = 0 \tag{32}$$

and

$$r^2 \frac{R''}{R} + r \frac{R'}{R} - \frac{r^2}{k} \frac{T'}{T} = p^2$$

or

$$\frac{R''}{R} + \frac{1}{r} \frac{R'}{R} - \frac{p^2}{r^2} = \frac{1}{k} \frac{T'}{T} \tag{33}$$

Now the left side of Eq. (33) is a function of $r$ alone, while the right side is a function of $t$. Hence, both sides of Eq. (33) are equal to a constant, say $-\lambda^2$. Thus,

$$T' + \lambda^2 k T = 0 \tag{34}$$

and

$$r^2 R'' + r R' + (\lambda^2 r^2 - p^2) R = 0. \tag{35}$$

The functions $\Theta(\theta)$ and $T(t)$ of the solution (29) are easily found by solving the simple differential equations (32) and (34). To find the function $R(r)$ of the solution (29), we must also solve Eq. (35). If we make the transformation $x = \lambda r$ and set $y(x) = R(r)$, we obtain

$$R' = \frac{dR}{dx} \cdot \frac{dx}{dr} = \lambda y' = \frac{x}{r} y' \quad \text{and} \quad R'' = \frac{d}{dx}(\lambda y') \frac{dx}{dr} = \lambda^2 y'' = \frac{x^2}{r^2} y''.$$

Substituting $R'$ and $R''$ in Eq. (35), we find Bessel's equation (27).

We now return to find solutions of Eq. (27) near the point $x_0 = 0$. Although $p$ could be a complex number, we assume for simplicity that $p \geq 0$. We have

$$a_2(x) = x^2, \quad a_1(x) = x, \quad a_0(x) = x^2 - p^2.$$

Since $a_2(0) = 0$, the point $x_0 = 0$ is a singular point. But

$$x \frac{a_1(x)}{a_2(x)} = 1 \quad \text{and} \quad x^2 \frac{a_0(x)}{a_2(x)} = -p^2 + x^2, \tag{36}$$

and so $x_0 = 0$ is a regular singular point with indicial equation $\lambda^2 - p^2 = 0$. The two roots of the indicial equation are

$$\lambda_1 = p \quad \text{and} \quad \lambda_2 = -p.$$

Hence, one solution of Bessel's equation is of the form

$$y_1(x) = |x|^p \sum_{n=0}^{\infty} a_n x^n \tag{37}$$

From Eqs. (36) and Theorem 1 the solutions of Eq. (27) near 0 are valid for $|x| > 0$, that is, for $x > 0$ or $x < 0$. The form of a second linearly independent solution $y_2(x)$ of Eq. (27) depends on the value of $\lambda_1 - \lambda_2 = 2p$ in accordance with Cases 1, 2, and 3 of Theorem 1. More precisely, $y_2(x)$ is given as follows.

**CASE 1** If $2p \neq$ integer, then

$$y_2(x) = |x|^{-p} \sum_{n=0}^{\infty} b_n x^n \tag{38}$$

In fact, this solution can be obtained from (37), replacing $p$ by $-p$.

**CASE 2** If $p = 0$, then

$$y_2(x) = y_1(x) \ln |x| + \sum_{n=0}^{\infty} b_n x^n. \tag{39}$$

**CASE 3** If $2p =$ positive integer, then

$$y_2(x) = Cy_1(x) \ln |x| + |x|^{-p} \sum_{n=0}^{\infty} b_n x^n. \tag{40}$$

Let us compute the coefficients $a_n$ of the solution (37). We have (for $x > 0$)

$$y_1(x) = \sum_{n=0}^{\infty} a_n x^{n+p}, \quad y_1'(x) = \sum_{n=0}^{\infty} (n+p)a_n x^{n+p-1},$$

and

$$y_1''(x) = \sum_{n=0}^{\infty} (n+p)(n+p-1)a_n x^{n+p-2}$$

$$\Rightarrow x^2 y_1'' = \sum_{n=0}^{\infty} (n+p)(n+p-1)a_n x^{n+p}$$

$$xy_1' = \sum_{n=0}^{\infty} (n+p)a_n x^{n+p}$$

$$x^2 y_1 = \sum_{n=0}^{\infty} a_n x^{n+p+2} = \sum_{n=2}^{\infty} a_{n-2} x^{n+p}$$

$$-p^2 y_1 = \sum_{n=0}^{\infty} - p^2 a_n x^{n+p}$$

$$0 = [p(p-1)a_0 + pa_0 - p^2a_0]x^p + [(1+p)pa_1 + (1+p)a_1 - p^2a_1]x^{1+p}$$
$$+ \sum_{n=2}^{\infty} [(n+p)(n+p-1)a_n + (n+p)a_n + a_{n-2} - p^2a_n)]x^{n+p}$$

$$\Rightarrow (1 + 2p)a_1 = 0 \quad \text{and}$$
$$(n+p)(n+p-1)a_n + (n+p)a_n + a_{n-2} - p^2a_n = 0, \quad n = 2, 3, \ldots$$

$$\Rightarrow a_1 = 0 \quad \text{and} \quad a_n = -\frac{1}{n(n+2p)}a_{n-2}, \quad n = 2, 3, \ldots$$

$$\Rightarrow a_1 = a_3 = a_5 = \cdots = 0 \quad \text{and}$$

$$a_{2n} = (-1)^n \frac{1}{2^{2n}n!(p+1)(p+2)\cdots(p+n)}a_0, \quad n = 1, 2, \ldots.$$

Hence, the solution (37) is given by

$$y_1(x) = a_0 |x|^p \left[1 + \sum_{n=1}^{\infty} \frac{(-1)^n}{2^{2n}n!(p+1)(p+2)\cdots(p+n)}x^{2n}\right], \quad (41)$$

and the series converges for all $x$. When $2p \neq$ integer, a second linearly independent solution of Eq. (27) can be found if we replace $p$ by $-p$ in Eq. (41). Thus,

$$y_2(x) = a_0 |x|^{-p} \left[1 + \sum_{n=1}^{\infty} \frac{(-1)^n}{2^{2n}n!(-p+1)(-p+2)\cdots(-p+n)}x^{2n}\right], \quad (42)$$

and the series converges for $x > 0$ or $x < 0$.

In the theory of Bessel functions the constant $a_0$ in the solution (41) is taken equal to

$$a_0 = \frac{1}{2^p \Gamma(p+1)}, \quad (43)$$

where $\Gamma$ is the *gamma function*, that is the function defined by the integral

$$\Gamma(p) = \int_0^\infty x^{p-1}e^{-x}dx. \quad (44)$$

With this choice of constant $a_0$, the solution (41) is called a *Bessel function of the first kind of order p* and is denoted by $J_p(x)$. If we use the identity

$$\Gamma(n+p+1) = (n+p)(n+p-1)\cdots(p+2)(p+1)\Gamma(p+1), \quad (45)$$

we obtain the formula

$$J_p(x) = \sum_{n=0}^{\infty} \frac{(-1)^n}{n!\Gamma(n+p+1)}\left(\frac{x}{2}\right)^{2n+p}. \quad (46)$$

The solution (42) with $a_0 = 1/2^{-p}\Gamma(-p + 1)$ is also a Bessel function of the first kind but of order $-p$ and is denoted by $J_{-p}(x)$. Clearly,

$$J_{-p}(x) = \sum_{n=0}^{\infty} \frac{(-1)^n}{n!\Gamma(n - p + 1)}\left(\frac{x}{2}\right)^{2n-p}. \tag{47}$$

Thus, $J_p(x)$ is always a solution of Bessel's equation (27). Furthermore, when $2p$ is not an integer, the functions $J_p(x)$ and $J_{-p}(x)$ are linearly independent solutions of Bessel's equation.

Setting $p = 0$ in Eq. (46) and observing that $\Gamma(n + 1) = n!$, we obtain

$$J_0(x) = \sum_{n=0}^{\infty} \frac{(-1)^n}{(n!)^2}\left(\frac{x}{2}\right)^{2n}. \tag{48}$$

The function $J_0(x)$ is a solution of Bessel's equation (27) for $p = 0$.

When $p = 0$ or when $2p$ is a positive integer, a second linearly independent solution of Bessel's equation is of the form (39) or (40), respectively, and can be obtained by direct substitution of the appropriate series into the differential equation. Such solutions of Eq. (27) with an appropriate choice of the constant $b_0$ are known as *Bessel functions of the second kind*.

**Laguerre's
Equation**

■ The differential equation

$$xy'' + (1 - x)y' + py = 0, \tag{49}$$

where $p$ is a constant, is *Laguerre's equation*. As we will see below, if the constant $p$ is a nonnegative integer, one of the solutions of the differential equation (49), near the point $x_0 = 0$, is a polynomial. When suitably normalized, these polynomial solutions are called *Laguerre polynomials*. The Laguerre polynomials are useful in the quantum mechanics of the hydrogen atom.

The point $x_0 = 0$ is a regular singular point of the differential equation (49) with indicial equation $\lambda^2 = 0$. Since $\lambda_1 = \lambda_2 = 0$, the differential equation has a solution of the form

$$y_1(x) = \sum_{n=0}^{\infty} a_n x^n \qquad \text{for } -\infty < x < \infty.$$

Choosing $a_0 = 1$, the reader can verify that

$$a_n = (-1)^n \frac{p(p - 1) \cdots (p - n + 1)}{(n!)^2} \qquad \text{for } n = 1, 2, \ldots. \tag{50}$$

Thus, one solution of Laguerre's equation is

$$y_1(x) = 1 + \sum_{n=1}^{\infty} (-1)^n \frac{p(p - 1) \cdots (p - n + 1)}{(n!)^2} x^n \tag{51}$$

and converges for all $x$.

From Eq. (50) we see that if $p$ is a nonnegative integer $k$, the coefficients $a_n$ vanish for $n \geq k + 1$. In this case the solution (51) is a polynomial of degree $k$. This polynomial multiplied by $k!$ is denoted by $L_k(x)$ and is called a *Laguerre polynomial*. For example, $1, 1 - x, 2 - 4x + x^2$, and $6 - 18x + 9x^2 - x^3$ are the Laguerre polynomials $L_0(x)$, $L_1(x)$, $L_2(x)$, and $L_3(x)$, respectively.

■ The differential equation

$$x(1 - x)y'' + [c - (a + b + 1)x]y' - aby = 0, \qquad (52)$$

where $a$, $b$, and $c$ are constants, is *Gauss's hypergeometric equation* or simply the hypergeometric equation. Equation (52) is of great theoretical and practical importance because many second-order linear differential equations are reducible to it and because many important special functions are closely related to its solutions.

Let us assume that $c$ is not equal to an integer. The indicial equation of (52) near the regular singular point $x_0 = 0$ is

$$\lambda^2 + (c - 1)\lambda = 0.$$

The two roots of the indicial equation are $0$ and $1 - c$. Since $c$ is not an integer, the two roots do not differ by an integer. Thus, by Theorem 1, Eq. (52) has two linearly independent solutions of the form

$$y_1(x) = \sum_{n=0}^{\infty} a_n x^n \quad \text{and} \quad y_2(x) = |x|^{1-c} \sum_{n=0}^{\infty} b_n x^n.$$

The reader can verify that

$$a_n = \frac{a(a + 1) \cdots (a + n - 1)b(b + 1) \cdots (b + n - 1)}{n!c(c + 1) \cdots (c + n - 1)} a_0,$$

$$n = 1, 2, 3, \ldots. \qquad (53)$$

Hence, choosing $a_0 = 1$, we obtain the solution

$$y_1(x) = 1 + \frac{ab}{1 \cdot c} x + \frac{a(a + 1)b(b + 1)}{1 \cdot 2c(c + 1)} x^2 + \cdots, \quad |x| < 1. \qquad (54)$$

The series solution (54) is called the *hypergeometric series* and is denoted by $F(a, b, c; x)$. Thus,

$$F(a, b, c; x) = 1 + \sum_{n=1}^{\infty} \frac{a(a + 1) \cdots (a + n - 1)b(b + 1) \cdots (b + n - 1)}{c(c + 1) \cdots (c + n - 1)} \frac{x^n}{n!}$$

$$(55)$$

and converges in the interval $|x| < 1$. It is important to note that many functions can be obtained from the hypergeometric functions for various values of the constants $a$, $b$, and $c$. For example,

$$F(a, b, c; 0) = 1$$

$$\lim_{a \to \infty} F\left(a, b, b; \frac{x}{a}\right) = e^x$$

$$F(-a, b, b; -x) = (1 + x)^a$$

$F(k + 1, -k, 1; x)$ is a polynomial of degree $k$ for $k$ a nonnegative integer

$$xF(1, 1, 2; -x) = \ln(1 + x).$$

**EXERCISES**

In Exercises 1 through 11, find the form of two linearly independent solutions near the point $x_0 = 0$.

**1.** $x^2y'' + xy' + (x^2 - \frac{1}{9})y = 0$     **2.** $x^2y'' + xy' + (x^2 - 4)y = 0$

**3.** $x^2y'' + xy' + x^2y = 0$     **4.** $x(1 - x)y'' + (\frac{1}{2} - 3x)y' - y = 0$

**5.** $xy'' + (1 - x)y' + y = 0$     **6.** $x^2y'' + 4x(1 - x)y' + 2y = 0$

**7.** $x^2y'' + x(1 - x)y' - \frac{1}{16}y = 0$     **8.** $x^2y'' + 3x(1 - x)y' + y = 0$

**9.** $xy'' + (1 - x)y' + 2y = 0$     **10.** $x^2y'' + xy' + (x^2 - 1)y = 0$

**11.** $xy'' + 3y' + 4x^3y = 0$

In Exercises 12 through 17, compute a nontrivial solution, free of logarithms, near the point $x_0 = 0$.

**12.** $x^2y'' + xy' + (x^2 - \frac{1}{9})y = 0$     **13.** $x^2y'' + xy' + (x^2 - 4)y = 0$

**14.** $x^2y'' + xy' + x^2y = 0$     **15.** $x(1 - x)y'' + (\frac{1}{2} - 3x)y' - y = 0$

**16.** $xy'' + (1 - x)y' + y = 0$     **17.** $xy'' + (1 - x)y' + 2y = 0$

In Exercises 18 and 19, compute two linearly independent solutions near the point $x_0 = 0$.

**18.** $x^2y'' + xy' + (x^2 - \frac{1}{9})y = 0$     **19.** $xy'' + (1 - x)y' + y = 0$

In Exercises 20 through 28, answer true or false.

**20.** The indicial equation of the differential equation

$$x^2y'' + xy' + (x^2 - p^2)y = 0$$

at $x_0 = 0$ is

$$\lambda^2 - p^2 = 0.$$

**21.** The indicial equation of the differential equation

$$xy'' + (1 - x)y' + 3y = 0$$

at $x_0 = 0$ is

$$\lambda^2 = 0.$$

**22.** The differential equation

$$xy'' + (1 - x)y' + 3y = 0$$

has two linearly independent solutions near $x_0 = 0$ of the form

$$y_1(x) = \sum_{n=0}^{\infty} a_n x^n \quad \text{and} \quad y_2(x) = \sum_{n=0}^{\infty} b_n x^n.$$

**23.** One of the solutions of the differential equation

$$x^2 y'' + xy' + (x^2 - \tfrac{1}{16})y = 0$$

near the point $x_0 = 0$ involves a logarithm.

**24.** The indicial equation of the differential equation

$$x^2 y'' - (x^2 + x)y' + y = 0$$

at $x_0 = 0$ is

$$\lambda^2 - 2\lambda + 1 = 0.$$

**25.** The indicial equation of the differential equation

$$x^2 y'' - (x^3 + x^2 + x)y' + (4x + 1)y = 0$$

at $x_0 = 0$ is

$$\lambda^2 - 2\lambda + 1 = 0.$$

**26.** The differential equation

$$x^2 y'' + xy' + (x^2 - \tfrac{1}{9})y = 0$$

has two linearly independent solutions near $x = 1$ of the form

$$\sum_{n=0}^{\infty} a_n(x - 1)^n.$$

**27.** The differential equation

$$x(1 - x)y'' + (\tfrac{2}{3} - 3x)y' - y = 0$$

has two linearly independent solutions near $x_0 = 0$ of the form

$$y_1(x) = \sum_{n=0}^{\infty} a_n x^n, \qquad y_2(x) = y_1(x) \ln |x| + \sum_{n=0}^{\infty} b_n x^n.$$

**28.** The differential equation

$$(x - 1)y'' + (2 - x)y' + y = 0$$

has two linearly independent solutions near $x = 1$ of the form

$$y_1(x) = \sum_{n=0}^{\infty} a_n(x - 1)^n, \qquad y_2(x) = y_1(x) \ln |x - 1| + \sum_{n=0}^{\infty} b_n(x - 1)^n.$$

**29.** Show that the point $x_0 = 0$ is a regular singular point for the differential equation

$$xy'' - y' + 4x^3 y = 0.$$

In this case, the roots of the indicial equation differ by an integer. Find the two linearly independent solutions and in the process verify that $C = 0$ in formula (8).

In Exercises 30 through 33, obtain a recurrence formula for the coefficients of the Frobenius-series solution [Eq. (5)] near the indicated point $x_0$.

**30.** $(x + 1)^2 y'' - (x + 3)y = 0$, near $x_0 = -1$

**31.** $(x - 1)^2 y'' - (x + 1)y = 0$, near $x_0 = 1$

**32.** $x^2 y'' - (x^2 + x)y' + y = 0$, near $x_0 = 0$

**33.** $2(x + 3)^2 y'' - (x^2 + 5x + 6)y' - y = 0$, near $x_0 = -3$

**34.** Verify the recurrence formula (50).

**35.** Verify the recurrence formula (53).

In Exercises 36 through 38, compute the Bessel functions of the first kind corresponding to Bessel's equation.

**36.** $x^2 y'' + xy' + (x^2 - \frac{1}{9})y = 0$          **37.** $x^2 y'' + xy' + x^2 y = 0$

**38.** $x^2 y'' + xy' + (x^2 - \frac{1}{4})y = 0$

In Exercises 39 through 41, compute the Laguerre polynomial corresponding to Laguerre's equation.

**39.** $xy'' + (1 - x)y' + y = 0$          **40.** $xy'' + (1 - x)y' + 2y = 0$

**41.** $xy'' - (x - 1)y' + 3y = 0$

In Exercises 42 and 43, compute the hypergeometric series corresponding to Gauss's equation.

**42.** $x(1 - x)y'' + (\frac{1}{2} - 3x)y' - y = 0$

**43.** $x(1 - x)y'' + (\frac{3}{4} - 4x)y' - 2y = 0$

**44.** Show that any differential equation of the form

$$(x - A)(x - B)y'' + (Cx + D)y' + Ey = 0,$$

where $A$, $B$, $C$, and $D$ are constants and $A \neq B$ can be transformed into Gauss's equation (52) by means of the transformation $x = A + (B - A)t$.

In Exercises 45 and 46, reduce the given differential equation to Gauss's hypergeometric equation by means of the transformation described in Exercise 44.

**45.** $(x - 1)(x + 2)y'' + (x + \frac{1}{2})y' + 2y = 0$

**46.** $(x^2 - \frac{1}{4})y'' + 2y' - 6y = 0$

## REVIEW EXERCISES

In Exercises 1 through 4, solve the IVP by the method of power series about the indicated initial point.

**1.** $y'' - xy = 0$
$\quad y(0) = -1$
$\quad y'(0) = 0$

**2.** $y'' - 2xy' + 6y = 0$
$\quad y(0) = 0$
$\quad y'(0) = -12$

**3.** $(1 - x^2)y'' - xy' + 9y = 0$
$y(0) = 0$
$y'(0) = -3$

**4.** $(1 - x^2)y'' - 2xy' + 12y = 0$
$y(0) = 0$
$y'(0) = -3$

In Exercises 5 through 8, compute the first four coefficients of the power-series solution about the initial point.

**5.** $x^2y'' + xy' + x^2y = 0$
$y(1) = 1$
$y'(1) = 0$

**6.** $(1 - x^2)y'' + xy' + y = 0$
$y(6) = 4$
$y'(6) = -1$

**7.** $(1 - x^2)y'' - xy' + y = 0$
$y(-6) = 4$
$y'(-6) = 1$

**8.** $xy'' + (1 - x)y' + 2y = 0$
$y(2) = 0$
$y'(2) = 0$

In Exercises 9 through 14, find the form of two linearly independent solutions about or near the indicated point $x_0$. What can you say about the radius of convergence of the solutions without solving the equation?

**9.** $y'' - xy = 0; x_0 = 0$

**10.** $x^2y'' + xy' + (x^2 - \frac{1}{4})y = 0; x_0 = 0$

**11.** $x^2y'' + xy' + (x^2 - \frac{1}{4})y = 0; x_0 = 1$

**12.** $(1 - x^2)y'' - xy' + 16y = 0; x_0 = 0$

**13.** $(1 - x^2)y'' - xy' + 16y = 0; x_0 = -1$

**14.** $x(3 - x)y'' + (1 - 3x)y' - y = 0; x_0 = 8$

In Exercises 15 through 18, compute the Frobenius-series solution near the point $x_0 = 0$.

**15.** $x^2y'' + xy' + (x^2 - 9)y = 0$

**16.** $x^2y'' + xy' + (x^2 - \frac{1}{4})y = 0$

**17.** $xy'' + (1 - x)y' + 3y = 0$

**18.** $x(1 - x)y'' + (\frac{1}{3} - 2x)y' + 2y = 0$

**19.** Derive the Taylor series for the function $\cos x$ by solving the IVP

$$y'' + y = 0, \qquad y(0) = 1, \qquad y'(0) = 0$$

by the method of power series.

**20.** Derive the Taylor series for the function $e^{-x}$ by solving the IVP

$$y'' - y = 0, \qquad y(0) = 1, \qquad y'(0) = -1$$

by the method of power series.

**21.** Show that the solutions of the Euler differential equation

$$x^2y'' + A_0xy' + B_0y = 0$$

as found in Section 2.7 are as described by Theorem 1, Section 5.5. The above can be used as a motivation for series solutions about a regular singular point.[1]

---

[1]See also *Amer. Math. Monthly* 67(1960): 278–79.

# CHAPTER 6

# Boundary Value Problems

## 6.1 INTRODUCTION AND SOLUTION OF BOUNDARY VALUE PROBLEMS

In this chapter we present a brief introduction to boundary value problems for linear second-order differential equations. There is a large body of theoretical material that has been developed for boundary value problems. Our presentation will bring out some essential features of the basic theory by means of simple yet instructive examples. Consider the second-order linear differential equation

$$a_2(x)y'' + a_1(x)y' + a_0(x)y = f(x), \tag{1}$$

where the coefficients $a_2(x)$, $a_1(x)$, $a_0(x)$ and the function $f(x)$ are continuous in some interval $a \le x \le b$ with $a_2(x) \ne 0$ in this interval. Until now we were mainly concerned with finding a solution $y(x)$ of the differential equation (1) which at some point $x = x_0$ in the interval $a \le x \le b$ satisfied two given initial conditions

$$y(x_0) = y_0 \quad \text{and} \quad y'(x_0) = y_1. \tag{2}$$

The differential equation (1), together with the initial conditions (2), constitutes an initial value problem (IVP). In most IVPs the independent variable $x$ of the differential equation usually represents time, $x_0$ being the initial time and $y_0$ and $y_1$ the initial conditions. However, when the independent variable $x$ represents a space variable (as in the diffusion applications in Section 2.1), we usually want to find a solution $y(x)$ of the differential equation (1) that satisfies a condition at each end point of the interval $a \le x \le b$. For example,

$$y(a) = A \quad \text{and} \quad y(b) = B, \tag{3}$$

where $A$ and $B$ are two constants. The conditions (3) which are given at the end points (or boundary points) of the interval $a \le x \le b$ are called *boundary conditions*. The differential equation (1), together with the boundary conditions (3), constitutes a *boundary value problem (BVP)*. The form of the boundary conditions at the end points $a$ and $b$ may vary widely, as we shall see in the examples and exercises below.

The procedure for solving a BVP is similar to the procedure used in an IVP. First, we find the general solution of the differential equation and then we use the boundary conditions to determine the arbitrary constants in the general solution. However, the striking difference between an IVP and a BVP is that while the IVP (1) and (2) has exactly one solution (as we know from Theorem 1 of Section 2.3), a BVP may have one, none, or infinitely many solutions. The following examples illustrate this point.

**EXAMPLE 1**  Solve the BVP

$$y'' + y = x \qquad \text{for} \qquad 0 \le x \le \frac{\pi}{2} \tag{4}$$

$$y(0) = 2, \qquad y\left(\frac{\pi}{2}\right) = 1. \tag{5}$$

**Solution**  First, we compute the general solution of the differential equation (4). The homogeneous solution is $y_h = c_1 \cos x + c_2 \sin x$. Using the method of undetermined coefficients, we find that a particular solution of the differential equation (4) is of the form $y_p = Ax + B$. Substituting $y_p$ into the differential equation (4), we find $Ax + B = x$. Thus, $A = 1$ and $B = 0$ and a particular solution is $y_p = x$. Hence, the general solution of the differential equation (4) is

$$y(x) = c_1 \cos x + c_2 \sin x + x. \tag{6}$$

Using the boundary conditions (5), we find that

$$y(0) = 2 \Rightarrow c_1 = 2$$

and

$$y\left(\frac{\pi}{2}\right) = 1 \Rightarrow c_2 + \frac{\pi}{2} = 1 \Rightarrow c_2 = 1 - \frac{\pi}{2}.$$

Thus, the BVP (4)–(5) has the *unique* solution

$$y(x) = 2 \cos x + \left(1 - \frac{\pi}{2}\right) \sin x + x. \tag{7}$$

**EXAMPLE 2**  Solve the BVP

$$y'' + y = x \qquad \text{for} \qquad 0 \le x \le \pi \tag{8}$$

$$y(0) = 2, \qquad y(\pi) = 1. \tag{9}$$

**Solution**  The general solution of the differential equation (8) was computed in Example 1 and is

$$y(x) = c_1 \cos x + c_2 \sin x + x.$$

Next, we examine for what values of the constants $c_1$ and $c_2$ the boundary conditions (9) are satisfied. We have

$$y(0) = 2 \Rightarrow c_1 = 2$$

and

$$y(\pi) = 1 \Rightarrow -c_1 + \pi = 1 \Rightarrow c_1 = \pi - 1.$$

Clearly, these conditions are inconsistent, since they imply that the constant $c_1$ has the values 2 and $\pi - 1$ simultaneously. Thus, we conclude in this case that the BVP (8)–(9) has *no* solution.

**EXAMPLE 3** Solve the BVP

$$y'' + y = x \qquad \text{for} \qquad 0 \le x \le \pi \tag{10}$$

$$y(0) = 2, \qquad y(\pi) = \pi - 2. \tag{11}$$

**Solution** The general solution of the differential equation (10) is

$$y(x) = c_1 \cos x + c_2 \sin x + x.$$

Now,

$$y(0) = 2 \Rightarrow c_1 = 2$$

and

$$y(\pi) = \pi - 2 \Rightarrow -c_1 + \pi = \pi - 2 \Rightarrow c_1 = 2.$$

Thus, $c_1 = 2$ while $c_2$ remains arbitrary. In this case, the BVP (10)–(11) has *infinitely many* solutions:

$$y(x) = 2 \cos x + c_2 \sin x + x. \tag{12}$$

The boundary conditions at the end points $a$ and $b$ need not always be of the type considered in the previous examples. They can contain combinations of $y$ and its derivatives at the points $a$ and $b$. We give a simple example.

**EXAMPLE 4** Solve the BVP

$$y'' + 4y = 0 \qquad \text{for} \qquad 0 \le x \le \pi \tag{13}$$

$$y(0) - 2y'(0) = -2. \tag{14}$$

$$y(\pi) + 3y'(\pi) = 3. \tag{15}$$

**Solution** The general solution of the differential equation (13) is

$$y(x) = c_1 \cos 2x + c_2 \sin 2x.$$

We have $y'(x) = -2c_1 \sin 2x + 2c_2 \cos 2x$, and using the boundary conditions, we find that

$$y(0) - 2y'(0) = -2 \Rightarrow c_1 - 4c_2 = -2$$

and

$$y(\pi) + 3y'(\pi) = 3 \Rightarrow c_1 + 6c_2 = 3.$$

Solving this system we obtain $c_1 = 0$ and $c_2 = \frac{1}{2}$. Therefore, the BVP (13)–(15) has the unique solution

$$y(x) = \tfrac{1}{2} \sin 2x.$$

Of course, it is very important to know under what conditions a BVP has a unique solution, no solution, or infinitely many solutions. This problem, in general, is very difficult and beyond our scope. We only develop here a simple but interesting result about the BVP consisting of the differential equation (1) and the simple boundary conditions (3).

Let $y_1(x)$ and $y_2(x)$ be two linearly independent solutions of the homogeneous differential equation corresponding to the differential equation (1), and let $y_p(x)$ be a particular solution of the differential equation (1). Then the general solution of the differential equation (1) is given by

$$y(x) = c_1 y_1(x) + c_2 y_2(x) + y_p(x). \tag{16}$$

In view of the boundary conditions (3), the constants $c_1$ and $c_2$ of the general solution should satisfy the following linear system of algebraic equations:

$$\left.\begin{array}{l} c_1 y_1(a) + c_2 y_2(a) + y_p(a) = A \\ c_1 y_1(b) + c_2 y_2(b) + y_p(b) = B \end{array}\right\}$$

or

$$\left.\begin{array}{l} c_1 y_1(a) + c_2 y_2(a) = A - y_p(a) \\ c_1 y_1(b) + c_2 y_2(b) = B - y_p(b) \end{array}\right\}. \tag{17}$$

Now it is clear that the BVP (1) and (3) has as many solutions as the system (17). We recall (Appendix A) that a linear system of the form

$$\left.\begin{array}{l} a_{11} x + a_{12}\, y = b_1 \\ a_{21} x + a_{22}\, y = b_2 \end{array}\right\} \tag{18}$$

has exactly one solution if the determinant of the coefficients satisfies

$$\begin{vmatrix} a_{11} & a_{12} \\ a_{21} & a_{22} \end{vmatrix} \neq 0. \tag{19}$$

On the other hand, if the determinant in (19) is equal to zero, the system (18) has no solution when

$$\begin{vmatrix} a_{11} & b_1 \\ a_{21} & b_2 \end{vmatrix} \neq 0 \tag{20}$$

and has infinitely many solutions when the determinant in (20) is equal to zero. Summarizing the facts above, we have the following.

### THEOREM 1

*Let $y_1(x)$ and $y_2(x)$ be two linearly independent solutions of the homogeneous differential equation corresponding to the differential equation (1) and let $y_p(x)$ be a particular solution of the differential equation (1). Then, the following statements are valid:*

*(a) If*

$$\begin{vmatrix} y_1(a) & y_2(a) \\ y_1(b) & y_2(b) \end{vmatrix} \neq 0, \tag{21}$$

the BVP (1) and (3) *has one and only one solution in the interval* $a \le x \le b$.

(b) *If the determinant in* (21) *is equal to zero, the BVP* (1) *and* (3) *has no solution or infinitely many solutions in the interval* $a \le x \le b$, *depending on whether the determinant*

$$\begin{vmatrix} y_1(a) & A - y_p(a) \\ y_1(b) & B - y_p(b) \end{vmatrix} \tag{22}$$

*is* $\ne 0$ *or equal to zero, respectively.*

Let us apply Theorem 1 to the BVPs in Examples 1, 2, and 3, where $y_1(x) = \cos x$, $y_2(x) = \sin x$, and $y_p(x) = x$. The BVP (4)–(5) has a unique solution because

$$\begin{vmatrix} y_1(a) & y_2(a) \\ y_1(b) & y_2(b) \end{vmatrix} = \begin{vmatrix} \cos 0 & \sin 0 \\ \cos \dfrac{\pi}{2} & \sin \dfrac{\pi}{2} \end{vmatrix} = \begin{vmatrix} 1 & 0 \\ 0 & 1 \end{vmatrix} = 1 \ne 0.$$

The BVP (8)–(9) has no solution, because

$$\begin{vmatrix} y_1(a) & y_2(b) \\ y_1(b) & y_2(b) \end{vmatrix} = \begin{vmatrix} \cos 0 & \sin 0 \\ \cos \pi & \sin \pi \end{vmatrix} = \begin{vmatrix} 1 & 0 \\ -1 & 0 \end{vmatrix} = 0$$

and

$$\begin{vmatrix} y_1(a) & A - y_p(a) \\ y_1(b) & B - y_p(b) \end{vmatrix} = \begin{vmatrix} \cos 0 & 2 - 0 \\ \cos \pi & 1 - \pi \end{vmatrix} = \begin{vmatrix} 1 & 2 \\ -1 & 1 - \pi \end{vmatrix} \ne 0.$$

Finally, the BVP (10)–(11) has infinitely many solutions, because

$$\begin{vmatrix} y_1(a) & y_2(a) \\ y_1(b) & y_2(b) \end{vmatrix} = \begin{vmatrix} \cos 0 & \sin 0 \\ \cos \pi & \sin \pi \end{vmatrix} = \begin{vmatrix} 1 & 0 \\ -1 & 0 \end{vmatrix} = 0$$

and

$$\begin{vmatrix} y_1(a) & A - y_p(a) \\ y_1(b) & B - y_p(b) \end{vmatrix} = \begin{vmatrix} \cos 0 & 2 - 0 \\ \cos \pi & \pi - 2 - \pi \end{vmatrix} = \begin{vmatrix} 1 & 2 \\ -1 & -2 \end{vmatrix} = 0.$$

**APPLICATIONS 6.1.1**

■ Suppose that a gas diffuses into a liquid in a long, narrow pipe (Figure 6.1). **Diffusion** Suppose that this process takes place for such a long period of time that the concentration $y(x)$ of gas in the pipe depends only on the distance $x$ from some

**Figure 6.1**

initial point 0 (and is independent of time). Then, as we proved in Section 2.1.1, $y(x)$ satisfies the BVP

$$y'' - \frac{k}{D}y = 0 \tag{23}$$

$$y(0) = A, \quad y(l) = 0. \tag{24}$$

In Exercise 16 the reader is asked to solve the BVP (23)–(24).

**Deflection of a Taut String** ■ A taut string, lying on the $xy$ plane, is stretched, under a tension $T$, between two fixed points $x = 0$ and $x = L$ on the $x$ axis. A transverse (perpendicular to the $x$ axis) load $f(x)$ is applied to the string and produces a deflection $u(x)$. Then it can be shown that the deflection $u(x)$ satisfies the differential equation

$$-Tu'' = f(x). \tag{25}$$

Since the ends of the string are kept fixed at the points $x = 0$ and $x = L$ of the $x$ axis, $u(x)$ must also satisfy the boundary conditions

$$u(0) = 0, \quad u(L) = 0. \tag{26}$$

**EXAMPLE 5**  Solve the BVP (25)–(26) when $f(x) \equiv f_0$ is constant.

**Solution**  The general solution of the differential equation (25) is

$$u(x) = c_1 + c_2 x - \frac{f_0}{2T}x^2.$$

But

$$u(0) = 0 \Rightarrow c_1 = 0$$

and

$$u(L) = 0 \Rightarrow c_1 + c_2 L - \frac{f_0}{2T}L^2 = 0.$$

Hence, $c_1 = 0$, $c_2 = f_0 L/2T$, and the BVP (25)–(26) has the unique solution

$$u(x) = \frac{f_0 L}{2T}x - \frac{f_0}{2T}x^2 = \frac{f_0}{2T}x(L - x).$$

**EXERCISES**

In Exercises 1 through 6, answer true or false.

**1.** $y(x) = \cos 3x$ is a solution of the BVP

$$y'' + 9y = 0$$

$$y(0) = 1, \quad y(\pi) = -1.$$

**2.** $y(x) = \cos 3x - \sin 3x$ is a solution of the BVP

$$y'' + 9y = 0$$

$$y(0) = 1, \quad y(\pi) = -1.$$

**3.** The BVP
$$y'' + 9y = 0$$
$$y(0) = 1, \quad y(\pi) = -1$$

has a unique solution.

**4.** The BVP
$$y'' + 9y = 0$$
$$y(0) = 1, \quad y(\pi) = B$$

has no solution if $B \neq -1$.

**5.** $y(x) = \cos 3x - \sin 3x$ is a solution of the BVP
$$y'' + 9y = 0$$
$$y(0) = 1, \quad y\left(\frac{\pi}{2}\right) = -1.$$

**6.** The BVP
$$y'' + 9y = 0$$
$$y(0) = 1, \quad y\left(\frac{\pi}{2}\right) = -1$$

has a unique solution.

In Exercises 7 through 11, solve the BVP.

**7.** $y'' + 9y = 0, 0 \le x \le \pi$
$y(0) = 1, y(\pi) = -1$

**8.** $y'' + 9y = 0, 0 \le x \le \pi$
$y(0) = 1, y(\pi) = 2$

**9.** $y'' + 9y = 0, 0 \le x \le \dfrac{\pi}{2}$
$y(0) = 1, y\left(\dfrac{\pi}{2}\right) = -1$

**10.** $y'' - 3y' + 2y = e^x, 0 \le x \le 1$
$y'(0) = 0, y(1) = 0$

**11.** $y'' - 3y' + 2y = 0, 0 \le x \le 1$
$y(0) = 0, y'(1) = 0$

**12.** Show that the BVP
$$y'' + 16y = 0, \quad 0 \le x \le \pi$$
$$4y(0) - y'(0) = A, \quad 4y(\pi) - y'(\pi) = B$$

has infinitely many solutions if $A = B$, and no solution if $A \neq B$.

**13.** Solve the BVP
$$y'' - y = 2e^x$$
$$y(0) - 2y'(0) = -2, \quad 3y(1) - y'(1) = e.$$

**14.** Solve the BVP

$$x^2y'' - 3xy' + 3y = 0, \qquad 1 \le x \le 2$$
$$y(1) - y'(1) = 2, \qquad y(2) - 2y'(2) = 4.$$

**15.** Show that the BVP

$$x^2y'' - 3xy' + 3y = \ln x, \qquad 1 \le x \le 2$$
$$y(1) = A, \qquad y(2) = B$$

has a unique solution for all values of $A$ and $B$.

**16.** Solve the BVP consisting of the differential equation (23) and the boundary conditions (24).

**17.** A (one-dimensional and steady-state) diffusion process in a medium bounded by the planes $x = 0$ and $x = l$ which are maintained at constant concentrations $c_1$ and $c_2$, respectively, is governed by the BVP

$$y''(x) = 0$$
$$y(0) = c_1, \qquad y(l) = c_2$$

(provided that the diffusion coefficient $D$ is constant). Solve this BVP and show that the concentration $y(x)$ changes linearly from $c_1$ to $c_2$ through the medium.

**18.** Solve the BVP (25)–(26) when

$$f(x) = x, \qquad T = 2, \qquad L = 1.$$

---

## 6.2  EIGENVALUES AND EIGENFUNCTIONS

---

From the point of view of applications, especially problems in mathematical physics, there is a very important class of BVPs in which the differential equation is of the form

$$(p(x)y')' + (q(x) + \lambda)y = 0, \tag{1}$$

where $p(x)$ ($> 0$) and $q(x)$ are continuous in some interval $a \le x \le b$, $\lambda$ is a real parameter, and the two boundary conditions at the end points $a$ and $b$ are of the form

$$\alpha_1 y(a) + \alpha_2 y'(a) = 0 \tag{2}$$
$$\beta_1 y(b) + \beta_2 y'(b) = 0.$$

The BVP (1)–(2) is always satisfied by the trivial solution $y(x) \equiv 0$. However, the important issue in this type of BVP is to determine values of $\lambda$ for which the BVP (1)–(2) has a nontrivial solution. These values of $\lambda$ are called the *eigenvalues* of the BVP (1)–(2). The corresponding nontrivial solutions are called *eigenfunctions*. The BVP (1)–(2) is also called an *eigenvalue problem* (*EVP*).

In this section we shall compute the eigenvalues and eigenfunctions of some simple but representative EVPs and state some of their properties.

**EXAMPLE 1** Compute the eigenvalues and eigenfunctions of the EVP

$$y'' + \lambda y = 0 \quad \text{for} \quad 0 \le x \le \pi \tag{3}$$

$$y(0) = 0, \quad y(\pi) = 0. \tag{4}$$

**Solution** Our problem here is to find all the values of the real parameter $\lambda$ for which the EVP (3)–(4) has a nontrivial solution and to find the corresponding nontrivial solutions. The characteristic roots of the differential equation (3) are $\pm \sqrt{-\lambda}$. Hence, the form of the solutions of the differential equation (3) depends on whether $\lambda$ is negative, zero, or positive. So in solving the eigenvalue problem (3)–(4) we have to examine three cases.

**CASE 1** $\lambda < 0$ Set $\lambda = -k^2$, where $k > 0$. Then the differential equation (3) becomes $y'' - k^2 y = 0$, and its general solution is

$$y(x) = c_1 e^{-kx} + c_2 e^{kx}. \tag{5}$$

Next, we use the boundary conditions (4) to determine the constants $c_1$ and $c_2$. Remember that we want to find solutions that are not identically equal to zero. We have

$$y(0) = 0 \Rightarrow c_1 + c_2 = 0 \tag{6}$$

and

$$y(\pi) = 0 \Rightarrow c_1 e^{-k\pi} + c_2 e^{k\pi} = 0. \tag{7}$$

Solving the system of simultaneous equations for $c_1$ and $c_2$, we find $c_1 = c_2 = 0$. In fact, from Eq. (6) we obtain $c_2 = -c_1$, and inserting this value into Eq. (7), we find that $c_1(e^{-k\pi} - e^{k\pi}) = 0$. Since $k > 0$, $e^{-k\pi} \ne e^{k\pi}$, and so $c_1 = 0$. Then $c_2 = 0$ and the solution (5) is identically zero. This implies that the EVP (3)–(4) does not have negative eigenvalues.

**CASE 2** $\lambda = 0$ In this case the differential equation (3) becomes $y'' = 0$, and its general solution is

$$y(x) = c_1 + c_2 x. \tag{8}$$

Using the boundary conditions (4), we see that

$$y(0) = 0 \Rightarrow c_1 = 0$$

and

$$y(\pi) = 0 \Rightarrow c_1 + c_2 \pi = 0 \Rightarrow c_2 \pi = 0 \Rightarrow c_2 = 0.$$

Thus, the solution (8) is identically zero, which implies that $\lambda = 0$ is not an eigenvalue.

**CASE 3**  $\lambda > 0$  Set $\lambda = k^2$, where $k > 0$. Then the differential equation (3) becomes $y'' + k^2y = 0$, and its general solution is

$$y(x) = c_1 \cos kx + c_2 \sin kx. \tag{9}$$

We now employ the boundary conditions

$$y(0) = 0 \Rightarrow c_1 = 0$$

and

$$y(\pi) = 0 \Rightarrow c_1 \cos k\pi + c_2 \sin k\pi = 0 \Rightarrow c_2 \sin k\pi = 0.$$

Thus, $c_1 = 0$ and $c_2 \sin k\pi = 0$. Recall once more that we try to compute nontrivial solutions of the differential equation (3). In this case, $c_1$ must be zero; hence, a nontrivial solution is possible only if $c_2$ can be chosen different from zero. Indeed, $c_2 \sin k\pi = 0$ if $\sin k\pi = 0$ or $k\pi = n\pi$ for $n = 1, 2, \ldots$ ; that is, $k = n$ for $n = 1, 2, \ldots$ . Now $\lambda = k^2$, and therefore the eigenvalues are given by

$$\lambda = n^2 \quad \text{for} \quad n = 1, 2, \ldots. \tag{10}$$

Equation (10) gives all the eigenvalues of the EVP (3)–(4). To find the eigenfunctions corresponding to these eigenvalues, we use (9) with $c_1 = 0$, $c_2 \neq 0$, and $k = n$. Then $y = c_2 \sin nx$. Thus, up to a nonzero constant multiple (that is, $c_2$), the eigenfunction of the EVP (3)–(4) that corresponds to the eigenvalue $\lambda = n^2$ is given by

$$y = \sin nx \quad \text{for} \quad n = 1, 2, \ldots. \tag{11}$$

Hence, the eigenvalues and eigenfunctions of the EVP (3)–(4) are given by

$$\lambda_n = n^2 \quad \text{for} \quad n = 1, 2, \ldots \tag{12}$$

and

$$y_n = \sin nx \quad \text{for} \quad n = 1, 2, \ldots, \tag{13}$$

where the subscript $n$ is attached to indicate the correspondence between eigenvalues and eigenfunctions.

**EXAMPLE 2**  Determine the eigenvalues and eigenfunctions of the EVP

$$y'' + (-4 + \lambda)y = 0 \quad \text{for} \quad 0 \le x \le 1 \tag{14}$$

$$y'(0) = 0, \quad y'(1) = 0. \tag{15}$$

**Solution**  The characteristic roots of the differential equation (14) are $\pm\sqrt{4 - \lambda}$. Hence, the form of the solutions of the differential equation (14) depends on whether the quantity $4 - \lambda$ is negative, zero, or positive.

**CASE 1** $4 - \lambda < 0$ Set $4 - \lambda = -k^2$, where $k > 0$. Then the differential equation (14) becomes $y'' + k^2 y = 0$, and its general solution is

$$y(x) = c_1 \cos kx + c_2 \sin kx.$$

We have $y'(x) = -c_1 k \sin kx + c_2 k \cos kx$ and, using the boundary conditions (15), we find that

$$y'(0) = 0 \Rightarrow c_2 k = 0 \Rightarrow c_2 = 0$$

and

$$y'(1) = 0 \Rightarrow -c_1 k \sin k = 0 \Rightarrow \sin k = 0 \Rightarrow k = n\pi, \quad \text{for} \quad n = 1, 2, \dots.$$

Here $\lambda = 4 + k^2$, and therefore

$$\lambda_n = 4 + n^2 \pi^2 \quad \text{for} \quad n = 1, 2, \dots \tag{16}$$

and

$$y_n = \cos n\pi x \quad \text{for} \quad n = 1, 2, \dots \tag{17}$$

are eigenvalues and eigenfunctions of the EVP (14)–(15).

**CASE 2** $4 - \lambda = 0$ Then the differential equation (14) becomes $y'' = 0$, and its general solution is

$$y(x) = c_1 + c_2 x.$$

Using the boundary conditions (15), we find that

$$y'(0) = 0 \Rightarrow c_2 = 0$$

and

$$y'(1) = 0 \Rightarrow c_2 = 0.$$

Hence, in this case we obtained the nontrivial solution $y(x) = c_1$ or up to a constant multiple $y(x) = 1$. The value $\lambda = 4$ is therefore an eigenvalue of the EVP (14)–(15) with corresponding eigenfunction $y(x) = 1$.

**CASE 3** $4 - \lambda > 0$ Set $4 - \lambda = k^2$, where $k > 0$. Then the differential equation (14) becomes $y'' - k^2 y = 0$, and its general solution is

$$y(x) = c_1 e^{-kx} + c_2 e^{kx}.$$

Using the boundary conditions (15), we find that

$$y'(0) = 0 \Rightarrow -c_1 k + c_2 k = 0 \Rightarrow -c_1 + c_2 = 0$$

and

$$y'(1) = 0 \Rightarrow -c_1 k e^{-k} + c_2 k e^{k} = 0 \Rightarrow -c_1 e^{-k} + c_2 e^{k} = 0.$$

Thus, $c_1 = c_2 = 0$, and there is no eigenvalue in this case.

If we observe that Eqs. (16) and (17) for $n = 0$ give us $\lambda_0 = 4$ and $y_0 = 1$, that is, the eigenvalue and eigenfunction found in Case 2, we conclude that the eigenvalues and eigenfunctions of the EVP (14)–(15) are given by the equations

$$\lambda_n = 4 + n^2\pi^2 \quad \text{for} \quad n = 0, 1, 2, \ldots \tag{18}$$

and

$$y_n = \cos n\pi x \quad \text{for} \quad n = 0, 1, 2, \ldots . \tag{19}$$

Finally, we state a theorem that summarizes some important properties of the eigenvalues and eigenfunctions of the EVP (1)–(2). To this end we first define the concepts of inner product and orthogonal functions.

## DEFINITION 1

*Let f and g be two continuous functions defined in the interval $a \leq x \leq b$. The inner product of the functions f and g is denoted by $(f, g)$ and is defined by*

$$(f, g) = \int_a^b f(x)g(x)dx.$$

*When $(f, g) = 0$, we say that the functions f and g are orthogonal.*

## THEOREM 1

*The EVP (1)–(2) possesses an infinite sequence of eigenvalues*

$$\lambda_1, \lambda_2, \ldots, \lambda_n, \ldots \to +\infty \tag{20}$$

*and a corresponding sequence of eigenfunctions*

$$y_1, y_2, \ldots, y_n, \ldots$$

*with the property that in the interval $a \leq x \leq b$,*

$$(y_n, y_m) = \begin{cases} 0, & \text{if } n \neq m \\ 1, & \text{if } n = m. \end{cases} \tag{21}$$

*Furthermore, any function $f(x)$ that is twice continuously differentiable and satisfies the boundary conditions (2) has the following eigenfunction expansion:*

$$f(x) = \sum_{n=1}^{\infty} (f, y_n)y_n, \tag{22}$$

*where series (22) converges uniformly in the interval $a \leq x \leq b$.*

In Example 1 it is clear that property (20) is satisfied. Let us verify (21). In the EVP (3)–(4) we found that [see Eq. (13)]

$$y_n = \sin nx \quad \text{for} \quad n = 1, 2, \ldots . \tag{13}$$

Computing the inner product $(y_n, y_m)$ we obtain that

$$(y_n, y_m) = \int_0^\pi \sin nx \sin mx \, dx = \begin{cases} 0, & \text{if} \quad n \neq m \\ \dfrac{\pi}{2}, & \text{if} \quad n = m. \end{cases}$$

In order that Eq. (21) be satisfied, we have to multiply each eigenfunction in (13) by $(2/\pi)^{1/2}$, which is acceptable since the eigenfunctions are defined up to a nonzero constant multiple. Thus, for Theorem 1 we have to replace the eigenfunction (13) of the EVP (3)–(4) by

$$y_n = \left(\frac{2}{\pi}\right)^{1/2} \sin nx \qquad \text{for} \quad n = 1, 2, \dots . \tag{13'}$$

With respect to these eigenfunctions, property (22) indicates that if $f(x)$ is a function that is twice continuously differentiable (in other words, $f''$ exists and is continuous) and satisfies the boundary conditions (4), that is, $f(0) = f(\pi) = 0$, then

$$f(x) = \frac{2}{\pi} \sum_{n=1}^\infty (f(x), \sin nx) \sin nx = \frac{2}{\pi} \sum_{n=1}^\infty \left(\int_a^b f(x) \sin nx \, dx\right) \sin nx, \tag{23}$$

and this series converges uniformly in the interval $0 \leq x \leq \pi$. The series (23) is known as the *Fourier sine series*[1] of the function $f(x)$ in the interval $0 \leq x \leq \pi$.

## APPLICATION 6.2.1

Eigenvalue problems occur frequently in problems of mathematical physics. In particular, they appear in the process of solving, by the method of separation of variables, initial–boundary value problems which involve one of the following partial differential equations:

$$u_t = \alpha^2 u_{xx} \qquad \text{(heat equation)}, \tag{24}$$

$$u_{tt} - c^2 u_{xx} = 0 \qquad \text{(wave equation)}, \tag{25}$$

and

$$u_{xx} + u_{yy} = 0 \qquad \text{(Laplace's equation)}. \tag{26}$$

Let us demonstrate this in the case of Eq. (24).

■ Equation (24) appears in the study of heat conduction and many other diffusion processes. For example, consider an insulated uniform bar of length $l$. Assume that the bar is oriented so that the $x$ axis coincides with the axis of the bar. Let $u(x, t)$ denote the temperature in the bar at the point $x$ at time $t$. Then

**The Heat Equation**

---

[1]See R. V. Churchill, *Fourier Series and Boundary Value Problems,* 2nd ed. (New York: McGraw-Hill, 1963).

$u$ satisfies Eq. (24) for $0 < x < l$. The constant $\alpha^2$ is called the *thermal diffusivity* and depends only on the material of which the bar is made. To determine the flow of heat $u(x, t)$ in the bar, we need to know some initial and boundary conditions. For example, assume that the initial temperature $f(x)$ in the bar is a known function of $x$, that is,

$$u(x, 0) = f(x), \qquad 0 \le x \le l, \tag{27}$$

and that the ends of the bar are kept at zero temperature, that is,

$$u(0, t) = u(l, t) = 0. \tag{28}$$

Using the method of separation of variables (see Section 5.5.1), we look for a solution $u(x, t)$ of the initial–boundary value problem (24), (27), and (28) of the form

$$u(x, t) = X(x)T(t). \tag{29}$$

We have

$$u_t = XT' \quad \text{and} \quad u_{xx} = X''T,$$

and Eq. (24) becomes

$$XT' = \alpha^2 X''T$$

or

$$\frac{X''}{X} = \frac{1}{\alpha^2}\frac{T'}{T}. \tag{30}$$

The left-hand side of Eq. (30) is a function of $x$ alone, and its right-hand side is a function of $t$ alone. The only way that this can happen is if both sides of Eq. (30) are equal to some constant $\lambda$. That is,

$$\frac{X''}{X} = \lambda$$

and

$$\frac{1}{\alpha^2}\frac{T'}{T} = \lambda$$

or

$$X'' - \lambda X = 0 \tag{31}$$

and

$$T' - \alpha^2 \lambda T = 0. \tag{32}$$

Using the boundary conditions (28), we find

$$0 = u(0, t) = X(0)T(t) \quad \text{and} \quad 0 = u(l, t) = X(l)T(t),$$

which imply that

$$X(0) = X(l) = 0. \tag{33}$$

Thus, *the function $X(x)$ of the solution (29) satisfies the EVP consisting of the differential equation (31) and the boundary conditions (33)*. We leave it to the student to show that the eigenvalues and eigenfunctions of the EVP (31), (33) are given by

$$\lambda_n = -\frac{n^2\pi^2}{l^2}, \qquad n = 1, 2, \ldots \tag{34}$$

and

$$X_n(x) = \sin\frac{n\pi x}{l}, \qquad n = 1, 2, \ldots \tag{35}$$

For the sake of completeness, we now indicate what remains to be done to find the solution $u(x, t)$ to the initial boundary value problem (24), (27), and (28). Since $\lambda$ has one of the values given by Eq. (34), the solution of Eq. (32), up to a constant multiple, is given by

$$T_n(t) = \exp(-\alpha^2 n^2\pi^2 l^{-2}t), \qquad n = 1, 2, \ldots \tag{36}$$

In view of the equations (29), (35), and (36), each of the functions

$$u_n(x, t) = \exp(-\alpha^2 n^2\pi^2 l^{-2}t)\sin\frac{n\pi x}{l}, \qquad n = 1, 2, \ldots \tag{37}$$

is a solution of the heat equation (24) and satisfies the boundary conditions (28). Furthermore, it can be shown that any linear combination

$$\sum_{n=1}^{\infty} c_n u_n(x, t) \tag{38}$$

of the functions in (37) is a solution of (24) and (28) provided that the series (38) converges and can be differentiated term by term a sufficient number of times with respect to $t$ and $x$. All that we have to do now is to choose the constants $c_n$, $n = 1, 2, \ldots$, in such a way that the series (38) satisfies the initial condition (27). That is,

$$f(x) = \sum_{n=1}^{\infty} c_n u_n(x, 0) = \sum_{n=1}^{\infty} c_n \sin\frac{n\pi x}{l}, \qquad 0 \le x \le l, \tag{39}$$

which means that the constants $c_n$ must be the coefficients of the Fourier sine series of the function $f(x)$ in the interval $0 \le x \le l$.

### EXERCISES

In Exercises 1 through 8, compute the eigenvalues and eigenfunctions of the EVP.

**1.** $y'' + \lambda y = 0$ for $0 \le x \le \frac{1}{2}$
$y(0) = 0$, $y(\frac{1}{2}) = 0$

**2.** $y'' + \lambda y = 0$ for $0 \le x \le 1$
$y'(0) = 0$, $y(1) = 0$

**3.** $y'' + \lambda y = 0$ for $-\frac{1}{2} \le x \le \frac{1}{2}$
$y(-\frac{1}{2}) = 0$, $y(\frac{1}{2}) = 0$

**4.** $y'' + \lambda y = 0$ for $0 \le x \le \pi$
$y(0) = y'(0)$, $y(\pi) = y'(\pi)$

**5.** $y'' + 2y + \lambda y = 0$ for $0 \le x \le 2$
$y(0) = 0$, $y(2) = 0$

**6.** $x^2 y'' + 3xy' + \lambda y = 0$ for $1 \le x \le e^2$
$y(1) = 0$, $y(e^2) = 0$

[*Hint:* Set $x = e^t$.]

**7.** $y'' + (2 + \lambda)y = 0$ for $0 \le x \le 1$
$y(0) = 0$, $y(1) = 0$

**8.** $y'' + (-2 + \lambda)y = 0$ for $0 \le x \le 1$
$y(0) = 0$, $y(1) = 0$

**9.** Show that the eigenfunctions of the EVP (14)–(15) are pairwise-orthogonal in the interval $0 \le x \le 1$; that is, $(y_n, y_m) = 0$ for $n \ne m$.

**10.** Show that the eigenvalues and eigenfunctions of the EVP (31), (33) are given by (34) and (35), respectively.

**11. The Wave Equation** Equation (25) appears in the study of phenomena involving the propagation of waves (for example, water waves, acoustic waves, electromagnetic waves). Consider the initial boundary value problem

$$u_{tt} - c^2 u_{xx} = 0, \qquad 0 < x < l, \qquad t > 0$$
$$u(x, 0) = f(x), \qquad u_t(x, 0) = g(x), \qquad 0 \le x \le l$$
$$u(0, t) = u(l, t) = 0, \qquad t \ge 0,$$

where $c$ is a constant sometimes called the *wave speed*. Assume that this problem has a solution of the form

$$u(x, t) = X(x)T(t).$$

Show that the function $X$ satisfies the EVP

$$X'' - \lambda X = 0$$
$$X(0) = X(l) = 0$$

and that the function $T$ satisfies the differential equation

$$T'' - \lambda c^2 T = 0.$$

Indicate a formal procedure for obtaining the solution of this problem.

**12. Laplace's Equation** Equation (26) appears in the study of steady-state heat flows, in potential theory, and so on. Consider the BVP

$$u_{xx} + u_{yy} = 0, \qquad 0 < x < A, \qquad 0 < y < B$$
$$u(x, 0) = u(x, B) = 0, \qquad 0 < x < A$$
$$u(0, y) = 0, \qquad u(A, y) = f(y), \qquad 0 \le y \le B.$$

Assume that this problem has a solution of the form

$$u(x, y) = X(x)Y(y).$$

Show that the function $Y$ satisfies the EVP

$$Y'' + \lambda Y = 0$$

$$Y(0) = Y(B) = 0,$$

and the function $X$ satisfies the differential equation

$$X'' - \lambda X = 0$$

and the initial condition

$$X(0) = 0.$$

Indicate a formal procedure for obtaining the solution of this problem.

## REVIEW EXERCISES

In Exercises 1 through 4, solve the BVP.

**1.** $y'' + 16y = 32x, \qquad 0 \le x \le \pi$
$y(0) = 0, \, y(\pi) = 1$

**2.** $y'' + 16y = 32x, \qquad 0 \le x \le \pi$
$y(0) = 0, \, y(\pi) = 2\pi$

**3.** $y'' + 16y = 32x, \qquad 0 \le x \le \dfrac{\pi}{8}$
$y(0) = 0, \, y\left(\dfrac{\pi}{8}\right) = 0$

**4.** $y'' + 16y = 32x, \qquad 0 \le x \le \pi$
$y(0) - y'(0) = 1, \, y'(\pi) = 0$

**5.** Under what conditions on $A$ and $B$ does the BVP

$$y'' + 16y = 32x, \qquad 0 \le x \le \pi$$
$$y(0) = A, \qquad y(\pi) = B$$

have exactly one solution, no solution, infinitely many solutions?

**6.** Show that the BVP

$$y'' + 16y = 32x, \qquad 0 \le x \le \pi$$
$$y(0) = A, \qquad y'(\pi) = B$$

has exactly one solution, no matter what the values of $A$ and $B$ are. Find the solution.

In Exercises 7 through 9, compute the eigenvalues and eigenfunctions of the EVP.

**7.** $y'' - \lambda y = 0, \qquad 0 \le x \le \pi$
$y(0) = 0, \, y(\pi) = 0$

**8.** $y'' + \lambda y = 0, \qquad 0 \le x \le \pi$
$y(0) = 0, \, y'(\pi) = 0$

**9.** $y'' + \lambda y = 0, \qquad 0 \le x \le \pi$
$y'(0) = 0, \, y'(\pi) = 0$

# CHAPTER 7

# Numerical Solutions of Differential Equations

---

## 7.1 INTRODUCTION

---

Up to now our treatment of differential equations has centered on the determination of solutions of the differential equation. More specifically, our methods have provided us with formulas for the solution. For example, the differential equation $y'' + y = 0$ has $y = c_1 \sin x + c_2 \cos x$ as a formula for its solutions, the differential equation $y'' - y = 0$ has $y = c_1 e^x + c_2 e^{-x}$ as a formula for its solutions, and the differential equation $y'' - xy = 0$ has

$$y = c_1 \left[ 1 + \sum_{n=1}^{\infty} \frac{1 \cdot 4 \cdot 7 \cdots (3n - 2)}{(3n)!} x^{3n} \right]$$

$$+ c_2 \left[ x + \sum_{n=1}^{\infty} \frac{2 \cdot 5 \cdot 8 \cdots (3n - 1)}{(3n + 1)!} x^{3n+1} \right]$$

as a formula for its solutions. It is natural to adopt the attitude that these formulas supply all the information about the solution that we need. For instance, if we want the value of $y$ corresponding to a specific value of $x$ in the cases above, we simply substitute the given value of $x$ into the appropriate formula and compute the corresponding value of $y$. Unfortunately, matters are not always that simple. Even in those cases where the solution is represented by such relatively simple functions as $e^x$, $\sin x$, or $\cos x$, the accuracy desired in the numerical value of $y$ may cause some computational problems. Furthermore, the formula for $y$ may itself impose computational difficulties. This is the case in the description of the solution of $y'' - xy = 0$ above. Of course, with the present-day proliferation of pocket calculators and the widespread availability of digital computers, such computational difficulties can be easily overcome if the accuracy desired is within the capability of the device used.

Our description of determining the numerical value of the solution of a differential equation, although it does point out one aspect of the problem, is *not* what is meant by numerical solutions of differential equations. *Rather, this terminology applies to methods associated with the determination of approximate numerical values of the solution when it is generally impossible to obtain a formula for the solution.* The reader perhaps recalls from calculus that many functions do not have antiderivatives, and hence it is impossible to represent the indefinite integral of these functions by a formula in terms of elementary functions. In this same context there are many differential equations for which it is impossible to obtain a formula for the solution. In such cases we "solve" the differential equation by applying certain *numerical methods.*

There are at present a large number of numerical methods for solving differential equations. Since our treatment is admittedly introductory and brief, we touch upon only a few of the elementary methods.

Any numerical method involves approximations of one sort or another. Consequently, questions arise concerning the accuracy of the results. In other words, how much error is involved in the approximation? Although we present a few definitions and results concerning the error involved, any reasonable attempt at answering such questions here would take us far beyond our goal of presenting a brief introduction to numerical methods.[1]

In our subsequent discussion, we assume that the differential equation in question is of the form $y' = f(x, y)$. We note that all differential equations can be written in this form or as a system of differential equations of this form. Hence, from one point of view there is no loss of generality in making this assumption. Furthermore, to avoid cumbersome treatment of integration constants, we concentrate on initial value problems. Thus we investigate the IVP

$$y' = f(x,y) \tag{1}$$

$$y(x_0) = y_0. \tag{2}$$

To obtain a feeling for the goal of numerical methods, let us imagine that the curve in Figure 7.1 represents the solution to the IVP (1)–(2). Suppose that we want to know the coordinates of the point $(x_1, y_1)$. In the case that $y$ is given explicitly by a formula involving $x$, it is simply a matter of knowing $x_1$.

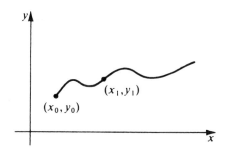

**Figure 7.1**

If $y$ is not known in terms of $x$ and the solution curve of Figure 7.1 exists only in theory (from the existence-uniqueness theorem), is it still possible to determine $y_1$ knowing $x_1$? The answer to this question is the basic goal of any numerical method.

---

[1]For some answers to questions about the accuracy of the methods, the interested reader is referred to E. Isaacson and H. B. Keller, *The Analysis of Numerical Methods* (New York: John Wiley & Sons, 1966).

We will say that we have solved a differential equation numerically on the interval $[x_0, b]$ when we have obtained a set of points $\{(x_0, y_0), (x_1, y_1), \ldots, (x_n, y_n)\}$ with $x_n = b$ and $y_i$ an approximation of the solution when $x$ has the value $x_i$.

# 7.2  EULER METHOD

Referring to Figure 7.2, we can consider that in transferring from the initial point $(x_0, y_0)$ to the new point $(x_1, y_1)$ we have induced changes in the coordinates. These changes are denoted by $\Delta x$ and $\Delta y$, where $\Delta x = x_1 - x_0$ and $\Delta y = y_1 - y_0$. Equivalently, we can write $x_1 = x_0 + \Delta x$ and $y_1 = y_0 + \Delta y$. We can then approximate $y_1$ by approximating $\Delta y$. Naturally, one considers $x_1$ and hence $\Delta x$ to be known exactly. The *Euler method* is to approximate $\Delta y$ by the differential $dy$. Now $dy = (dy/dx)dx = f(x,y)dx \approx f(x,y)\Delta x$. Of course, if this approximation is to be reasonably accurate, then $\Delta x$ should be quite small (see Figure 7.2). Given the IVP (1)–(2) of Section 7.1 and a specific value of $x$, say $x_1$, then $y_1$ is given by the Euler method to be $y_0 + f(x_0, y_0)\Delta x = y_0 + f(x_0, y_0)(x_1 - x_0)$. Thus, one can calculate $y_1$ directly from the information given. One can then use $(x_1, y_1)$ to approximate $y_2$ for some given $x_2$ ($> x_1$). In this case, $y_2 = y_1 + f(x_1, y_1)\Delta x = y_1 + f(x_1, y_1)(x_2 - x_1)$. This process can then be repeated to obtain additional points. Usually, this process is carried out in a specified interval with a prescribed spacing between the $x$ coordinates. The spacing is normally taken to be uniform (in which case the spacing is often called the "mesh width" or simply "mesh") and is denoted by $h$. If it is desired to obtain $n$ points in addition to the initial point, then $h$ is given by the formula $(b - x_0)/n$. The sequence of points $\{(x_i, y_i)\}_{i=0}^n$, when plotted, hopefully lie very close to the solution curve.

The Euler method can be summarized by the following formulas of the coordinates of the points $(x_i, y_i)$:

$$x_i = x_{i-1} + h, \qquad\qquad i = 1, 2, 3, \ldots, n \qquad\qquad (1)$$

$$y_i = y_{i-1} + f(x_{i-1}, y_{i-1})h, \qquad i = 1, 2, 3, \ldots, n. \qquad (2)$$

**Figure 7.2**

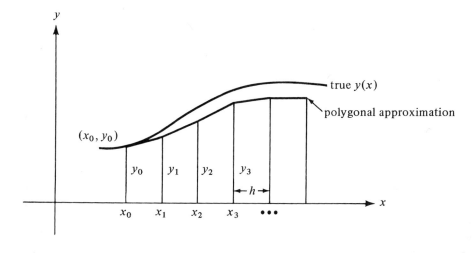

**Figure 7.3**

The "solution" to the IVP is then the sequence of points $\{(x_i, y_i)\}_{i=0}^n$. This solution is usually displayed in tabular form, although it could be displayed geometrically by plotting each of the points. (See Figure 7.3.)

**EXAMPLE 1**  Use the Euler method to solve numerically the IVP

$$y' = x^2 + y^2 \tag{3}$$

$$y(0) = 0 \tag{4}$$

on the interval $0 \le x \le 1$.

**Solution**  We were not told how many points to obtain; therefore, the choice is at our disposal. The more points we ask for, the smaller the value of $h$ will be and so the more accurate our results will be. The price to be paid, of course, is that the smaller $h$ is, the more calculations we require. We choose $n = 5$. Thus, $h = (1 - 0)/5 = 0.2$. The results are displayed in Table 7.1. Even though we are only interested in $x_i$ and $y_i$, we present the intermediate calculations to assist the reader in following the details. Where necessary we rounded numbers off to six decimal places. (There are problems of accuracy associated with round-off also, but again we skirt the issue since it would take us beyond our treatment.)

Thus, the collection of points $\{(0,0), (0.2,0), (0.4,0.008), (0.6,0.040013), (0.8,0.112333), (1,0.242857)\}$ is our numerical solution (by the Euler method with $n = 5$) of the IVP (3)–(4) in the interval $0 \le x \le 1$. Here and in subsequent applications, it is to be considered that if one were to plot the points of our numerical solution and connect these points with straight lines, the resulting polygonal curve would be an approximation to the graph of the solution on the interval $0 \le x \le 1$.

**Table 7.1** Numerical Solution of the IVP (3)–(4) by the Euler Method

| $i$ | $x_{i-1}$ | $y_{i-1}$ | $f(x_{i-1}, y_{i-1})$ | $f(x_{i-1}, y_{i-1})h$ | $x_i$ | $y_i$ |
|---|---|---|---|---|---|---|
| 1 | 0 | 0 | $0^2 + 0^2 = 0$ | $0(0.2) = 0$ | 0.2 | 0 |
| 2 | 0.2 | 0 | $(0.2)^2 + 0^2 = 0.04$ | $(0.04)(0.2) = 0.008$ | 0.4 | 0.008 |
| 3 | 0.4 | 0.008 | $(0.4)^2 + (0.008)^2 = 0.160064$ | $(0.160064)(0.2) = 0.032013$ | 0.6 | 0.040013 |
| 4 | 0.6 | 0.040013 | $(0.6)^2 + (0.040013)^2 = 0.361601$ | $(0.361601)(0.2) = 0.072320$ | 0.8 | 0.112333 |
| 5 | 0.8 | 0.112333 | $(0.8)^2 + (0.112333)^2 = 0.652619$ | $(0.652619)(0.2) = 0.130524$ | 1.0 | 0.242857 |

The calculations presented in Table 7.1 can be carried out with the aid of a computer or programmable calculator.[2] A flowchart of the steps involved is shown in Figure 7.4. A program for Example 1 would use the initial values $x = 0$, $y = 0$, $b = 1$, and $n = 5$. Once a program has been written it is a simple matter to produce a solution with any other set of initial values. In particular, if we increase the value of $n$, we decrease the value of $h = \Delta x$ and generally improve the accuracy of the method.

The error resulting from the use of a numerical method, such as the Euler method, to approximate a true solution is called *truncation* or *discretization error*. Generally, increasing $n$ (decreasing the step size) reduces the truncation error. However, another type of error, called *round-off error*, tends to become more significant as $h$ gets smaller. This is due to the limitation of all computers that can carry only a specified number of digits. This implies that the final digit in any number represented in the computer may be in error due to round-off. As $n$ increases, the accumulated round-off error can more than offset the gains made in improving the accuracy of the method. Numerical analysts have spent and continue to spend much time searching for the proper balance between reducing the truncation error (increasing $n$) and the round-off error (decreasing $n$).

One fairly straightforward procedure to suggest a possible best solution for a given computer is to write a program that repeats the sequence of steps illustrated in the flowchart several times, each time doubling the size of $n$, until the computed results [that is, $(x_n, y_n)$] no longer differ in the number of decimal places required. While this is no guarantee of the accuracy of the results, many programmers find this practical rule very useful for a first guess.

Of course, one way to obtain a better approximation of the solution is to use another, more accurate method with fewer steps. We shall do this in the remaining sections.

With regard to the error in the Euler approximation, we state the following theorem without proof.

**THEOREM 1.**

*Suppose that the functions $f(x, y)$, $\dfrac{\partial f}{\partial x}$ and $\dfrac{\partial f}{\partial y}$ are defined and continuous on the rectangle $R = \{(x, y); x_0 \le x \le x_0 + a, y_0 - b \le y \le y_0 + b\}$. Let $M = \max\limits_{(x,y)\in R} |f(x, y)|$ and $\gamma = \min(a, \dfrac{b}{M})$. Let $K$ and $L$ represent two positive numbers such that*

$$\max_{(x,y)\in R} \left| \frac{\partial f}{\partial y} \right| \le K, \qquad \max_{(x,y)\in R} \left| \frac{\partial f}{\partial x} + f \frac{\partial f}{\partial y} \right| \le L.$$

[2]The authors wish to thank Thomas M. Green for the development of the flowcharts and the computer and calculator exercises.

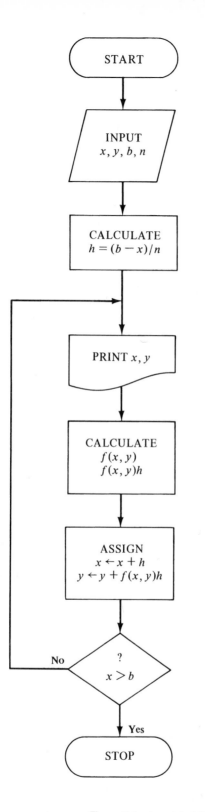

1. Input initial values:
$$x \leftarrow x_0$$
$$y \leftarrow y_0$$

2. $h = \Delta x$

3. Print $\{(x_i, y_i)\}_{i}^{n} = 0$

4. $y' = f(x, y)$
$$dy = f(x, y)h$$

5. Increment the values of $x$ and $y$:
$$x_i = x_{i-1} + h$$
$$y_i = y_{i-1} + (dy)_{i-1}$$
$$i = 1, 2, 3, \ldots, n$$

6. Test value of $x$.

**Figure 7.4** Annotated Flowchart for the Euler Method.

*If $E_i$ represents the error made in the ith step by approximating the actual solution $y(x_i)$ by the Euler approximation $y_i = y_{i-1} + hf(x_{i-1}, y_{i-1})$, then*

$$E_i \leq \alpha h, \quad i = 1, 2, \ldots, n$$

*where $\alpha = L(e^x K - 1)/2K$.*

## EXERCISES

In all problems, maintain four-decimal-place accuracy. In Exercises 1 through 6, use the Euler method with $h = 0.2$ to solve each of the IVPs on $0 \leq x \leq 1$.

**1.** $y' = x - 2y$
$y(0) = 3$

**2.** $y' = e^{-y} - y + 1$
$y(0) = 1$

**3.** $y' = 3x^2 - 2y$
$y(0) = 0$

**4.** $y' = \sin y + x$
$y(0) = 0$

**5.** $y' = y^2 - 2x^2 + 1$
$y(0) = 0$

**6.** $y' = 2xy^{-1}$
$y(0) = 1$

**7.** Write a computer or calculator program to repeat Exercise 1 using (a) $h = 0.1$ and (b) $h = 0.01$ and compare with the results of Exercise 1.

**8.** Write a computer or calculator program to repeat Exercise 3 using (a) $h = 0.1$ and (b) $h = 0.01$ and compare with the results of Exercise 3.

**9.** Write a computer or calculator program to repeat Exercise 5 using (a) $h = 0.1$ and (b) $h = 0.01$ and compare with the results of Exercise 5.

**10.** Solve Exercise 6 exactly and compare the actual set of points $(x_i, y_i)$, $i = 1, 2, 3, 4, 5$ with the approximate set obtained in Exercise 6.

**11.** Given the IVP $y' = 3xy^2$, $y(\frac{1}{10}) = 0.2$, use the Euler method with $h = 0.25$ to approximate $y(0.85)$. Compare your result with the actual value.

**12.** Solve the IVP $y' = x + y$, $y(0) = 0$, on the interval $0 \leq x \leq 1$. Use the Euler method with $h = 0.2$. Compare with the exact results.

**13.** Repeat Exercise 12 with $h = 0.1$.

**14.** Write a computer or calculator program to repeat Exercise 12 with $h = 0.5, 0.25, 0.125, 0.0625, \ldots$, allowing $h$ to decrease until there is no change in the first three decimal places of the approximations for $y(1)$. Adjust the program so that only $(x_n, y_n)$ is printed and not the other points on the interval for $i < n$.

# 7.3 TAYLOR-SERIES METHOD

In Section 2.9 and in Chapter 5 we demonstrated how the Taylor series can be used to generate series representations of solutions of a differential equation. In some cases it is possible to generate the complete series, and in other cases it is not possible (or perhaps not feasible) to obtain the complete series. The

Taylor series can be used in another way to obtain an approximation to the value of the solution of an IVP at specific values of $x$. Given the IVP

$$y' = f(x,y), \qquad x_0 \le x \le b \tag{1}$$

$$y(x_0) = y_0, \tag{2}$$

we again wish to obtain a sequence of points $\{(x_i, y_i)\}_{i=0}^{n}$ that lie close to the solution curve on the interval $[x_0, b]$. We recall that the Taylor-series representation of the solution $y$ about $x_0$ is

$$y(x) = \sum_{n=0}^{\infty} \frac{y^{(n)}(x_0)}{n!} (x - x_0)^n.$$

Setting $x = x_1 = x_0 + h$, we have

$$y(x_1) = y(x_0 + h) = \sum_{n=0}^{\infty} \frac{y^{(n)}(x_0)}{n!} h^n$$

$$= y(x_0) + y'(x_0)h + \frac{y''(x_0)}{2!} h^2 + \frac{y'''(x_0)}{3!} h^3 + \cdots$$

$$= y_0 + f(x_0, y_0)h + \frac{y''(x_0)}{2!} h^2 + \frac{y'''(x_0)}{3!} h^3 + \cdots$$

The Euler method can be thought of as an approximation of the Taylor series by retaining only the first two terms of this series. In general, if the Taylor series is approximated by

$$y(x_0) + y'(x_0)h + \cdots + \frac{y^{(l)}(x_0)}{l!} h^l,$$

we say that the approximation is a *Taylor approximation of order l*. In this language the Euler method is a Taylor approximation of order 1.

If we assume that all partial derivatives of $f(x,y)$ of any order exist at $(x_0, y_0)$, then the differential equation (1) and the initial condition (2) can be used to determine each $y^{(k)}(x_0)$, $k = 2, 3, \ldots$, and consequently $y_1$. For example,

$$y'(x_0) = y'(x) \Big|_{\substack{x=x_0 \\ y=y_0}} = f(x_0, y_0)$$

$$y''(x_0) = \frac{d}{dx}(y'(x)) \Big|_{\substack{x=x_0 \\ y=y_0}} = \frac{d}{dx}(f(x, y)) \Big|_{\substack{x=x_0 \\ y=y_0}}$$

$$= \left[ \frac{\partial f}{\partial x} + \frac{\partial f}{\partial y}\frac{dy}{dx} \right] \Big|_{\substack{x=x_0 \\ y=y_0}} = f_x(x_0, y_0) + f_y(x_0, y_0) \cdot f(x_0, y_0).$$

Subsequent derivatives can be obtained in like fashion. Knowing $(x_1, y_1)$ and $x_2$, we can then calculate $y_2$ in a similar manner. Thus,

$$y_2 = y(x_2) = y(x_1 + h) = \sum_{n=0}^{\infty} \frac{y^{(n)}(x_1)}{n!} h^n,$$

where $y(x_1) = y_1$, $y'(x_1) = f(x_1, y_1)$, $y''(x_1) = f_x(x_1, y_1) + f_y(x_1, y_1) \cdot f(x_1, y_1)$, and so on. Additional points are obtained in the same way.

In any application of the Taylor method, one must specify the order $l$ of the approximation. The larger $l$ is, the more accurate we anticipate the results to be. Once again the computations involved increase with increasing $l$. We illustrate the method by repeating Example 1 of Section 7.2.

**EXAMPLE 1**  Use the Taylor approximation of order 2 to obtain a numerical solution of the IVP

$$y' = x^2 + y^2 \tag{3}$$

$$y(0) = 0 \tag{4}$$

on $0 \le x \le 1$.

**Solution**  We choose $h = 0.2$. From the differential equation (3), we obtain

$$y' = x^2 + y^2$$

$$y'' = 2x + 2yy' = 2x + 2y(x^2 + y^2)$$

$$= 2x + 2x^2 y + 2y^3.$$

In general, we have

$$y_i = y(x_i) = y(x_{i-1} + h) \approx y(x_{i-1}) + y'(x_{i-1})h + \frac{y''(x_{i-1})}{2}h^2$$

$$\approx y_{i-1} + f(x_{i-1}, y_{i-1})h + \{f_x(x_{i-1}, y_{i-1})$$

$$+ f_y(x_{i-1}, y_{i-1}) \cdot f(x_{i-1}, y_{i-1})\} \frac{h^2}{2}$$

$$\approx y_{i-1} + (x_{i-1}^2 + y_{i-1}^2)h$$

$$+ (2x_{i-1} + 2x_{i-1}^2 y_{i-1} + 2y_{i-1}^3) \frac{h^2}{2}.$$

We present the results in Table 7.2, incorporating the intermediate calculations.

Thus, our numerical solution of the IVP (3)–(4), by the Taylor approximation of order 2 with $h = 0.2$, on the interval $0 \le x \le 1$, is the set

$\{(0,0), (0.2,0), (0.4,0.016), (0.6, 0.064154), (0.8, 0.161911), (1, 0.331469)\}$.

*For the Taylor approximation of order $l$ it can be shown that the error is proportional to $h^l$. Thus we anticipate that the Taylor approximation of order*

**Table 7.2** Numerical Solution of the IVP (3)–(4) by the Taylor Approximation of Order 2

| $i$ | $x_i$ | $y_i$ | $[x_i^2 + y_i^2]h$ | $[2x_i + 2x_i^2 y_i + 2y_i^3]\dfrac{h^2}{2}$ | $x_{i+1}$ | $y_{i+1}$ |
|---|---|---|---|---|---|---|
| 0 | 0 | 0 | $[0^2 + 0^2](0.2) = 0$ | $[2(0) + 2(0)^2(0) + 2(0)^3]\dfrac{(0.2)^2}{2} = 0$ | 0.2 | 0 |
| 1 | 0.2 | 0 | $[(0.2)^2 + 0^2](0.2) = 0.008$ | $[2(0.2) + 2(0.2)^2(0) + 2(0)^3]\dfrac{(0.2)^2}{2} = 0.008$ | 0.4 | 0.016 |
| 2 | 0.4 | 0.016 | $[(0.4)^2 + (0.016)^2](0.2)$ $= 0.032051$ | $[2(0.4) + 2(0.4)^2(0.016) + 2(0.016)^3]\dfrac{(0.2)^2}{2}$ $= 0.016103$ | 0.6 | 0.064154 |
| 3 | 0.6 | 0.064154 | $[(0.6)^2 + (0.064154)^2](0.2)$ $= 0.072823$ | $[2(0.6) + 2(0.6)^2(0.064154) + 2(0.064154)^3]\dfrac{(0.2)^2}{2}$ $= 0.024934$ | 0.8 | 0.161911 |
| 4 | 0.8 | 0.161911 | $[(0.8)^2 + (0.161911)^2](0.2)$ $= 0.133243$ | $[2(0.8) + 2(0.8)^2(0.161911) + 2(0.161911)^3]\dfrac{(0.2)^3}{2}$ $= 0.036315$ | 1.0 | 0.331469 |

2 is more accurate than the Taylor approximation of order 1. Of course a higher-order Taylor approximation would be even more accurate. Naturally, the higher the order, the more laborious the calculations. Special forms of "higher-order methods" are presented in the next section. The forms presented are those commonly called the Runge–Kutta methods.

The reader will be asked to compare the various methods of this chapter in Review Exercises. Some exercises involve IVPs that can be solved exactly, and therefore the numerical results of each method can be compared to the actual results as well as comparisons with the other methods.

We have only to make minor modifications in our flowchart for the Euler method to produce a flowchart for the Taylor-series method. Namely, we change steps 4 and 5 as in Figure 7.5.

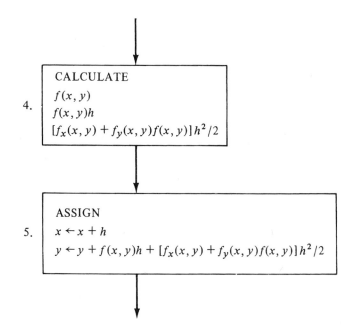

**Figure 7.5.** Changes in Flowchart Presented in Section 7.2 to Produce Flowchart for Taylor-Series Approximation of Order 2.

## EXERCISES

In all problems, maintain four-decimal-place accuracy. In Exercises 1 through 8, use the Taylor approximation of order 2 with $h = 0.2$ to solve the IVPs on $0 \le x \le 1$.

**1.** $y' = x^2 + y$
$y(0) = 0$

**2.** $y' = x^2 + e^y$
$y(0) = 0$

**3.** $y' = x - y$
$y(0) = 1$

**4.** $y' = \sin y + x$
$y(0) = 0$

**5.** $y' = xy$
$y(0) = 1$

**6.** $y' = 3x + 2y$
$y(0) = -1$

**7.** $y' = x^2y$
$y(0) = 2$

**8.** $y' = e^y \sin y$
$y(0) = 0.1$

**9.** Use the Taylor approximation of order 2 with $h = 0.2$ to solve the IVP

$$y' = x + y$$

$$y(0) = 0$$

on $0 \le x \le 1$. Compare with the exact results. Also compare with the Euler method (Exercise 12 in Section 7.2).

**10.** Use the Taylor approximation of order 3 with $h = 0.2$ to solve Exercise 9. Compare with the results of Exercise 9 and with the exact results. [*Hint:* Calculate $y''$ and $y'''$ directly from the differential equation.]

**11.** Repeat Exercise 9 with $h = 0.1$ and compare with the corresponding results of Exercise 9 and with the exact values. For instance, $y(0.2)$ of this exercise is to be compared with $y(0.2)$ of Exercise 9 and with the exact value of $y(0.2)$.

**12.** Repeat Exercise 10 with $h = 0.1$ and compare with the corresponding results of Exercise 10 and with the exact results (see Exercise 11).

In Exercises 13 through 15 write a computer or calculator program for the Taylor approximation of order 2.

**13.** Repeat Exercise 1 with (a) $h = 0.1$ and (b) $h = 0.01$.

**14.** Repeat Exercise 3 with (a) $h = 0.1$ and (b) $h = 0.01$.

**15.** Repeat Exercise 6 with (a) $h = 0.1$ and (b) $h = 0.01$.

# 7.4  RUNGE–KUTTA METHODS

In the description of the Taylor approximation of order 2, we wrote

$$y(x_0 + h) \approx y(x_0) + f(x_0, y_0)h + [f_x(x_0, y_0) + f_y(x_0, y_0) \cdot f(x_0, y_0)]\frac{h^2}{2}.$$

The Taylor approximation of order 3 would have the additional term

$$\{f_{xx}(x_0, y_0) + 2f_{xy}(x_0, y_0) \cdot f(x_0, y_0) + f_{yy}(x_0, y_0)[f(x_0, y_0)]^2$$

$$+ f_x(x_0, y_0)f_y(x_0, y_0) + [f_y(x_0, y_0)]^2 f(x_0, y_0)\} \frac{h^3}{6}.$$

Based on our knowledge of the Taylor series (or convergent series in general) from calculus, we know that the accuracy of our approximation improves with the number of terms retained. On the other hand, the description of the higher-order terms gets more and more complicated and the associated calculations more profuse. The basic idea of the *Runge–Kutta methods*[3] is to preserve the order of a Taylor approximation (in the sense of the error involved) while eliminating the necessity of calculating the various partial derivatives of $f$ that are involved. The alternative proposed by these methods involves evaluating the function $f$ at certain judicious points rather than evaluating the specific partial derivatives. We present here the Runge–Kutta formulas of order 3 and order 4.[4]

The Runge–Kutta approximation of order 3 is defined by the following collection of formulas:

$$k_1 = hf(x_i, y_i) \tag{1}$$

$$k_2 = hf(x_i + \tfrac{1}{2}h, y_i + \tfrac{1}{2}k_1) \tag{2}$$

$$k_3 = hf(x_i + h, y_i + 2k_2 - k_1) \tag{3}$$

$$y_{i+1} = y_i + \tfrac{1}{6}(k_1 + 4k_2 + k_3). \tag{4}$$

The scheme then is to obtain $y_1$ from formulas (1)–(4) by setting $i = 0$. With this value of $y_1$ and with $x_1$ we obtain $y_2$ from formulas (1)–(4) by setting $i = 1$. This process is repeated until we have the desired number of points.

**EXAMPLE 1**  Use the Runge–Kutta approximation of order 3 with $h = 0.2$ to solve the IVP

$$y' = x^2 + y^2 \tag{5}$$

$$y(0) = 0 \tag{6}$$

on $0 \le x \le 1$.

**Solution**  As before, it is convenient to display our solution (see Table 7.3). Once again we show the intermediate calculations, in the hope that they help the reader to follow the development.

Our numerical solution using the Runge–Kutta approximation of order 3 with $h = 0.2$ of the IVP (5)–(6) on $0 \le x \le 1$ is given by the set

$\{(0, 0), (0.2, 0.002667), (0.4, 0.021371), (0.6, 0.072509),$

$$(0.8, 0.174273), (1, 0.350721)\}.$$

---

[3]Runge first developed the approximation of order 3, and Kutta generalized and refined the approximations of orders 3 and 4.

[4]For a development of these formulas and an analysis of their accuracy, the reader is referred to K. S. Kunz, *Numerical Analysis* (New York: McGraw-Hill, 1957).

**Table 7.3**  Numerical Solution of the IVP (5)–(6) Using the Runge–Kutta Method of Order 3

| $i$ | $x_i$ | $y_i$ | $k_1$ | $k_2$ | $k_3$ | $\frac{1}{6}(k_1 + 4k_2 + k_3)$ | $x_{i+1}$ | $y_{i+1}$ |
|---|---|---|---|---|---|---|---|---|
| 0 | 0 | 0 | $(0.2)[0^2 + 0^2] = 0$ | $(0.2)[(.1)^2 + 0^2]$ $= 0.002$ | $(0.2)[(.2)^2 + (.004)^2]$ $= 0.008003$ | $\frac{1}{6}[0 + 4(.002) + .008003]$ $= 0.002667$ | 0.2 | 0.002667 |
| 1 | 0.2 | 0.002667 | $0.2[(.2)^2 + (.002667)^2]$ $= 0.008001$ | $0.2[(.3)^2 + (.006668)^2]$ $= 0.018009$ | $0.2[(.4)^2 + (.030684)^2]$ $= 0.032188$ | $\frac{1}{6}[.008001 + 4(.018009) + .032188]$ $= 0.018704$ | 0.4 | 0.021371 |
| 2 | 0.4 | 0.021371 | $0.2[(.4)^2 + (.021371)^2]$ $= 0.032091$ | $0.2[(.5)^2 + (.037417)^2]$ $= 0.050280$ | $0.2[(.6)^2 + (.089840)^2]$ $= 0.073614$ | $\frac{1}{6}[.032091 + 4(.050280) + .073614]$ $= 0.051138$ | 0.6 | 0.072509 |
| 3 | 0.6 | 0.072509 | $0.2[(.6)^2 + (.072509)^2]$ $= 0.073052$ | $0.2[(.7)^2 + (.109035)^2]$ $= 0.100378$ | $0.2[(.8)^2 + (.200213)^2]$ $= 0.136017$ | $\frac{1}{6}[.073052 + 4(.100378) + .136017]$ $= 0.101764$ | 0.8 | 0.174273 |
| 4 | 0.8 | 0.174273 | $0.2[(.8)^2 + (.174273)^2]$ $= 0.134074$ | $0.2[(.9)^2 + (.241388)^2]$ $= 0.173646$ | $0.2[(1)^2 + (.387491)^2]$ $= 0.230030$ | $\frac{1}{6}[.134074 + 4(.173646) + .230030]$ $= 0.176448$ | 1.0 | 0.350721 |

The Runge–Kutta approximation of order 4 is defined by the following formulas:

$$l_1 = hf(x_i, y_i) \tag{7}$$

$$l_2 = hf(x_i + \tfrac{1}{2}h, y_i + \tfrac{1}{2}l_1) \tag{8}$$

$$l_3 = hf(x_i + \tfrac{1}{2}h, y_i + \tfrac{1}{2}l_2) \tag{9}$$

$$l_4 = hf(x_i + h, y_i + l_3) \tag{10}$$

$$y_{i+1} = y_i + \tfrac{1}{6}(l_1 + 2l_2 + 2l_3 + l_4). \tag{11}$$

We employ formulas (7)–(11) in exactly the same manner as the Runge–Kutta formulas of order 3. We illustrate with an example.

**EXAMPLE 2** Solve the IVP

$$y' = x^2 + y^2 \tag{12}$$

$$y(0) = 0 \tag{13}$$

and $0 \le x \le 1$. Use the Runge–Kutta approximation of order 4 with $h = 0.2$.

**Solution** As in the previous example, we demonstrate our results in a table (see Table 7.4).

Our solution of the IVP (12)–(13) on $0 \le x \le 1$ using the Runge–Kutta approximation of order 4 is the set

$\{(0, 0), (0.2, 0.002667), (0.4, 0.021360), (0.6, 0.072451),$

$(0.8, 0.174090), (1, 0.350257)\}.$

Steps 4 and 5 of our flowchart given in Section 7.2 would be changed to those in Figure 7.6 in order to produce a flowchart for the Runge–Kutta method of order 4.

The Runge–Kutta method is highly accurate (small truncation error) even when $n$ is relatively small (small round-off error). For these reasons it is a widely used method.

It can be shown that the error involved in the Runge–Kutta approximation of order $k(k = 3, 4)$ is proportional to $h^k$. Thus the Runge–Kutta method is more accurate than the Euler method and although it has the same order of accuracy as the Taylor approximation of order $k$, it has the feature that the calculations depend on knowledge of the function $f$ only and not on derivatives of $f$ as in the case of the Taylor method.

**EXERCISES**

In all problems, maintain four-decimal-place accuracy. In Exercises 1 through 8, use the Runge–Kutta approximation of order 3 with $h = 0.2$ to solve the IVPs on $0 \le x \le 1$.

**1.** $y' = 3x + 2y$      **2.** $y' = y^2 + 2y - x$      **3.** $y' = x - 4y$
      $y(0) = 1$            $y(0) = 0$             $y(0) = 0$

**4.** $y' = 2xy - x^2$      **5.** $y' = y^2 - 2x$      **6.** $y' = 2yx^2 + 4$
      $y(0) = 0$            $y(0) = 0$             $y(0) = 1$

**7.** $y' = 3y + 2x^2 - x$      **8.** $y' = y + y^2$
      $y(0) = 0$            $y(0) = -1$

**9.** Use the Runge–Kutta approximation of order 4 with $h = 0.2$ to solve the IVP of Exercise 1 on $0 \le x \le 1$.

**10.** Use the Runge–Kutta approximation of order 4 with $h = 0.2$ to solve the IVP of Exercise 3 on $0 \le x \le 1$.

**11.** Use the Runge–Kutta approximation of order 4 with $h = 0.2$ to solve the IVP of Exercise 5 on $0 \le x \le 1$.

**Table 7.4**   Numerical Solution of the IVP (12)–(13) Using the Runge–Kutta Method of Order 4

| $i$ | $x_i$ | $y_i$ | $l_1$ | $l_2$ | $l_3$ |
|---|---|---|---|---|---|
| 0 | 0 | 0 | $0.2[0^2 + 0^2] = 0$ | $0.2[(.1)^2 + 0^2] = 0.002$ | $0.2[(.1)^2 + (.001)^2]$ $= 0.002$ |
| 1 | 0.2 | 0.002667 | $0.2[(.2)^2 + (.002667)^2]$ $= 0.008001$ | $0.2[(.3)^2 + (.006668)^2]$ $= 0.018009$ | $\frac{1}{5}[(.3)^2 + (.011672)^2]$ $= 0.018027$ |
| 2 | 0.4 | 0.021360 | $\frac{1}{5}[(.4)^2 + (.021360)^2]$ $= 0.032091$ | $\frac{1}{5}[(.5)^2 + (.037406)^2]$ $= 0.050280$ | $\frac{1}{5}[(.5)^2 + (.046500)^2]$ $= 0.050432$ |
| 3 | 0.6 | 0.072451 | $\frac{1}{5}[(.6)^2 + (.072451)^2]$ $= 0.073050$ | $\frac{1}{5}[(.7)^2 + (.108976)^2]$ $= 0.100375$ | $\frac{1}{5}[(.7)^2 + (.122639)^2]$ $= 0.101008$ |
| 4 | 0.8 | 0.174090 | $\frac{1}{5}[(.8)^2 + (.174090)^2]$ $= 0.134061$ | $\frac{1}{5}[(.9)^2 + (.241121)^2]$ $= 0.173628$ | $\frac{1}{5}[(.9)^2 + (.260904)^2]$ $= 0.175614$ |

| $l_4$ | $\frac{1}{6}(l_1 + 2l_2 + 2l_3 + l_4)$ | $x_{i+1}$ | $x_{i+1}$ |
|---|---|---|---|
| $0.2[(.2)^2 + (.002)^2]$ $= 0.008001$ | $\frac{1}{6}[(0 + 2(.002) + 2(.002) + 0.008001]$ $= 0.002667$ | 0.2 | 0.002667 |
| $\frac{1}{5}[(.4)^2 + (.020694)^2]$ $= 0.032086$ | $\frac{1}{6}[(.008001 + 2(.018009)$ $+ 2(.018027) + 0.032086] = 0.0018693$ | 0.4 | 0.021360 |
| $\frac{1}{5}[(.6)^2 + (.071792)^2]$ $= 0.073031$ | $\frac{1}{6}[(0.032091 + 2(.050280)$ $+ 2(.050432) + 0.073031] = 0.051091$ | 0.6 | 0.072451 |
| $\frac{1}{5}[(.8)^2 + (.173459)^2]$ $= 0.134018$ | $\frac{1}{6}[.073050 + 2(.100375)$ $+ 2(.101008) + 0.134018] = 0.101639$ | 0.8 | 0.174090 |
| $\frac{1}{5}[(1)^2 + (.349704)^2]$ $= 0.224459$ | $\frac{1}{6}[.134061 + 2(.173629)$ $+ 2(.175614) + 0.224459] = 0.176167$ | 1.0 | 0.350257 |

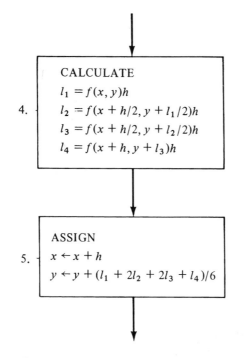

**Figure 7.6** Changes in Flowchart Presented in Section 7.2 to Produce Flowchart for the Runge–Kutta Approximation of Order 4.

**12.** Use the Runge–Kutta approximation of order 4 with $h = 0.2$ to solve the IVP of Exercise 4 on $0 \leq x \leq 1$.

**13.** Use the Runge–Kutta approximation of order 3 with $h = 0.2$ to solve the IVP

$$y' = x + y$$

$$y(0) = 0$$

on $0 \leq x \leq 1$. Compare with the exact results.

**14.** Use the Runge–Kutta approximation of order 4 with $h = 0.2$ to solve the IVP

$$y' = x + y$$

$$y(0) = 0$$

on $0 \leq x \leq 1$. Compare with the exact results and compare with the results of Exercise 13. Also compare with the Euler method (Exercise 12 in Section 7.2).

In Exercises 15 through 18 write a computer or calculator program for the Runge–Kutta approximation.

**15.** Repeat Exercise 13 with (a) $h = 0.1$ and (b) $h = 0.01$.

**16.** Repeat Exercise 14 with (a) $h = 0.1$ and (b) $h = 0.01$.

**17.** Repeat Exercise 1 with (a) $h = 0.1$ and (b) $h = 0.01$.

**18.** Repeat Exercise 3 with (a) $h = 0.1$ and (b) $h = 0.01$.

# 7.5 SYSTEMS OF FIRST-ORDER DIFFERENTIAL EQUATIONS

The methods presented in Sections 7.2 through 7.4 can also be applied to initial value problems for a system of first-order differential equations. We restrict our discussion to systems with two unknown functions, but the methods apply to systems with any number of unknown functions.

Thus, we consider the initial value problem

$$x' = f_1(t, x, y)$$

$$y' = f_2(t, x, y)$$

$$(1)$$

$$x(t_0) = x_0, \qquad y(t_0) = y_0. \tag{2}$$

In this system $x$ and $y$ are the unknown functions and $t$ is the independent variable. By a numerical solution of the initial value problem (1)–(2) on the interval $t_0 \leq t \leq b$, we mean a set of points $\{(t_i, x_i, y_i)\}_{i=0}^n$ where $t_n = b$, and $x_i$, $y_i$ are approximations to the solution of the system (1) when $t = t_i$. We assume that $\Delta t_i$ is the same for all $i$; thus, $\Delta t_i = h = (b - t_0)/n$.

For the initial value problem (1)–(2), the Euler formulas are as follows:

$$x_i = x_{i-1} + f_1(t_{i-1}, x_{i-1}, y_{i-1})h, \qquad i = 1, 2, \ldots, n \tag{3}$$

$$y_i = y_{i-1} + f_2(t_{i-1}, x_{i-1}, y_{i-1})h, \qquad i = 1, 2, \ldots, n. \tag{4}$$

**EXAMPLE 1** Assume $1 \leq t \leq 2$. Use the Euler method with $h = 0.2$ to solve the initial value problem

$$x' = t + x^2 - y^2 \tag{5}$$

$$y' = t^2 - 2x^2 + 3y^2 \tag{6}$$

$$x(1) = 0, \qquad y(1) = 0. \tag{7}$$

**Solution**　We exhibit the details of the solution in tabular form (Table 7.5). Thus, the numerical solution is the set

$\{(1, 0, 0), (1.2, 0.2, 0.2), (1.4, 0.44, 0.496), (1.6, 0.7095, 0.9582),$

$(1.8, 0.9466, 1.8197), (2.0, 0.8236, 4.0959)\}.$

**Table 7.5** Numerical Solution of the IVP (5)–(7) Using the Euler Method

| $t_i$ | $x_i$ | $y_i$ | $f_1(t_i, x_i, y_i)$ | $f_2(t_i, x_i, y_i)$ | $x_i + f_1(t_i, x_i, y_i)h$ | $y_i + f_2(t_i, x_i, y_i)h$ |
|---|---|---|---|---|---|---|
| 1 | 0 | 0 | $1 + 0^2 - 0^2 = 1$ | $1^2 - 2(0)^2 + 3(0)^2 = 1$ | $0 + 1(0.2) = 0.2$ | $0 + 1(0.2) = 0.2$ |
| 1.2 | 0.2 | 0.2 | $1.2 + (0.2)^2 - (0.2)^2 = 1.2$ | $(1.2)^2 - 2(0.2)^2 + 3(0.2)^2 = 1.48$ | $0.2 + (1.2)(0.2) = 0.44$ | $0.2 + (1.48)(.2) = 0.496$ |
| 1.4 | 0.44 | 0.496 | $1.4 + (0.44)^2 - (.496)^2 = 1.3476$ | $(1.4)^2 - 2(0.44)^2 + 3(.496)^2 = 2.3108$ | $0.44 + (1.3476)(0.2) = 0.7095$ | $0.496 + (2.3108)(.2) = 0.9582$ |
| 1.6 | 0.7095 | 0.9582 | $1.6 + (.7095)^2 - (.9582)^2 = 1.8533$ | $(1.6)^2 - 2(.7095)^2 + 3(.9582)^2 = 4.3074$ | $0.7095 + (1.8533)(0.2) = 0.9466$ | $0.9582 + (4.3074)(.2) = 1.8197$ |
| 1.8 | 0.9466 | 1.8197 | $1.8 + (.9466)^2 - (1.8197)^2 = -0.6151$ | $(1.8)^2 - 2(.9466)^2 + 3(1.8197)^2 = 11.3814$ | $0.9466 + (-0.6151)(0.2) = 0.8236$ | $1.8197 + (11.3814)(.2) = 4.0959$ |
| 2 | 0.8236 | 4.0959 | | | | |

The Taylor formulas corresponding to systems are rather complicated in general. For example, suppose we desire $d^2y/dt^2 \big|_{t=t_0}$, then

$$\frac{d^2y}{dt^2}\bigg|_{t=t_0} = \frac{d}{dt}[y'(t)]\bigg|_{\substack{t=t_0\\x=x_0\\y=y_0}} = \frac{d}{dt}[f_2(t, x, y)]\bigg|_{\substack{t=t_0\\x=x_0\\y=y_0}}$$

$$= \left[\frac{\partial f_2}{\partial t} + \frac{\partial f_2}{\partial x}\frac{dx}{dt} + \frac{\partial f_2}{\partial y}\frac{dy}{dt}\right]\bigg|_{\substack{t=t_0\\x=x_0\\y=y_0}}$$

$$= \frac{\partial f_2}{\partial t}(t_0, x_0, y_0) + \frac{\partial f_2}{\partial x}(t_0, x_0, y_0) \cdot f_1(t_0, x_0, y_0)$$

$$+ \frac{\partial f_2}{\partial y}(t_0, x_0, y_0) \cdot f_2(t_0, x_0, y_0).$$

Naturally, the higher the order of the Taylor approximation, the more complex the calculations become. We give here the formulas for a Taylor expansion of order 2.

$$x_{i+1} = x_i + f_1(t_i, x_i, y_i)h + \left[\frac{\partial f_1}{\partial t}(t_i, x_i, y_i) + \frac{\partial f_1}{\partial x}(t_i, x_i, y_i)\right.$$

$$\left. \cdot f_1(t_i, x_i, y_i) + \frac{\partial f_1}{\partial y}(t_i, x_i, y_i) \cdot f_2(t_i, x_i, y_i)\right]\frac{h^2}{2}$$

$$y_{i+1} = y_i + f_2(t_i, x_i, y_i)h + \left[\frac{\partial f_2}{\partial t}(t_i, x_i, y_i) + \frac{\partial f_2}{\partial x}(t_i, x_i, y_i)\right.$$

$$\left. \cdot f_1(t_i, x_i, y_i) + \frac{\partial f_2}{\partial y}(t_i, x_i, y_i) \cdot f_2(t_i, x_i, y_i)\right]\frac{h^2}{2}.$$

For the Runge–Kutta method, the generalization for systems is more straightforward. For example, the Runge–Kutta approximation of order 3 would be as follows.

$$\alpha_1 = hf_1(t_i, x_i, y_i)$$

$$\beta_1 = hf_2(t_i, x_i, y_i)$$

$$\alpha_2 = hf_1(t_i + \tfrac{1}{2}h, x_i + \tfrac{1}{2}\alpha_1, y_i + \tfrac{1}{2}\beta_1)$$

$$\beta_2 = hf_2(t_i + \tfrac{1}{2}h, x_i + \tfrac{1}{2}\alpha_1, y_i + \tfrac{1}{2}\beta_1)$$

$$\alpha_3 = hf_1(t_i + h, x_i + 2\alpha_2 - \alpha_1, y_i + 2\beta_2 - \beta_1)$$

$$\beta_3 = hf_2(t_i + h, x_i + 2\alpha_2 - \alpha_1, y_i + 2\beta_2 - \beta_1)$$

$$x_{i+1} = x_i + \tfrac{1}{6}(\alpha_1 + 4\alpha_2 + \alpha_3)$$

$$y_{i+1} = y_i + \tfrac{1}{6}(\beta_1 + 4\beta_2 + \beta_3).$$

The Runge–Kutta approximations of order 4 follow in a similar fashion and are left as an exercise.

## EXERCISES

Maintain four-decimal-place accuracy. In each exercise it is assumed that $0 \le t \le 1$.

1. Use the Euler method with $h = 0.2$ to obtain a numerical solution to the initial value problem

$$x' = 1 + t + y$$
$$y' = y + x^2$$
$$x(0) = 0, \quad y(0) = 0.$$

2. Use the Taylor approximation of order 2 with $h = 0.2$ to obtain a numerical solution of the initial value problem of Exercise 1.

3. Use the Runge–Kutta approximation of order 3 with $h = 0.2$ to solve the initial value problem of Exercise 1.

4. Write the Runge–Kutta formulas of order 4 for the initial value problem (1)–(2).

For each of the initial value problems in Exercises 5 through 10, use the Euler method with $h = 0.2$ to obtain a numerical solution on $0 \le t \le 1$.

5. $x' = t - y + x$
   $y' = x + 2y$
   $x(0) = 1, y(0) = 0$

6. $x' = t^2 - 3x^2 + y$
   $y' = 2t + x - y$
   $x(0) = 0, y(0) = 1$

7. $x' = x + y^2$
   $y' = -t + x - y$
   $x(0) = 1, y(0) = 1$

8. $x' = 5t + 6xy$
   $y' = -t - y + 3x$
   $x(0) = 1, y(0) = 0$

9. $x' = t^3 - y^2$
   $y' = t + x - xy$
   $x(0) = 0, y(0) = 1$

10. $x' = y$
    $y' = -2x + 3y$
    $x(0) = 2, y(0) = 3$

11. Exercise 10 can be solved explicitly (see Example 1, Section 3.2). Use this explicit solution to evaluate $x(1)$ and $y(1)$. Compare with the results of Exercise 10.

12. Solve Exercise 10 by the Taylor approximation of order 2 with $h = 0.2$, and compare the results with the actual values (see Exercise 11).

13. Solve Exercise 10 by the Runge–Kutta approximation of order 3 with $h = 0.2$ and compare the results with the actual values (see Exercise 11).

14. Write a computer or calculator program for the Runge–Kutta approximation of order 3 with $h = 0.1$ to solve Exercise 10.

## 7.6  APPLICATIONS

**Flood Waves**  ■ In his study of the behavior of flood waves in rivers, Stoker,[5] derives the following differential equation:

$$(v_0 + c_0)^2 \frac{dc_1}{d\xi} - 3c_1^2 + c_1 gS \left( \frac{1}{v_0} - \frac{2}{3c_0} \frac{gB}{gB + 2c_0^2} \right) = 0. \tag{1}$$

In this equation $g$ represents the acceleration due to gravity ($g = 32.16$ ft/sec$^2$), $B$ represents the width of the river (any cross section of the river perpendicular to the direction of the river flow is assumed to be a rectangle of constant width $B$ but variable depth $y$), $S$ represents the slope of the river bed, $v_0$ is the initial velocity of the wave, $c_0$ is related to the initial depth $y_0$ by $c_0 = (gy_0)^{1/2}$ (note that $c_0$ has units of velocity), $c_1$ is an unknown function that is related to the depth $y$ of the river by the approximation $c_1(\xi) \approx (\sqrt{gy} - c_0)/\tau$. Furthermore, $\xi = x$, $\tau = (v_0 + c_0)t - x$, and $t$ is time, $x$ is a space variable whose axis is parallel to the direction of flow of the river.

Equation (1) is a Bernoulli equation and hence could be solved by the method associated with that type of differential equation (see Exercise 19). The results are rather cumbersome, however, so numerical methods might be more appropriate as a means of solution (see Exercises 1 through 4).

**An Outstanding Long Jump**  ■ A novel and entertaining application of differential equations is presented in an article by Brearley.[6]

During the 1968 Olympic Games in Mexico City, R. Beamon of the United States set a world record of 29 feet $2\frac{1}{2}$ inches in the long jump. This improved the existing world mark by an astounding 2 feet $7\frac{5}{8}$ inches. Many critics attributed the length of the jump to the thinness of the air at Mexico City. In his article Brearley attempts to disclaim these critics. We present here Brearley's model only and recommend that the reader consult the article for Brearley's complete argument.

After takeoff, the center of mass $G$ of the long jumper describes a path that one can study as a projectile moving through a resisting medium. The equation of motion in vector form is

$$M \frac{d\mathbf{V}}{dt} = M\mathbf{g} + \mathbf{D}, \tag{2}$$

where $M$ is the mass of the long jumper, $\mathbf{V}$ is the velocity vector of $G$ at any instant, $\mathbf{g}$ is the acceleration vector due to gravity, and $\mathbf{D}$ is the air drag on the long jumper.

The direction of $\mathbf{D}$ is opposite to that of $\mathbf{V}$, and the magnitude of $\mathbf{D}$ is given experimentally as $k\rho V^2$, where $\rho$ is air density, $V = |\mathbf{V}|$, and $k$ is constant for fixed body posture. Hence, $\mathbf{D} = -\rho k V^2 \mathbf{U}$, where $\mathbf{U}$ is a unit vector in the direction of $\mathbf{V}$.

[5]J. J. Stoker, *Water Waves* (New York: Interscience, 1957).
[6]M. N. Brearley, "The Long Jump Miracle of Mexico City," *Math. Magazine* 45 (1972).

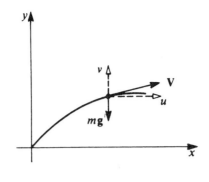

**Figure 7.7**

The path of $G$ is considered relative to a two-dimensional coordinate system $(x, y)$ with the origin located at the position of $G$ at $t = 0$ (takeoff). At time $t$ after takeoff, $u$ and $v$ represent the velocity components of $G$ parallel to the $x$ and $y$ axes, respectively (see Figure 7.7).

Equation (2) can be written as the following system

$$M \frac{du}{dt} = -\rho k u (u^2 + v^2)^{1/2} \tag{3}$$

$$M \frac{dv}{dt} = -Mg - \rho k v (u^2 + v^2)^{1/2}. \tag{4}$$

We also have the initial conditions $u(0) = u_0$, $v(0) = v_0$. Brearley gives the values

$$u_0 = 9.45 \text{ m/sec} \tag{5}$$

$$v_0 = 4.15 \text{ m/sec} \tag{6}$$

$$k = 0.182 \text{ m}^2 \tag{7}$$

$$\rho = 0.984 \text{ kg/m}^3 \quad \text{(at Mexico City)} \tag{8}$$

$$M = 80 \text{ kg} \quad \text{(approximate mass of R. Beamon)}. \tag{9}$$

Equations (3) and (4) are difficult to solve in their present form and hence can be studied numerically. Equations (3) and (4), treated as a system, can more conveniently be solved numerically as indicated in Section 7.5. However, we shall modify the problem by making an assumption that Brearley did not make. (See Exercises 5 through 8.)

**EXERCISES**

For the differential equation (1) applied to the Pawcatuck River in Rhode Island, take the following values: $B = 500$ ft, $S = 0.4$ ft/mile, $y_0 = 15$ ft, $v_0 = 2.19$

mph, and $c_0 = 15.6$ mph. With this information and $c_1(0) = 22.3$, solve Eq. (1) numerically on the interval $0 \leq \xi \leq 1$ using $h = 0.2$ by:

**1.** the Euler method;

**2.** the Taylor approximation of order 2;

**3.** the Runge–Kutta method of order 3;

**4.** the Runge–Kutta method of order 4.

In Eq. (3) we assume that $v = \frac{1}{2}(u - 1.15)$, so we obtain

$$M \frac{du}{dt} = -\rho k u[u^2 + \tfrac{1}{4}(u - 1.15)^2]^{1/2}, \tag{10}$$

which is a differential equation involving only the unknown function $u$. Using the values given in (5), (7), (8), and (9), solve the differential equation (10) numerically on the interval $0 \leq t \leq 1$ with $h = 0.2$ by:

**5.** the Euler method;

**6.** the Taylor approximation of order 2;

**7.** the Runge–Kutta method of order 3;

**8.** the Runge–Kutta method of order 4.

**Heat Loss**  Consider a mass $M$ of water of specific heat unity that is mixed so well that its temperature $T$ is the same throughout. $T_1$ represents the temperature of the medium surrounding the mass. The rate of loss of heat from the mass is given by the differential equation

$$M \frac{dT}{dt} = -A(T^4 - T_1^4) \tag{11}$$

Assuming that $M = 100$, $A = 3 \times 10^{-4}$, $T_1 = 10$, and $T_0 = T(0) = 30$, solve Eq. (11) numerically on $0 \leq t \leq 1$ with $h = 0.2$ by:

**9.** the Euler method;

**10.** the Taylor approximation of order 2;

**11.** the Runge–Kutta method of order 3;

**12.** the Runge–Kutta method of order 4.

**13.** The differential equation (11) can be solved by separation of variables and partial fractions. Solve Eq. (11) by these techniques. Observe that the expression involved in the solution is not convenient for obtaining numerical values for $T$.

**14. Gas Ionization**  A gas is to be ionized in such a way that the number of electrons per unit volume equals the number of positive ions per unit volume. Positive ions and electrons recombine to form neutral molecules at a rate equal to $kn^2$, where the constant $k$ is called the *constant of recombination*. We assume that the gas is initially ($t = 0$) un-ionized, and for $t > 0$, $A$ (a constant) ions per unit volume are produced. This leads to the IVP

$$\frac{dn}{dt} = A - kn^2 \tag{12}$$

$$n(0) = 0. \tag{13}$$

If we take $A = 100{,}000$ and $k = 5 \times 10^{-6}$, solve the IVP (12)–(13) numerically on $0 \leq t \leq 1$ with $h = 0.2$ by Euler's method.

**15.** Solve the IVP (12)–(13) exactly by separation of variables. Using the values of $A$ and $k$ given in Exercise 14, compare the numerical results of Exercise 14 with the actual results.

**16. Physics** In the study of the simple pendulum (see Chapter 8), the following differential equation occurs:

$$\frac{d\theta}{dt} = -\sqrt{\frac{g}{l}}(2 \cos \theta - 2 \cos \alpha)^{1/2}, \tag{14}$$

where $\theta$ is the angle that the pendulum makes with the vertical, $g$ is the acceleration due to gravity, $l$ is the length of the pendulum, and $\theta(0) = \theta_0 = \alpha$. If we assume that $\sqrt{g/l} = 2$ and $\alpha = \pi/6$, solve Eq. (14) numerically on $0 \leq t \leq 1$ with $h = 0.2$ using the Taylor approximation of order 2.

**17. Astronomy** If one takes into account the general theory of relativity when studying planetary orbits, the following differential equation occurs[7]:

$$\frac{d^2u}{d\theta^2} + u = \frac{GM}{h^2}(1 + \epsilon u^2). \tag{15}$$

In this differential equation $r$, $\theta$ are polar coordinates, the equation of the orbit is assumed in the form $r = r(\theta)$, $u = r^{-1}$, $M$ is the mass of the sun, $G$ is a universal constant, $h$ is twice the area swept out by the radius vector in unit time, and $\epsilon$ is a small parameter. If we set $v = du/d\theta$, we can write $d^2u/d\theta^2 = v(dv/du)$ and the differential equation (15) takes the form

$$\frac{dv}{du} = \frac{(GM/h^2)(1 + \epsilon u^2) - u}{v}. \tag{16}$$

For simplicity we take $GM/h^2 = 1$, $\epsilon = 0.01$ and $v(1) = 1$. Solve Eq. (16) numerically on $1 \leq u \leq 2$ with $h = 0.2$ by the Runge–Kutta method of order 3.

---

[7]C. C. Lin and L.A. Segel, *Mathematics Applied to Deterministic Problems in the Natural Sciences* (New York: Macmillan, 1974).

**18.** The following differential equation occurs in nonlinear oscillation theory in the consideration of limit cycles[8]:

$$\frac{dv}{d\xi} = \alpha^2 \frac{(v - v^2/3) - \xi}{v}, \tag{17}$$

where $v$ and $\xi$ are space variables and $\alpha$ is a parameter. Choose $\alpha = 0.1$, $v(0) = 1$, and solve Eq. (17) on $0 \leq \xi \leq 1$ with $h = 0.2$ using the Runge–Kutta method of order 4.

**19.** Equation (1) can be written in the form $(dc_1/d\xi) + ac_1 = bc_1^2$, where the constants $a$ and $b$ are given by

$$a = \frac{gS}{(v_0 + c_0)^2} \left( \frac{1}{v_0} - \frac{2}{3c_0} \frac{gB}{gB + 2c_0^2} \right)$$

$$b = \frac{3}{(v_0 + c_0)^2} \, .$$

Solve the differential equation $(dc_1/d\xi) + ac_1 = bc_1^2$ by the method associated with Bernoulli's differential equation leaving the answer in terms of $a$ and $b$. (See Exercise 27, Section 1.4.)

### REVIEW EXERCISES

Exercises 1 through 5 relate to the following initial value problem:

$$\frac{dy}{dx} = x + y + y^2 \tag{A}$$

$$y(1) = 0. \tag{B}$$

**1.** Solve the initial value problem (A)–(B) on the interval $1 \leq x \leq 2$ by the Euler method with $h = 0.2$.

**2.** Solve the initial value problem (A)–(B) on the interval $1 \leq x \leq 2$ by the Taylor approximation of order 2 with $h = 0.2$.

**3.** Solve the initial value problem (A)–(B) on the interval $1 \leq x \leq 2$ by the Taylor approximation of order 3 with $h = 0.2$.

**4.** Solve the initial value problem (A)–(B) on the interval $1 \leq x \leq 2$ by the Runge–Kutta approximation of order 3 with $h = 0.2$.

**5.** Solve the initial value problem (A)–(B) on the interval $1 \leq x \leq 2$ by the Runge–Kutta approximation of order 4 with $h = 0.2$.

Exercises 6 through 10 relate to the following initial value problem:

$$\dot{x} = -2xy, \qquad \dot{y} = x^2 + y^2 \tag{C}$$

$$x(1) = 1, \qquad y(1) = -1. \tag{D}$$

[8]J. J. Stoker, *Nonlinear Vibrations in Mechanical and Electrical Systems* (New York: Interscience, 1950).

**6.** Solve the initial value problem (C)–(D) on the interval $1 \leq t \leq 2$ by the Euler method with $h = 0.2$.

**7.** Solve the initial value problem (C)–(D) on the interval $1 \leq t \leq 2$ by the Runge–Kutta approximation of order 3 with $h = 0.2$.

**8.** Solve the initial value problem (C)–(D) on the interval $1 \leq t \leq 2$ by the Taylor approximation of order 2 with $h = 0.2$.

**9.** Solve the initial value problem (C)–(D) on the interval $1 \leq t \leq 2$ by the Runge–Kutta approximation of order 4 with $h = 0.2$.

**10.** If $(x(t),y(t))$ is a solution of the initial value problem (C)–(D), show that

$$y = -\sqrt{\frac{4-x^3}{3x}}.$$

**11.** Use the results of either Exercise 6, 7, 8, or 9 to determine whether or not the relationship in Exercise 10 is valid for all $t \geq 1$.

If we account[9] for the nonlinear dissipative effects in the neck of a Helmholtz cavity resonator, an approximate equation for free oscillations of such a resonator is of the form

$$m\frac{d^2x}{dt^2} + R\frac{dx}{dt}\left[1 + \beta\left(\frac{dx}{dt}\right)^2\right] + Kx = 0. \tag{E}$$

**12.** Setting $v = \frac{dx}{dt}$ and $v\frac{dv}{dx} = \frac{d^2x}{dt^2}$ in Eq. (E), show that this equation can be rewritten in the form

$$\frac{dv}{dx} = -\frac{\left(\frac{K}{R}\right)x + v(1 + \beta v^2)}{\left(\frac{m}{R}\right)v}.$$

Taking $\dfrac{m}{R} = 900/(K/R)$, $K/R = 100$, and $\beta = 2 \cdot 10^{-5}$, solve this equation on the interval $0 \leq x \leq 1$ by the Runge–Kutta approximation of order 3 with $h = 0.2$ if it is given that $v = 700$ cm/sec when $x = 0$.

Exercises 13 through 15 relate to the following initial value problem (see Example 2, Section 1.5):

$$\frac{dy}{dx} = \frac{(4-y)[(x\cos x)\ln(2y-8) + 1]}{x\sin x}, \tag{F}$$

$$y(1) = \frac{9}{2}. \tag{G}$$

---

[9]This is Exercise 3 in Philip M. Morse and K. Uno Ingard, *Theoretical Acoustics* (New York: McGraw-Hill, 1968), p. 883. Reprinted by permission of McGraw-Hill Book Company.

**13.** Solve the initial value problem (F)–(G) on the interval $1 \leq x \leq 2$ by the Euler method with $h = 0.2$ and compare the approximate results with the exact results.

**14.** Solve the initial value problem (F)–(G) on the interval $1 \leq x \leq 2$ by the Taylor approximation of order 2 with $h = 0.2$ and compare the approximate results with the exact results.

**15.** Solve the initial value problem (F)–(G) on the interval $1 \leq x \leq 2$ by the Runge–Kutta approximation of order 3 with $h = 0.2$ and compare the approximate results with the exact results.

# CHAPTER 8

# Nonlinear Differential Equations and Systems

## 8.1 INTRODUCTION

Nonlinear differential equations and systems of nonlinear differential equations occur frequently in applications. However, only a few types of nonlinear differential equations (for example, separable, homogeneous, exact) can be solved explicitly. The same is true for nonlinear systems. Fortunately, the situation is not as hopeless as it seems. We are often primarily interested in certain properties of the solutions and not the solutions themselves. For example, are the solutions bounded for all time? Are they periodic? Does the limit of the solution exist as $t \to +\infty$? A very effective technique in studying nonlinear differential equations and systems is to "linearize" them, that is, to approximate them by linear differential equations. Also, the coefficients and initial conditions of differential equations are often determined approximately. It is therefore very important to know what effect "small" changes in the coefficients of differential equations and their initial conditions have on their exact solutions. It is often possible to answer some of the questions above by utilizing the form of the differential equation and properties of its coefficients without the luxury of knowing the exact solution.

The questions raised above belong to that branch of differential equations known as *qualitative theory*. Our aim in this chapter is to present some elementary qualitative results for nonlinear differential equations and systems.

## 8.2 EXISTENCE AND UNIQUENESS THEOREMS

Here we state an existence and uniqueness theorem for systems of two differential equations of the first order with two unknown functions of the form

$$\begin{aligned}
\dot{x}_1 &= f_1(t, x_1, x_2) \\
\dot{x}_2 &= f_2(t, x_1, x_2),
\end{aligned} \tag{1}$$

subject to the initial conditions

$$x_1(t_0) = x_1^0, \qquad x_2(t_0) = x_2^0. \tag{2}$$

The same theorem applies to second-order differential equations of the form

$$\ddot{y} = f(t, y, \dot{y}), \tag{3}$$

subject to the initial conditions

$$y(t_0) = y^0, \qquad \dot{y}(t_0) = y^1. \tag{4}$$

In fact, if we set

$$y = x_1 \quad \text{and} \quad \dot{y} = x_2, \tag{5}$$

the differential equation (3) is equivalent to the system

$$\dot{x}_1 = x_2$$
$$\dot{x}_2 = f(t, x_1, x_2), \tag{6}$$

and the initial conditions (4) are equivalent to

$$x_1(t_0) = y^0, \qquad x_2(t_0) = y^1. \tag{7}$$

Before we state the existence and uniqueness theorem we explain what is meant by a *Lipschitz condition*.

### DEFINITION 1

*Let $f(t, x_1, x_2)$ be a function defined in a region $\mathcal{R}$ of the three-dimensional Euclidean space $R^3$. We say that $f(t, x_1, x_2)$ satisfies a Lipschitz condition (with respect to $x_1$ and $x_2$) in $\mathcal{R}$ if there exist constants $L_1$ and $L_2$ such that*

$$|f(t, x_1, x_2) - f(t, y_1, y_2)| \le L_1 |x_1 - y_1| + L_2 |x_2 - y_2|$$

*for all points $(t, x_1, x_2)$, $(t, y_1, y_2)$ in $\mathcal{R}$. The constants $L_1$ and $L_2$ are called Lipschitz constants. When the partial derivatives $\partial f / \partial x_1$ and $\partial f / \partial x_2$ exist and are continuous, it can be shown that $f$ satisfies a Lipschitz condition.*

### THEOREM 1   (Existence and Uniqueness)

*Let each of the functions $f_1(t, x_1, x_2)$ and $f_2(t, x_1, x_2)$ be continuous and satisfy a Lipschitz condition (with respect to $x_1$ and $x_2$) in the region*

$$\mathcal{R} = \left\{ (t, x_1, x_2): \begin{array}{l} |t - t_0| \le A \\ |x_1 - x_1^0| \le B \\ |x_2 - x_2^0| \le C \end{array} \right\}. \tag{8}$$

*Then the initial value problem (1)–(2) has a unique solution defined in some interval $a < t < b$ about the point $t_0$.*

### COROLLARY 1

*Let the function $f(t, x_1, x_2)$ be continuous and satisfy a Lipschitz condition (with respect to $x_1$ and $x_2$) in the region $\mathcal{R}$ defined by Eq. (8) with $x_1^0 = y^0$ and $x_2^0 = y^1$. Then the IVP (3)–(4) has a unique solution defined in some interval $a < t < b$ about the point $t_0$.*

Theorem 1 and its corollary can be easily extended to systems of $n$ differential equations with $n$ unknown functions and to differential equations of order $n$, respectively.

With the hypotheses of Theorem 1 it can be shown that the solutions of system (1) depend continuously on the initial conditions (2). Therefore, "small" changes in the initial conditions (2) will result in small changes in the solutions of system (1). This result is very important in physical applications, because we can never measure the initial conditions exactly.

### EXERCISES

**1.** Show that the function $f(t, x_1, x_2) = x_1^{1/2} + x_2$ does not satisfy a Lipschitz condition (with respect to $x_1$ and $x_2$) in the region

$$-1 \le t \le 1, \qquad 0 \le x_1 \le 1, \qquad 0 \le x_2 \le 1.$$

How about in the region

$$-1 \le t \le 1, \qquad 3 \le x_1 \le 4, \qquad 0 \le x_2 \le 1?$$

**2.** For what points $(t_0, y^0, y^1)$ does Corollary 1 imply that the IVP

$$\ddot{y} = \dot{y} + y^{1/2}$$

$$y(t_0) = y^0$$

$$\dot{y}(t_0) = y^1$$

has a unique solution?

**3.** Suppose that we make an error of magnitude $10^{-3}$ in measuring the initial conditions $A$ and $B$ in the IVP

$$\ddot{y} + y = 0$$

$$y(0) = A$$

$$\dot{y}(0) = B.$$

What is the largest error we make in evaluating the solution $y(t)$?

# 8.3  SOLUTIONS AND TRAJECTORIES OF AUTONOMOUS SYSTEMS

A system of differential equations of the form

$$\dot{x} = f(x, y), \qquad \dot{y} = g(x, y), \tag{1}$$

where the functions $f$ and $g$ are not time-dependent is called an *autonomous system*. Second-order differential equations of the form

$$\ddot{x} = F(x, \dot{x}) \tag{2}$$

can be also written as an autonomous system by setting

$$\dot{x} = y, \qquad \dot{y} = F(x, y).$$

For the sake of existence and uniqueness of solutions, we assume that each of the functions $f$ and $g$ is continuous and satisfies a Lipschitz condition (with respect to $x$ and $y$) in some region $\mathfrak{R}$ of the $xy$ plane. Then by Theorem 1 of Section 8.2, if $(x_0, y_0)$ is any point in $\mathfrak{R}$ and if $t_0$ is any real number, there exists a unique solution

$$x = x(t), \qquad y = y(t) \tag{3}$$

of system (1), defined in some interval $a < t < b$ which contains the point $t_0$, and satisfying the initial conditions

$$x(t_0) = x_0, \qquad y(t_0) = y_0. \tag{4}$$

Clearly, the solution (3) defines a curve in the three-dimensional space $t$, $x$, $y$. If we regard $t$ as a parameter, then as $t$ varies in the interval $a < t < b$, the points $(x(t), y(t))$ trace out a curve in the $xy$ plane called the *trajectory* or *orbit* of the solution (3). In the study of physical systems, the pair $(x, y)$ is called a *phase* of the system and, therefore, the $xy$ plane is usually called the *phase plane*. Let us emphasize again that a solution of system (1) is a curve in the three-dimensional space $t$, $x$, $y$, while a trajectory is a curve in the phase plane that is described parametrically by a solution of system (1).

For example, $x = \cos t$, $y = \sin t$, $-\infty < t < \infty$ is a solution of the system

$$\dot{x} = -y, \qquad \dot{y} = x. \tag{5}$$

This solution is a helix in the three-dimensional space $t$, $x$, $y$. And, as $t$ varies in the interval $-\infty < t < \infty$, the points $(\cos t, \sin t)$ trace out the unit circle $x^2 + y^2 = 1$ an infinite number of times. Thus, the trajectory of this solution is the circle $x^2 + y^2 = 1$ in the $xy$ plane. Another solution of system (5) is $x = \sin t$, $y = -\cos t$. The trajectory of this solution is again the unit circle $x^2 + y^2 = 1$. As this example shows, different solutions of a system may have the same trajectory.

With the assumptions on $f$ and $g$ made above, it can be shown that system (1) has exactly one trajectory through any point $(x_0, y_0) \in \mathfrak{R}$. If we know a solution (3) of system (1), we can find the corresponding trajectory by eliminating, if possible, the parameter $t$ from the two equations in (3). This was actually done in the example above. However, we can get a lot of information about the trajectories of a system even when we cannot find the solutions explicitly. In fact, in some cases we can find the trajectories explicitly, while an explicit solution of the system is impossible. To see this, assume that in some region $D$, $f(x, y) \neq 0$. Then, using the two equations in (1) and the chain rule, we obtain the first-order differential equation

$$\frac{dy}{dx} = \frac{dy}{dt} \cdot \frac{dt}{dx} = \frac{g(x, y)}{f(x, y)}. \tag{6}$$

The solutions of Eq. (6) give the trajectories of system (1) through the points of $D$. If we can solve Eq. (6), then we obtain explicitly the trajectories. Even if we are unable to find explicit solutions, we can use Eq. (6) to compute the slope of any trajectory through a point of $D$. Similarly, if $g(x, y) \neq 0$ in some region $S$, we can utilize the first-order differential equation

$$\frac{dx}{dy} = \frac{f(x, y)}{g(x, y)} \tag{7}$$

to compute the trajectories through the points of $S$.

The points $(x_0, y_0)$ in the $xy$ plane (the phase plane), where both functions $f$ and $g$ are zero, are of special interest. For such points neither Eq. (6) nor Eq. (7) is valid. Such points are called *critical points* (or *equilibrium points*) of system (1). Clearly, if $(x_0, y_0)$ is a critical point of system (1), then $x(t) = x_0$, $y(t) = y_0$ is a solution for all $t$. The trajectory corresponding to this solution is the single point $(x_0, y_0)$ in the phase plane.

As we mentioned above, Eq. (2) can be easily put in the form of system (1) by setting $\dot{x} = y$. Then

$$\dot{x} = y, \qquad \dot{y} = F(x, y). \tag{8}$$

If $(x_0, y_0)$ is a critical point of (8), it follows that $y_0 = 0$, and hence

$$x(t) = x_0, \qquad y(t) = 0 \tag{9}$$

is a solution of (8) for all $t$. Thus, if $x$ represents the position and $y$ the velocity of a particle moving according to the differential equation (2), it follows that $x = x_0$ is a position of equilibrium of the motion. With such an interpretation, we see that critical points are identified with positions of equilibrium.

**EXAMPLE 1**   Find the critical points and the trajectories of the following system of differential equations:

$$\dot{x} = -y, \qquad \dot{y} = x. \tag{10}$$

In particular, draw the trajectory corresponding to the solution of (10) with initial conditions

$$x(0) = 1, \qquad y(0) = \sqrt{3}. \tag{11}$$

**Solution**   The critical points of (10) are determined by the two equations $-y = 0$, $x = 0$. Thus, $(0, 0)$ is the only critical point of (10). Using Eq. (6), we see that the trajectories of (10) for $y \neq 0$ are the solution curves of the differential equation

$$\frac{dy}{dx} = -\frac{x}{y}, \qquad y \neq 0.$$

The general solution of this separable differential equation is

$$x^2 + y^2 = c^2. \tag{12}$$

For $y = 0$ we see from (10) that $x = 0$. Thus, $(0, 0)$ is also a trajectory that is contained in Eq. (12) for $c = 0$. Hence, Eq. (12), which is a one-parameter family of circles centered at the origin, gives all the trajectories of system (10)

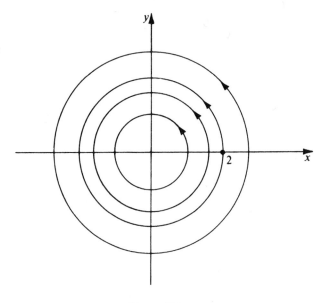

**Figure 8.1**

(see Figure 8.1). The arrows on the trajectories indicate the direction of increasing $t$ and can be found from system (10). For example, when $y > 0$, the first equation in (10) implies that $\dot{x} < 0$. This means that $x(t)$ decreases (and so it moves in the counterclockwise direction).

From Eq. (12) we see that the trajectory of the solution of the IVP (10)–(11) is the circle $x^2 + y^2 = 4$, centered at the origin with radius equal to 2.

Section 8.5 contains many worked-out examples in which we find the shape of all trajectories of linear systems.


**EXERCISES**

In Exercises 1 through 4, find the critical points and the trajectories of the system. Indicate the direction of increasing $t$.

**1.** $\dot{x} = x, \dot{y} = y$                          **2.** $\dot{x} = -x, \dot{y} = 2y$

**3.** $\dot{x} = y, \dot{y} = x$                          **4.** $\dot{x} = y, \dot{y} = -x$

In Exercises 5 through 10, compute the trajectory of the solution of the IVP. Indicate the direction of increasing $t$.

**5.** $\dot{x} = x, \dot{y} = 2y$,                $x(0) = 1, \quad y(0) = -1$

**6.** $\dot{x} = x, \dot{y} = -2y$,               $x(0) = 1, \quad y(0) = 1$

**7.** $\dot{x} = x, \dot{y} = x + y$,             $x(0) = 1, \quad y(0) = 0$

**8.** $\dot{x} = -x, \dot{y} = -x - y$,           $x(0) = 1, \quad y(0) = 0$

**9.** $\dot{x} = -x + y, \dot{y} = -x - y,$      $x(0) = -1, \ y(0) = 1$

**10.** $\dot{x} = -y, \dot{y} = x,$                         $x(0) = 1, \quad y(0) = -1$

In Exercises 11 through 14, find the critical points and the equations of the trajectories of the solutions of the system.

**11.** $\dot{x} = -x^2 + y^2, \dot{y} = 2xy$            **12.** $\dot{x} = y, \dot{y} = -\sin x$

**13.** $\dot{x} = y, \dot{y} = -4 \sin x$               **14.** $\dot{x} = x - xy, \dot{y} = -y + xy$

In Exercises 15 through 18, transform the differential equation into an equivalent system (by setting $\dot{x} = y$) and compute the equation of the trajectories.

**15.** $\ddot{x} + x = 0$   **16.** $\ddot{x} + \sin x = 0$   **17.** $\ddot{x} - x = 0$   **18.** $\ddot{x} - x + x^3 = 0$

**19.** (a) Show that the point $(0, 0)$ is the only critical point of the system

$$\dot{x} = ax + by, \quad \dot{y} = cx + dy$$

if and only if $ad - bc \neq 0.$

(b) Show that the system has a line of critical points if $ad - bc = 0.$

# 8.4   STABILITY OF CRITICAL POINTS OF AUTONOMOUS SYSTEMS

Consider again the autonomous system

$$\dot{x} = f(x, y), \quad \dot{y} = g(x, y). \tag{1}$$

We recall that a point $(x_0, y_0)$ is a critical point (or an equilibrium point) of system (1) if $f(x_0, y_0) = 0$ and $g(x_0, y_0) = 0$. Since the derivative of a constant is zero, it follows that if the point $(x_0, y_0)$ is a critical point of (1), then the pair of constant functions

$$x(t) \equiv x_0, \quad y(t) \equiv y_0 \tag{2}$$

is a solution of (1) for all $t$.

In many situations it is important to know whether every solution of (1) that starts sufficiently close to the solution (2) at $t = 0$ will remain close to (2) for all future time $t > 0$. If this is true, the solution (2), or the critical point $(x_0, y_0)$, is called stable. More precisely, we give the following definitions.

## DEFINITION 1

*The critical point $(x_0, y_0)$ (or the constant solution (2)) of system (1) is called stable if for every positive number $\epsilon$ there corresponds a positive number $\delta$ such that, every solution $(x(t), y(t))$ of (1) which at $t = 0$ satisfies*

$$[x(0) - x_0]^2 + [y(0) - y_0]^2 < \delta \tag{3}$$

*exists and satisfies*

$$[x(t) - x_0]^2 + [y(t) - y_0]^2 < \epsilon \tag{4}$$

*for all $t \geq 0$.*

**DEFINITION 2**

*A critical point* $(x_0, y_0)$ *[or the constant solution (2)] is called asymptotically stable if it is stable and if in addition there exists a positive number* $\delta_0$ *such that, every solution* $(x(t), y(t))$ *of (1) which at* $t = 0$ *satisfies*

$$[x(0) - x_0]^2 + [y(0) - y_0]^2 < \delta_0 \tag{5}$$

*exists for all* $t \geq 0$ *and satisfies*

$$\lim_{t \to \infty} x(t) = x_0, \qquad \lim_{t \to \infty} y(t) = y_0. \tag{6}$$

**DEFINITION 3**

*A critical point that is not stable is called unstable.*

Roughly speaking, stability means that a small change in the initial conditions causes only a small effect on the solution, asymptotic stability means that the effect of a small change tends to die out, while instability means that a small change in the initial conditions has a large effect on the solution.

The concepts of stable, asymptotically stable, and unstable critical point are illustrated in Figure 8.2(a), (b), and (c), respectively.

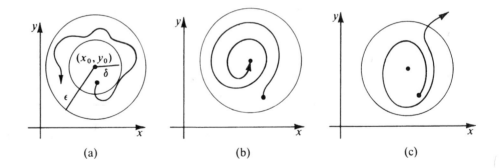

|     |     |     |
| :-: | :-: | :-: |
| (a) | (b) | (c) |

**Figure 8.2**

**EXAMPLE 1**  Show that the critical point $(0, 0)$ of the system

$$\dot{x} = -y, \qquad \dot{y} = x \tag{7}$$

is stable. Is it asymptotically stable?

**Solution**  We shall apply Definition 1. Let $\epsilon > 0$ be given. Choose $\delta = \epsilon$. Every solution of (7) is of the form

$$x(t) = c_1 \cos t + c_2 \sin t$$

$$y(t) = c_1 \sin t - c_2 \cos t,$$

where $c_1$ and $c_2$ are arbitrary constants. Here $x_0 = y_0 = 0$, $x(0) = c_1$, and $y(0) = -c_2$. We must show that (3) implies (4). In fact, $c_1^2 + c_2^2 < \delta$ implies that

$$(c_1 \cos t + c_2 \sin t)^2 + (c_1 \sin t - c_2 \cos t)^2 = c_1^2 + c_2^2 < \delta = \epsilon, \qquad (8)$$

and the proof that the critical point $(0, 0)$ of system (7) is stable is complete. From Eq. (8) we see that the trajectories of the system (7) are circles centered at the critical point $(0, 0)$. Hence, they do not approach the critical point $(0, 0)$ as $t \to \infty$. That is, Eq. (6) does not hold. Therefore, $(0, 0)$ is not an asymptotically stable critical point of (7).

**EXAMPLE 2**  Show that the critical point $(0, 0)$ of the system

$$\dot{x} = -x, \qquad \dot{y} = -y \qquad (9)$$

is asymptotically stable.

**Solution**  First, we must prove that $(0, 0)$ is stable. Let $\epsilon > 0$ be given. Choose $\delta = \epsilon$. The general solution of (9) is

$$x(t) = c_1 e^{-t}, \qquad y(t) = c_2 e^{-t},$$

where $c_1$ and $c_2$ are arbitrary constants. Here $x_0 = y_0 = 0$, $x(0) = c_1$, and $y(0) = c_2$. Again we must show that (3) implies (4). In fact, $c_1^2 + c_2^2 < \delta$ implies that

$$(c_1 e^{-t})^2 + (c_2 e^{-t})^2 = (c_1^2 + c_2^2)e^{-2t} \le c_1^2 + c_2^2 < \delta = \epsilon.$$

Thus, $(0, 0)$ is stable. Since (for any $c_1$ and $c_2$)

$$\lim_{t \to \infty} x(t) = \lim_{t \to \infty} c_1 e^{-t} = 0, \qquad \lim_{t \to \infty} y(t) = \lim_{t \to \infty} c_2 e^{-t} = 0,$$

it follows that the critical point $(0, 0)$ is asymptotically stable.

**EXAMPLE 3**  Show that the critical point $(0, 0)$ of the system

$$\dot{x} = -3x + 4y, \qquad \dot{y} = -2x + 3y \qquad (10)$$

is unstable.

**Solution**  If the critical point $(0, 0)$ were stable, then given $\epsilon > 0$ there should exist a $\delta > 0$ such that (3) implies (4). Consider the solution

$$x(t) = \frac{\sqrt{\delta}}{2} e^t, \qquad y(t) = \frac{\sqrt{\delta}}{2} e^t$$

of (10). Here $x_0 = y_0 = 0$, $x(0) = y(0) = \sqrt{\delta}/2$, and therefore Eq. (3) is satisfied. However, Eq. (4) becomes

$$\frac{\delta}{2} e^{2t} < \epsilon,$$

which cannot be true for all $t \geq 0$. Hence, $(0, 0)$ is an unstable critical point of the system (10).

When the autonomous system (1) is linear with constant coefficients, that is, when

$$\dot{x} = ax + by, \qquad \dot{y} = cx + dy, \tag{11}$$

where $a$, $b$, $c$, and $d$ are constants, we can obtain the solution explicitly. Therefore, it is not surprising that the stability character of the critical point $(0, 0)$ of system (11) is easy to study. We shall assume that $ad - bc \neq 0$. Then the point $(0, 0)$ is the only critical point of (11). [See Exercise 19(a) in Section 8.3.] As we explained in Section 3.3, the solutions of system (11) are of the form

$$x = Ae^{\lambda t}, \qquad y = Be^{\lambda t},$$

where $\lambda$ is a root of the characteristic equation

$$\lambda^2 - (a + d)\lambda + ad - bc = 0. \tag{12}$$

The stability character of the critical point $(0, 0)$ of system (11) depends almost entirely on the roots of Eq. (12). In fact, we have the following theorem.

**THEOREM 1**

*(a) The critical point $(0, 0)$ of system (11) is stable if, and only if, both roots of Eq. (12) are real and negative or have nonpositive real parts.*

*(b) The critical point $(0, 0)$ of system (11) is asymptotically stable if, and only if, both roots of Eq. (12) are real and negative or have negative real parts.*

*(c) The critical point $(0, 0)$ of system (11) is unstable if one (or both) of the roots of Eq. (12) is real and positive or if at least one of the roots has a positive real part.*

The three examples considered were all linear. In Example 1, the characteristic equation is (here $a = 0$, $b = -1$, $c = 1$, and $d = 0$) $\lambda^2 + 1 = 0$. Its roots are $\pm i$. Since they have zero real parts, using Theorem 1(a), it follows that the critical point $(0, 0)$ of system (7) is stable.

In Example 2, the characteristic equation is (here $a = -1$, $b = 0$, $c = 0$, and $d = -1$) $\lambda^2 + 2\lambda + 1 = 0$. Its roots are $\lambda_1 = \lambda_2 = -1$ (a double root). Since they are real and negative, using Theorem 1(b) we see that the critical point $(0, 0)$ of system (9) is asymptotically stable.

In Example 3, the characteristic equation is (here $a = -3$, $b = 4$, $c = -2$, and $d = 3$) $\lambda^2 - 1 = 0$. Its roots are $\lambda_1 = 1$ and $\lambda_2 = -1$. Since one of the roots is real and positive, using Theorem 1(c) it follows that the critical point $(0, 0)$ of system (10) is unstable.

Consider again the autonomous system (1) and assume that $(x_0, y_0)$ is a critical point of the system. Since a transformation of the form $X = x - x_0$, $Y = y - y_0$ transforms the autonomous system (1) into an equivalent system with $(0, 0)$ as a critical point, we assume, without loss of generality, that $(0, 0)$ is a critical point of system (1). An effective technique in studying the autonomous system (1) near the critical point $(0, 0)$ is to approximate it by a linear system of the form (11). In many cases one can prove that if the approximation is "good", the linear system (11) has solutions which themselves are "good" approximations to the solutions of system (1). The result that we are about to state is of this nature.

Assume that system (1) is of the form

$$\dot{x} = ax + by + F(x, y)$$
$$\dot{y} = cx + dy + G(x, y),$$
(13)

with $ad - bc \neq 0$ and $F(0, 0) = G(0, 0) = 0$. [Hence $(0, 0)$ is a critical point of (13).] Assume, furthermore, that near the origin $(0, 0)$, the functions $F$ and $G$ are continuous, have continuous first partial derivatives, and that

$$\lim_{\substack{x \to 0 \\ y \to 0}} \frac{F(x, y)}{\sqrt{x^2 + y^2}} = \lim_{\substack{x \to 0 \\ y \to 0}} \frac{G(x, y)}{\sqrt{x^2 + y^2}} = 0.$$
(14)

[Roughly speaking, the condition (14) means that the linear system (11) is a good approximation of system (13).] Then the following result holds.

**THEOREM 2**

(a) *The critical point* $(0, 0)$ *of the nonlinear system* (13) *is asymptotically stable if the critical point* $(0, 0)$ *of the "linearized" system* (11) *is asymptotically stable.*

(b) *The critical point* $(0, 0)$ *of the nonlinear system* (13) *is unstable if the critical point* $(0, 0)$ *of system* (11) *is unstable.*

This theorem offers no conclusion about system (13) when $(0, 0)$ is just a stable point of system (11).

**EXAMPLE 4** Show that the critical point $(0, 0)$ of the nonlinear system

$$\dot{x} = -x + y + (x^2 + y^2)$$
$$\dot{y} = -2y - (x^2 + y^2)^{3/2}$$
(15)

is asymptotically stable.

**Solution** Here $a = -1$, $b = 1$, $c = 0$, $d = -2$, and $ad - bc = 2 \neq 0$. $F(x, y) = (x^2 + y^2)$, $G(x, y) = -(x^2 + y^2)^{3/2}$, and $F(0, 0) = G(0, 0) = 0$. Clearly, the conditions (14) are satisfied. The linearized system is

$$\dot{x} = -x + y, \qquad \dot{y} = -2y.$$
(16)

The characteristic equation of system (16) is $\lambda^2 + 3\lambda + 2 = 0$. Its roots are $\lambda_1 = -1$ and $\lambda_2 = -2$. Since they are both negative, the point $(0, 0)$ is an asymptotically stable critical point of (16). Hence, by Theorem 2(a), the point $(0, 0)$ is also an asymptotically stable critical point of the nonlinear system (15).

**EXAMPLE 5**　Show that the critical point $(0, 0)$ of the nonlinear system

$$\dot{x} = -3x + 4y + x^2 - y^2$$
$$\dot{y} = -2x + 3y - xy \tag{17}$$

is unstable.

**Solution**　Here $a = -3$, $b = 4$, $c = -2$, and $d = 3$ with $ad - bc = -1 \neq 0$. $F(x, y) = x^2 - y^2$, $G(x, y) = -xy$, with $F(0, 0) = G(0, 0) = 0$. We express $x$ and $y$ in polar coordinates: $x = r \cos \theta$, $y = r \sin \theta$. Then $(x \to 0$ and $y \to 0$ is equivalent to $r \to 0)$

$$\frac{F(x, y)}{\sqrt{x^2 + y^2}} = \frac{r^2(\cos^2 \theta - \sin^2\theta)}{r} = r \cos 2\theta \to 0 \qquad \text{as } r \to 0$$

and

$$\frac{G(x, y)}{\sqrt{x^2 + y^2}} = -\frac{r^2 \cos \theta \sin \theta}{r} = -r \cos \theta \sin \theta \to 0 \qquad \text{as } r \to 0.$$

Hence, the conditions (14) are satisfied. The linearized system is

$$\dot{x} = -3x + 4y, \qquad \dot{y} = -2x + 3y. \tag{18}$$

The characteristic equation of system (18) is $\lambda^2 - 1 = 0$. Its roots are $\lambda_1 = 1$ and $\lambda_2 = -1$. Since one of them is positive, the point $(0, 0)$ is an unstable critical point of (18). By Theorem 2(b), the point $(0, 0)$ is also an unstable critical point of the nonlinear system (17).

**EXERCISES**

By using the definition, determine whether the critical point $(0, 0)$ of each of the systems in Exercises 1 through 3 is stable, asymptotically stable, or unstable.

**1.** $\dot{x} = y$, $\dot{y} = -x$　**2.** $\dot{x} = -y$, $\dot{y} = -x$　**3.** $\dot{x} = -x + y$, $\dot{y} = -2y$.

In Exercises 4 through 10, determine whether the critical point $(0, 0)$ of the system is stable, asymptotically stable, or unstable.

**4.** $\dot{x} = -x + y$, $\dot{y} = -x - y$　　　　**5.** $\dot{x} = x - y$, $\dot{y} = x + 3y$

**6.** $\dot{x} = x - y$, $\dot{y} = 5x - y$　　　　　**7.** $\dot{x} = -3x + 4y$, $\dot{y} = -2x + 3y$

**8.** $\dot{x} = 5x - 6y$, $\dot{y} = 6x - 7y$　　　**9.** $\dot{x} = -3x + 5y$, $\dot{y} = -x + y$

**10.** $\dot{x} = 3x - 2y$, $\dot{y} = 4x - y$

**11.** Assume that the system

$$\dot{x} = 4x - 6y, \qquad \dot{y} = 8x - 10y$$

represents two competing populations, where $x$ is the desirable population and $y$ is a parasite. Show that the critical point $(0, 0)$ of this system is asymptotically stable and therefore both populations are headed for extinction.

**12.** Show that the critical point $(1, 1)$ of the system

$$\dot{x} = 5x - 6y + 1, \quad \dot{y} = 6x - 7y + 1,$$

which appears in Turing's theory of morphogenesis is asymptotically stable. [*Hint:* Transform the critical point to $(0, 0)$ by means of $X = x - 1$, $Y = y - 1$.]

In Exercises 13 through 17, show that the point $(0, 0)$ is a critical point for the given system and determine, if possible, whether it is asymptotically stable or unstable.

**13.** $\dot{x} = 5x - 6y + xy, \ \dot{y} = 6x - 7y - xy$

**14.** $\dot{x} = 3x - 2y + (x^2 + y^2)^2, \ \dot{y} = 4x - y + (x^2 - y^2)^5$

**15.** $\dot{x} = y + x^2 - xy, \ \dot{y} = -2x + 3y + y^2$

**16.** $\dot{x} = x - xy, \ \dot{y} = -y + xy$

**17.** $\dot{x} = -x - x^2 + y^2, \ \dot{y} = -y + 2xy$

Transform the differential equation $\ddot{x} + p\dot{x} + qx = 0$ into an equivalent system by setting $\dot{x} = y$, and investigate the stability character of the resulting system at the critical point $(0, 0)$ in each of the following cases.

**18.** $p^2 > 4q, \ q > 0,$ and $p < 0$      **19.** $p^2 > 4q, \ q > 0,$ and $p > 0$

**20.** $p^2 > 4q$ and $q < 0$      **21.** $p^2 = 4q$ and $p > 0$

**22.** $p^2 = 4q$ and $p < 0$      **23.** $p^2 < 4q$ and $p > 0$

**24.** $p^2 < 4q$ and $p < 0$      **25.** $p = 0$ and $q > 0$

# 8.5   PHASE PORTRAITS OF AUTONOMOUS SYSTEMS

The picture of all trajectories of a system is called the *phase portrait* of the system. In this section we first present, by means of examples, all possible phase portraits of the linear autonomous system

$$\dot{x} = ax + by, \quad \dot{y} = cx + dy \quad (1)$$

near the critical point $x = 0, y = 0$. We shall assume that $(0, 0)$ is the only critical point of system (1), and this is equivalent to assuming that $ad - bc \neq 0$. (See Exercise 19, Section 8.3.) At the end of this section we state a theorem about the phase portraits of some nonlinear systems.

As we explained in Section 3.3, the solutions of system (1) are of the form

$$x = Ae^{\lambda t}, \quad y = Be^{\lambda t},$$

where $\lambda$ is an eigenvalue of the matrix

$$\begin{pmatrix} a & b \\ c & d \end{pmatrix};$$

that is, $\lambda$ is a root of the characteristic equation

$$\lambda^2 - (a + d)\lambda + ad - bc = 0. \tag{2}$$

The phase portrait of system (1) depends almost entirely on the roots $\lambda_1$ and $\lambda_2$ of Eq. (2). [The assumption $ad - bc \neq 0$ implies that $\lambda = 0$ is not a root of Eq. (2). ] There are five different cases that must be studied separately.

**CASE 1**    *Real and distinct roots of the same sign,* that is,

$$\lambda_2 < \lambda_1 < 0 \quad \text{or} \quad \lambda_1 > \lambda_2 > 0.$$

**CASE 2**    *Real roots with opposite sign,* that is,

$$\lambda_1 < 0 < \lambda_2.$$

**CASE 3**    *Equal roots,* that is,

$$\lambda_1 = \lambda_2 = \lambda \quad \text{with} \quad \lambda < 0 \quad \text{or} \quad \lambda > 0.$$

In this case we obtain two distinct portraits, depending on whether or not the matrix

$$\begin{pmatrix} a - \lambda & b \\ c & d - \lambda \end{pmatrix}$$

is equal to zero.

**CASE 4**    *Complex conjugate roots but not pure imaginary,* that is,

$$\lambda_1 = k + il, \quad \lambda_2 = k - il \quad \text{with} \quad k < 0 \quad \text{or} \quad k > 0.$$

**CASE 5**    *Pure imaginary roots,* that is,

$$\lambda_1 = il, \quad \lambda_2 = -il \quad \text{with} \quad l \neq 0.$$

It is sufficient to illustrate these cases by means of specific examples. This is true because any two systems falling into one and the same of these cases (and corresponding subcase) can be transformed into each other by means of a linear change of variables.

**EXAMPLE 1**    $(\lambda_2 < \lambda_1 < 0)$    Draw the phase portrait of the linear system

$$\dot{x} = -2x + y, \quad \dot{y} = x - 2y. \tag{3}$$

**Solution**   Here the characteristic roots are $\lambda_1 = -1$ and $\lambda_2 = -3$. The general solution of system (3) is

$$x = c_1 e^{-t} + c_2 e^{-3t}, \qquad y = c_1 e^{-t} - c_2 e^{-3t}, \tag{4}$$

where $c_1$ and $c_2$ are arbitrary constants. We want to find the trajectories of all the solutions given by (4) for all different values of the constants $c_1$ and $c_2$. When $c_1 = c_2 = 0$, we have the solution $x = 0$, $y = 0$, whose trajectory is the origin $(0, 0)$ in Figure 8.3. When $c_1 \neq 0$ and $c_2 = 0$, we find the solutions

$$x = c_1 e^{-t}, \qquad y = c_1 e^{-t}, \tag{5}$$

and when $c_1 = 0$ and $c_2 \neq 0$, we have the solutions

$$x = c_2 e^{-3t}, \qquad y = -c_2 e^{-3t}. \tag{6}$$

For $c_1 > 0$, all the solutions (5) have the same trajectory, $y = x > 0$. Similarly, for $c_1 < 0$, all the solutions (5) have the common trajectory, $y = x < 0$. From (6) with $c_2 > 0$ and $c_2 < 0$, we also find the trajectories $y = -x < 0$ and $y = -x > 0$, respectively. These four trajectories are half-lines shown in Figure 8.3. The arrows on the half-lines indicate the direction of motion on the trajectories as $t$ increases. To obtain the other trajectories explicitly, we must eliminate $t$ from the two equations in (4) and investigate the resulting curves for all nonzero values of the constants $c_1$ and $c_2$. When this approach is complicated (as in this example), we can still obtain a good picture of the phase portrait of the system as follows. First, from (4), it is clear that every trajectory of system (3) approaches the origin $(0, 0)$ as $t$ approaches infinity. Furthermore, for $c_1 \neq 0$ and $c_2 \neq 0$, we have

$$\frac{y}{x} = \frac{c_1 e^{-t} - c_2 e^{-3t}}{c_1 e^{-t} + c_2 e^{-3t}} = \frac{c_1 - c_2 e^{-2t}}{c_1 + c_2 e^{-2t}} \to 1 \qquad \text{as } t \to \infty.$$

Thus, all these trajectories approach the origin tangent to the line $y = x$. In fact, all the trajectories of system (3), except the pair $y = -x > 0$ and $y = -x < 0$, approach the origin tangent to the line $y = x$. Figure 8.3 shows a few

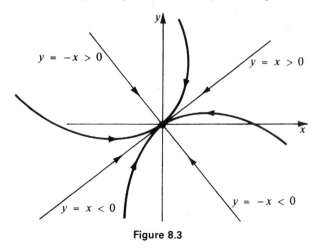

**Figure 8.3**

trajectories of the phase portrait of system (3). The arrows on the trajectories indicate the direction of increasing $t$. This type of critical point is called a *stable node*.

When $\lambda_1 > \lambda_2 > 0$, as in Exercise 1, the phase portrait is exactly the same, except that the direction of the arrows on the trajectories is reversed and the trajectories approach the origin as $t \to -\infty$. This type of critical point is called an *unstable node*. In the case of an unstable node, all the solutions, except the solution $x = 0$, $y = 0$, approach infinity as $t \to \infty$.

**EXAMPLE 2**  $(\lambda_1 < 0 < \lambda_2)$  Draw the phase portrait of the linear system

$$\dot{x} = 3x - 2y, \qquad \dot{y} = 2x - 2y. \tag{7}$$

**Solution**  Here the characteristic roots are $\lambda_1 = -1$ and $\lambda_2 = 2$. The general solution of (7) is

$$x = c_1 e^{-t} + c_2 e^{2t}, \qquad y = 2c_1 e^{-t} + \tfrac{1}{2} c_2 e^{2t}, \tag{8}$$

where $c_1$ and $c_2$ are arbitrary constants. When $c_1 \neq 0$ and $c_2 = 0$, we obtain the solutions

$$x = c_1 e^{-t}, \qquad y = 2c_1 e^{-t}, \tag{9}$$

and when $c_1 = 0$ and $c_2 \neq 0$, we have the solutions

$$x = c_2 e^{2t}, \qquad y = \tfrac{1}{2} c_2 e^{2t}. \tag{10}$$

For $c_1 > 0$, the trajectory of the solutions (9) is the half-line $y = 2x > 0$, while for $c_1 < 0$, the trajectory is the half-line $y = 2x < 0$ (see Figure 8.4). The arrows on these trajectories point toward the origin because both $x$ and $y$ approach zero as $t \to \infty$. This observation follows from (9).

For $c_2 > 0$, the trajectory of the solutions (10) is the half-line $y = \tfrac{1}{2}x > 0$, while for $c_2 < 0$, the trajectory is the half-line $y = \tfrac{1}{2}x < 0$ (see Figure 8.4). The arrows on these trajectories point away from the origin, since in this case both $x$ and $y$ tend to $\pm\infty$ as $t \to \infty$, as can be seen from (10).

For $c_1 \neq 0$ and $c_2 \neq 0$, we have

$$\frac{y}{x} = \frac{2c_1 e^{-t} + \tfrac{1}{2} c_2 e^{2t}}{c_1 e^{-t} + c_2 e^{2t}} = \frac{2c_1 e^{-3t} + \tfrac{1}{2} c_2}{c_1 e^{-3t} + c_2} \to \frac{1}{2} \qquad \text{as } t \to \infty.$$

Hence, all trajectories are asymptotic to the line $y = \tfrac{1}{2}x$ as $t$ tends to $\infty$.

On the other hand,

$$\frac{y}{x} = \frac{2c_1 e^{-t} + \tfrac{1}{2} c_2 e^{2t}}{c_1 e^{-t} + c_2 e^{2t}} = \frac{2c_1 + \tfrac{1}{2} c_2 e^{3t}}{c_1 + c_2 e^{3t}} \to 2 \qquad \text{as } t \to -\infty,$$

which implies that all trajectories are asymptotic to the line $y = 2x$ as $t$ tends to $-\infty$. Figure 8.4 shows a few trajectories of the phase portrait of system (7).

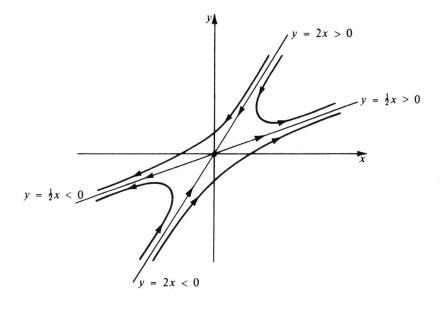

$y = 2x > 0$

$y = \frac{1}{2}x > 0$

$y = \frac{1}{2}x < 0$

$y = 2x < 0$

**Figure 8.4**

This type of critical point is called a *saddle point*. Since one of the roots of the characteristic equation is positive, it follows that a saddle point is an unstable point of the system. As we see from Figure 8.4, only two trajectories approach the origin; the rest tend away from it.

**EXAMPLE 3A**   $(\lambda_1 = \lambda_2 = \lambda < 0$ and the matrix $\begin{pmatrix} a - \lambda & b \\ c & d - \lambda \end{pmatrix}$ is zero$)$

Draw the phase portrait of the linear system

$$\dot{x} = -2x, \qquad \dot{y} = -2y. \tag{11}$$

**Solution**   Here the characteristic roots are $\lambda_1 = \lambda_2 = -2 < 0$, and the general solution of the system is

$$x = c_1 e^{-2t}, \qquad y = c_2 e^{-2t}, \tag{12}$$

where $c_1$ and $c_2$ are arbitrary constants. The trajectories of the solutions (12) are the half-lines $c_1 y = c_2 x$ with slope $c_2/c_1$. (Thus, all slopes are possible.) The phase portrait of system (11) has the form described in Figure 8.5. This type of critical point is called a *stable node* when $\lambda < 0$ and an unstable node when $\lambda > 0$. The phase portrait of an unstable node is as in Figure 8.5 except that the direction of the arrows on the trajectories is reversed.

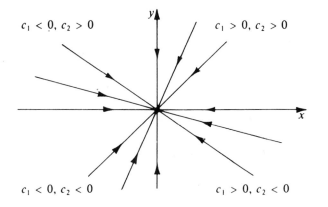

$c_1 < 0, c_2 > 0$          $c_1 > 0, c_2 > 0$

$c_1 < 0, c_2 < 0$          $c_1 > 0, c_2 < 0$

Figure 8.5

**EXAMPLE 3B**   ($\lambda_1 = \lambda_2 = \lambda < 0$ and the matrix $\begin{pmatrix} a - \lambda & b \\ c & d - \lambda \end{pmatrix}$ is not zero)

Draw the phase portrait of the system

$$\dot{x} = -x, \qquad \dot{y} = -x - y. \tag{13}$$

**Solution**   Here the characteristic roots are $\lambda_1 = \lambda_2 = -1 < 0$, and the general solution of the system is

$$x = c_1 e^{-t}, \qquad y = -c_1 t e^{-t} + c_2 e^{-t}, \tag{14}$$

where $c_1$ and $c_2$ are arbitrary constants. From (14) we see that all trajectories approach the origin as $t \to \infty$. When $c_1 = 0$, $c_2 \neq 0$, we find the solutions

$$x = 0, \qquad y = c_2 e^{-t}. \tag{15}$$

Clearly, the positive and negative $y$ axes are the trajectories of the solutions (15) (see Figure 8.6). For $c_1 \neq 0$, we can eliminate $t$ between the two equations in (14), obtaining

$$y = \frac{c_2}{c_1} x + x \ln \left| \frac{x}{c_1} \right|. \tag{16}$$

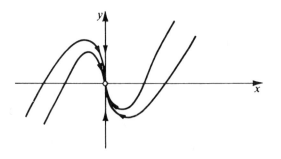

Figure 8.6

Equation (16) gives explicitly the trajectories of the solutions (14) for $c_1 \neq 0$. All these trajectories approach the origin tangent to the $y$ axis (see Figure 8.6). This type of critical point is called a *stable node* when $\lambda < 0$ and an *unstable node* when $\lambda > 0$. The phase portrait of the unstable node is as in Figure 8.6 except that the arrows on the trajectories are reversed.

**EXAMPLE 4** ($\lambda_1 = k + il$, $\lambda_2 = k - il$ with $k < 0$) Draw the phase portrait of the linear system

$$\dot{x} = -x + y$$
$$\dot{y} = -x - y. \qquad (17)$$

**Solution** Here the characteristic roots are $\lambda_1 = -1 + i$ and $\lambda_2 = -1 - i$. The general solution of the system is

$$x = c_1 e^{-t} \cos t + c_2 e^{-t} \sin t$$

$$y = -c_1 e^{-t} \sin t + c_2 e^{-t} \cos t.$$

It seems complicated to find the trajectories through this form of the solution. For this reason let us introduce polar coordinates in (17) by setting

$$x = r \cos \phi, \qquad y = r \sin \phi.$$

Then

$$\dot{x} = \dot{r} \cos \phi - (r \sin \phi)\dot{\phi} \qquad \text{and} \qquad \dot{y} = \dot{r} \sin \phi + (r \cos \phi)\dot{\phi}.$$

Substituting these values of $x$, $y$, $\dot{x}$, and $\dot{y}$ into (17) and solving for $\dot{r}$ and $\dot{\phi}$, we obtain the equivalent system

$$\dot{r} = -r, \qquad \dot{\phi} = -1. \qquad (18)$$

The general solution of (18) is

$$r = Ae^{-t}, \qquad \phi = -t + B. \qquad (19)$$

The trajectories are therefore logarithmic spirals approaching the origin as $t$ tends to infinity. The phase portrait of system (17) has the form described in Figure 8.7. This type of critical point is called a *stable focus* when $k < 0$ and an *unstable focus* when $k > 0$ (see Figure 8.8). As is expected, in the unstable focus the arrows have opposite direction.

**EXAMPLE 5** ($\lambda_1 = il$, $\lambda_2 = -il$ with $l \neq 0$) Draw the phase portrait of the linear system

$$\dot{x} = y, \qquad \dot{y} = -x. \qquad (20)$$

**Solution** Here the characteristic roots are $\lambda_1 = i$ and $\lambda_2 = -i$, and the general solution of the system is

$$x = c_1 \cos t + c_2 \sin t, \qquad y = -c_1 \sin t + c_2 \cos t, \qquad (21)$$

Figure 8.7

Figure 8.8

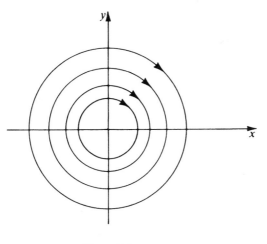

Figure 8.9

where $c_1$ and $c_2$ are arbitrary constants. From (21) we see that $x^2 + y^2 = c_1^2 + c_2^2$. Therefore, the trajectories of the system are circles centered at the origin with radius $\sqrt{c_1^2 + c_2^2}$. The phase portrait of system (20) has the form described in Figure 8.9. This type of critical point is called a *center*. The direction of the arrows in Figure 8.9 is found from system (20). From the first equation in (20) we see that $\dot{x} > 0$ when $y > 0$; that is, $x$ increased when $y$ is positive. This implies that the motion along the trajectories is in the clockwise direction.

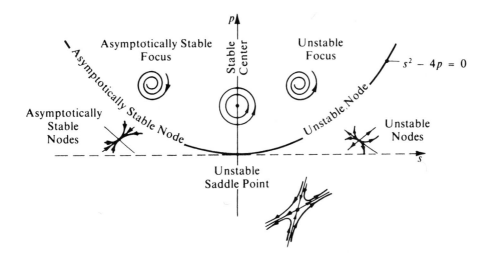

Figure 8.10 Summary of Portraits

**Summary**  In Figure 8.10 we summarize the nature and stability of the critical point $(0, 0)$ of the system

$$\dot{x} = ax + by, \qquad \dot{y} = cx + dy,$$

when $ad - bc \ne 0$. If $\lambda_1$ and $\lambda_2$ are the characteristic roots, then $\lambda_1 + \lambda_2 = a + d$ and $\lambda_1\lambda_2 = ad - bc$. We set $s = \lambda_1 + \lambda_2$, $p = \lambda_1\lambda_2$, and $D = s^2 - 4p$. In the $sp$ plane we have graphed the parabola $s^2 - 4p = 0$. Since we have assumed that $ad - bc \ne 0$, the entire $s$ axis is excluded from the diagram.

Finally, we state a theorem about the phase portrait, near a critical point $(0, 0)$, of a nonlinear system of the form

$$\dot{x} = ax + by + F(x, y)$$
$$\dot{y} = cx + dy + G(x, y), \tag{22}$$

where $a$, $b$, $c$, $d$, and the functions $F$ and $G$ satisfy the conditions mentioned before Theorem 2 in Section 8.4. Let $\lambda_1$ and $\lambda_2$ be the two roots of the characteristic equation (2) of the linearized system (1). Under these hypotheses the following result holds.

**THEOREM 1**

*The critical point $(0, 0)$ of the nonlinear system (22) is*
  *(a) a node if $\lambda_1$ and $\lambda_2$ are real, distinct, and of the same sign;*
  *(b) a saddle if $\lambda_1$ and $\lambda_2$ are real and of opposite sign;*
  *(c) a focus if $\lambda_1$ and $\lambda_2$ are complex conjugates but not pure imaginary;*
  *(d) a focus or center if $\lambda_1$ and $\lambda_2$ are pure imaginary.*

In Example 5 of Section 8.4 we verified that the functions $F(x, y) = x^2 - y^2$ and $G(x, y) = -xy$ satisfy the hypotheses of Theorem 1. Then from Theorem 1 and the Examples 1, 2, and 4, we conclude that the critical point $(0, 0)$ of the systems

  (a) $\dot{x} = -2x + y + x^2 - y^2, \dot{y} = x - 2y - xy,$
  (b) $\dot{x} = 3x - 2y + x^2 - y^2, \dot{y} = 2x - 2y - xy,$ and
  (c) $\dot{x} = -x + y + x^2 - y^2, \dot{y} = -x - y - xy$

is a node, a saddle, and a focus point, respectively.

**EXERCISES**

In Exercises 1 through 10, determine whether the critical point $(0, 0)$ is a node, saddle point, focus, or center, and draw the phase portrait.

**1.** $\dot{x} = x, \dot{y} = 3y$                    **2.** $\dot{x} = -x, \dot{y} = -3y$

**3.** $\dot{x} = x, \dot{y} = -y$                    **4.** $\dot{x} = -x, \dot{y} = y$

**5.** $\dot{x} = 2x, \dot{y} = 2y$                   **6.** $\dot{x} = -x, \dot{y} = -y$

**7.** $\dot{x} = -x, \dot{y} = x - y$        **8.** $\dot{x} = x, \dot{y} = -x + y$

**9.** $\dot{x} = -2x + y, \dot{y} = -x - 2y$      **10.** $\dot{x} = -y, \dot{y} = x.$

In Exercises 11 through 14, determine whether the critical point $(0, 0)$ is a node, saddle point, focus, or center, and graph a few trajectories.

**11.** $\dot{x} = -x + y, \dot{y} = -x - y$      **12.** $\dot{x} = -x + y, \dot{y} = -x - 3y$

**13.** $\dot{x} = 4x + 6y, \dot{y} = -7x - 9y$     **14.** $\dot{x} = 2x - y, \dot{y} = -x + 2y$

In Exercises 15 through 19, determine whether the critical point $(0, 0)$ is a node, a saddle, or a focus.

**15.** $\dot{x} = -2x + y - x^2 + 2y^2, \dot{y} = 3x + 2y + x^2y^2$

**16.** $\dot{x} = -x + x^2, \dot{y} = -3y + xy$

**17.** $\dot{x} = -x + xy, \dot{y} = y + (x^2 + y^2)^2$

**18.** $\dot{x} = 2x + y^2, \dot{y} = 3y - x^2$

**19.** $\dot{x} = x - xy, \dot{y} = -y + xy$

Transform the differential equation $\ddot{x} + p\dot{x} + qx = 0$ into an equivalent system by setting $\dot{x} = y$ and investigate the type of the phase portrait of the resulting system near the critical point $(0, 0)$ in each of the following cases:

**20.** $p^2 - 4q > 0, q > 0,$ and $p < 0$     **21.** $p^2 - 4q > 0, q > 0,$ and $p > 0$

**22.** $p^2 - 4q > 0$ and $q < 0$         **23.** $p^2 = 4q$ and $p > 0$

**24.** $p^2 = 4q$ and $p < 0$            **25.** $p^2 - 4q < 0$ and $p > 0$

**26.** $p = 0$ and $q > 0$

In Exercises 27 through 34, determine the nature and stability of the critical point $(0, 0)$ by using Figure 8.10.

**27.** $\dot{x} = x + y, \dot{y} = x - 3y$       **28.** $\dot{x} = 3x - 2y, \dot{y} = 2x + 3y$

**29.** $\dot{x} = -2x - 3y, \dot{y} = -3x + 2y$   **30.** $\dot{x} = -3x + 2y, \dot{y} = -2x - 3y$

**31.** $\dot{x} = x, \dot{y} = y$               **32.** $\dot{x} = -x, \dot{y} = -y$

**33.** $\dot{x} = -x, \dot{y} = -x - y$        **34.** $\dot{x} = -3x + 2y, \dot{y} = -2x + 2y$

# 8.6   APPLICATIONS

In this section we present a few applications of nonlinear differential equations and systems. In comparison to mathematical models described by linear equations and systems, nonlinear equations and systems describe situations that are closer to reality in many cases.

■ Consider two species and assume that one of them, called prey, has an abundant supply of food while the other species, called predator, feeds exclusively on the first. The mathematical study of such an ecosystem was initiated independently by Lotka and Volterra in the mid 1920s. Let us denote by $x(t)$ the number of prey and by $y(t)$ the number of predators at time $t$.

If the two species were in isolation from each other, they would vary at a rate proportional to their number present,

$$\dot{x} = ax \quad \text{and} \quad \dot{y} = -cy. \tag{1}$$

In Eq. (1), $a > 0$ because the prey population has an abundant supply of food and therefore increases, while $-c < 0$ because the predator population has no food and hence decreases.

However, we have assumed that the two species interact in such a way that the predator population eats the prey. It is reasonable to assume that the number of fatal encounters per unit time is proportional to $x$ and $y$, and therefore to $xy$. Then the prey will decrease while the predators will increase at rates proportional to $xy$. That is, the two interacting populations satisfy the nonlinear system

$$\dot{x} = ax - bxy, \quad \dot{y} = -cy + dxy, \tag{2}$$

where $b$ and $d$ are positive constants.

Although system (2) is nonlinear and there is no known way to solve it explicitly, it is nevertheless possible, using the qualitative theory of such systems, to obtain many properties of its solutions, and in turn to make useful predictions about the behavior of the two species.

Solving the system of simultaneous equations $ax - bxy = 0$ and $-cy + dxy = 0$, we find the critical points $(0, 0)$ and $(c/d, a/b)$ of system (2). Thus, system (2) has the two equilibrium solutions $x(t) = 0$, $y(t) = 0$ and $x(t) = c/d$, $y(t) = a/b$. Of course, only the second of these is of interest in this application.

Let us compute the trajectories of the solutions of (2). Clearly, $x(t) = 0$, $y(t) = y(0)e^{-ct}$ is a solution of (2) with trajectory the positive $y$ axis. Also, $y(t) = 0$, $x(t) = x(0)e^{at}$ is another solution of (2), with trajectory the positive $x$ axis. Because of the uniqueness of solutions, it follows that every solution of (2) which at $t = 0$ starts in the first quadrant cannot cross the $x$ or the $y$ axis, and therefore should remain in the first quadrant forever. The remaining trajectories of the solutions of (2) satisfy the differential equation

$$\frac{dy}{dx} = \frac{-cy + dxy}{ax - bxy} = \frac{(-c + dx)y}{(a - by)x},$$

which is a separable differential equation. Separating the variables $x$ and $y$, we obtain

$$\frac{a - by}{y} \, dy = \frac{-c + dx}{x} \, dx$$

or

$$\left(\frac{a}{y} - b\right) dy = \left(-\frac{c}{x} + d\right) dx. \tag{3}$$

Integrating both sides of (3), we obtain the general solution of (3),

$$a \ln y - by = -c \ln x + dx + k, \qquad (4)$$

where $k$ is an arbitrary constant. Equation (4) can be rewritten as follows:

$$\ln y^a + \ln x^c = by + dx + k$$

$$\Rightarrow \quad \ln y^a x^c = by + dx + k$$

$$\Rightarrow \quad y^a x^c = e^{by + dx + k}$$

$$\Rightarrow \quad \frac{y^a}{e^{by}} \cdot \frac{x^c}{e^{dx}} = K, \qquad (5)$$

where $K$ is $e^k$. If we allow $K$ to be equal to zero, Eq. (5) gives all the trajectories of system (2).

It can be shown that for each $K > 0$, the trajectory (5) is a closed curve, and therefore each solution $(x(t), y(t))$ of (2) with initial value $(x(0), y(0))$ in the first quadrant is a periodic function of time $t$ (see Figure 8.11). If $T$ is the period of a solution $(x(t), y(t))$, that is, if $(x(t + T), y(t + T)) = (x(t), y(t))$ for all $t \geq 0$, then the average values of the populations $x(t)$ and $y(t)$ are, by definition, given by the integrals

$$\bar{x} = \frac{1}{T} \int_0^T x(t)dt, \qquad \bar{y} = \frac{1}{T} \int_0^T y(t)dt.$$

Surprisingly, these average values can be computed directly for system (2) without knowing explicitly the solution and its period. In fact, from the second equation in (2), we obtain

$$\frac{\dot{y}}{y} = -c + dx.$$

Integrating both sides from 0 to $T$, we find that

$$\ln y(T) - \ln y(0) = -cT + d \int_0^T x(t)dt.$$

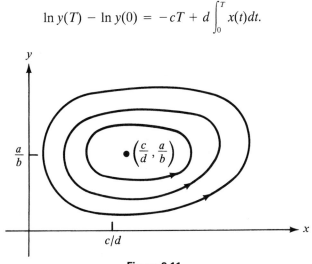

Figure 8.11

Since $y(T) = y(0)$, it follows that

$$-cT + d \int_0^T x(t)dt = 0 \qquad \text{or} \qquad \frac{1}{T}\int_0^T x(t)dt = \frac{c}{d}.$$

Therefore,

$$\bar{x} = \frac{c}{d}. \tag{6}$$

Similarly, from the first equation in (2), we obtain

$$\frac{\dot{x}}{x} = a - by.$$

Integrating both sides from 0 to $T$ and using the fact that $x(T) = x(0)$, we find

$$\bar{y} = \frac{a}{b}. \tag{7}$$

From (6) and (7) we can make the interesting prediction that the average sizes of two populations $x(t)$ and $y(t)$ which interact according to the mathematical model described by system (2) will be exactly their equilibrium values $x = c/d$ and $y = a/b$.

We can utilize this observation to make another interesting prediction. In addition to the hypotheses about the prey and predator populations which lead to the mathematical model described by system (2), assume that the prey population $x(t)$ is harvested in "moderate amounts." Then both the prey and predator populations will decrease at rates, say, $\epsilon x(t)$ and $\epsilon y(t)$, respectively. In this case, system (2) should be replaced by the system

$$\dot{x} = ax - bxy - \epsilon x, \qquad \dot{y} = -cy + dxy - \epsilon y$$

or

$$\dot{x} = (a - \epsilon)x - bxy, \qquad \dot{y} = -(c + \epsilon)y + dxy. \tag{8}$$

Applying equations (6) and (7) to system (8), we conclude, with surprise, that if the harvesting of the preys is such that $a > \epsilon$, the average size of the prey population will be

$$\bar{x} = \frac{c + \epsilon}{d}, \tag{6'}$$

which is somewhat higher than before there was any harvesting. On the other hand, the average size of the predator population will be

$$\bar{y} = \frac{a - \epsilon}{b}, \tag{7'}$$

which is somewhat smaller than before there was any harvesting.[1]

---

[1]For more details on this and other related applications, the reader is referred to U. D'Ancona, *The Struggle for Existence* (Leiden: Brill, 1954).

■ A simple pendulum consists of a bob $B$ of mass $m$ at the end of a very light and rigid rod of length $L$, pivoted at the top so that the system can swing in a vertical plane (see Figure 8.12). We pull the bob to one side and release it from rest at time $t = 0$. Let $\theta(t)$, in radians, be the angular displacement of the rod from its equilibrium position $OA$ at time $t$. The angle $\theta(t)$ is taken to be positive when the bob is to the right of the equilibrium and negative when it is to the left. We want to study $\theta(t)$ as the bob swings back and forth along the circular arc $CC'$. From this information we know already that

**Nonlinear Mechanics**
The Motion of a Simple Pendulum

$$\theta(0) = \theta_0 \quad \text{and} \quad \dot{\theta}(0) = 0, \tag{9}$$

where $\theta_0$ is the initial angular displacement of the rod and $\dot{\theta}(0) = 0$ because the bob was released from rest. There are two forces acting upon the bob $B$ at any time $t$. One is its weight $-mg$, where $m$ is the mass of the bob and $g$ is the acceleration of gravity. The other force is the tension $T$ from the rod.

The force $-mg$ is resolved into the two components $-mg \cos \theta$ and $-mg \sin \theta$, as shown in Figure 8.12. The force $-mg \cos \theta$ balances the tension $T$ in the rod, while the force $-mg \sin \theta$ moves the bob along the circular arc $BA$. By Newton's second law of motion, we obtain

$$m\frac{d^2s}{dt^2} = -mg \sin \theta, \tag{10}$$

where $s$ is the length of the arc $AB$, and $d^2s/dt^2$ is the acceleration along the arc. Since $L$ is the length of the rod, we have $s = L\theta$ and therefore

$$\frac{d^2s}{dt^2} = L\frac{d^2\theta}{dt^2}. \tag{11}$$

Using Eq. (11) in (10) and simplifying the resulting equation, we obtain

$$\ddot{\theta} + \frac{g}{L}\sin \theta = 0. \tag{12}$$

The nonlinear differential equation (12), together with the initial conditions (9), describes completely the motion (in a vacuum) of the simple pendulum.

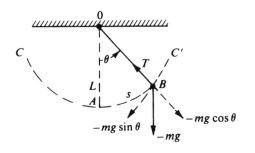

**Figure 8.12**

Let us denote for simplicity the constant $g/L$ by $\omega^2$. If we now set $\theta = x$ and $\dot\theta = y$, the differential equation (12) is equivalent to the system

$$\dot x = y, \qquad \dot y = -\omega^2 \sin x. \tag{13}$$

In what follows we present a phase-plane analysis of system (13).

The critical points of system (13) are the solutions of the simultaneous equations

$$y = 0 \qquad \text{and} \qquad -\omega^2 \sin x = 0.$$

Since $\sin x = 0$ for $x = 0, \pm\pi, \pm 2\pi, \ldots$, the critical points are

$$(0, 0), (\pi, 0), (-\pi, 0), (2\pi, 0), (-2\pi, 0), \ldots. \tag{14}$$

Clearly, all of them are located on the $x$ axis. Since $x = 0$ is the angular displacement of the rod from the equilibrium position $OA$ (see Figure 8.12), we conclude that when $x = 0, 2\pi, -2\pi, \ldots$, the pendulum is pointing vertically downward, and when $x = \pi, -\pi, 3\pi, -3\pi, \ldots$, the pendulum is pointing vertically upward. Because of the periodicity of the function $\sin x$ it suffices to study the nature of the critical points $(0, 0)$ and $(\pi, 0)$ only.

First, we study the critical point $(0, 0)$. Since

$$\sin x = x - \frac{x^3}{3!} + \frac{x^5}{5!} - \cdots,$$

we can approximate (13) by the linear system

$$\dot x = y, \qquad \dot y = -\omega^2 x. \tag{15}$$

The characteristic equation of (15) is $\lambda^2 + \omega^2 = 0$ with characteristic roots $\lambda = \pm\omega i$ (pure imaginary). Hence, the critical point $(0, 0)$ is stable and a center for system (15) (see Figure 8.13). The arrows on the trajectories point in the clockwise direction because, as we see from the first equation in (15), $x$ increases

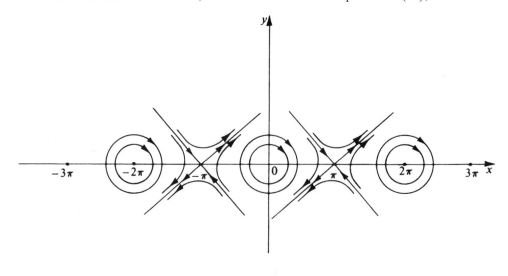

**Figure 8.13**

when $y$ is positive. By Theorem 1(d) of Section 8.5 (whose hypotheses are satisfied here), it follows that $(0, 0)$ is either a center or a focus point for the nonlinear system (13). In this case, however, it can be shown that it is a center (the simple pendulum is a conservative system, and hence there is no gain or loss of energy as in the case of a focus). System (13) also has a center at each of the critical points $(2\pi n, 0)$ for $n = \pm 1, \pm 2, \ldots$. The closed trajectories in Figure 8.14 correspond to the oscillatory motions of the pendulum about the stable points $(2\pi n, 0)$, $n = 0, \pm 1, \pm 2, \ldots$.

Next we study the critical point $(\pi, 0)$. The Taylor-series expansion of $\sin x$ about $x = \pi$ is given by

$$\sin x = -(x - \pi) + \frac{(x - \pi)^3}{3!} - \frac{(x - \pi)^5}{5!} + \cdots.$$

Hence, the linearized system is

$$\dot{x} = y, \qquad \dot{y} = \omega^2(x - \pi) \tag{16}$$

and has a critical point at $(\pi, 0)$. We can transform the critical point $(\pi, 0)$ to $(0, 0)$ by setting $v = x - \pi$ in (16). Then (16) becomes

$$\dot{v} = y, \qquad \dot{y} = \omega^2 v. \tag{17}$$

The characteristic equation of (17) is $\lambda^2 - \omega^2 = 0$ with characteristic roots $\pm \omega$. Since the roots are real with opposite sign, it follows that the critical point $(0, 0)$ of (17) is a saddle point and hence an unstable equilibrium point. Thus, $(\pi, 0)$ is also a saddle point and hence an unstable equilibrium point of the linearized system (16) (see Figure 8.13). Now, from Theorem 1(b) of Section 8.5, we conclude that $(\pi, 0)$ is a saddle point and therefore an unstable equilibrium point for system (13). System (13) also has a saddle point at each of the critical points $((2n + 1)\pi, 0)$ for $n = \pm 1, \pm 2, \ldots$. The trajectories in this case are the heavy loops shown in Figure 8.14.

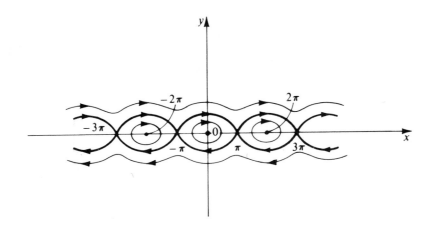

**Figure 8.14**

The equations of the trajectories of the solutions of system (13) can be found explicitly. In fact, from (13) we see that the trajectories satisfy the separable differential equation

$$\frac{dy}{dx} = -\omega^2 \frac{\sin x}{y}.$$

Separating the variables $x$ and $y$ and integrating, we obtain the family of trajectories

$$\tfrac{1}{2}y^2 - \omega^2 \cos x = c, \tag{18}$$

where $c$ is the integration constant. For the sake of describing the phase portrait, it is convenient to express $c$ in terms of initial conditions. Assume that $y = y_0$ when $x = 0$. Then from (18) we find that

$$c = \tfrac{1}{2}y_0^2 - \omega^2$$

$$\tfrac{1}{2}y^2 - \omega^2 \cos x = \tfrac{1}{2}y_0^2 - \omega^2$$

$$y^2 + 2\omega^2(1 - \cos x) = y_0^2$$

$$y^2 + 4\omega^2 \sin^2 \frac{x}{2} = y_0^2. \tag{19}$$

The three types of curves shown in Figure 8.14 correspond to the following three cases.

**CASE 1**  $|y_0| < 2\omega$  Then the maximum value of the angle $x$ (remember that $x = \theta$) is attained when $y = 0$ and

$$x_{\max} = 2 \arcsin \frac{y_0}{2\omega} < \pi.$$

In this case the pendulum oscillates between the extreme angles $\pm x_{\max}$. The trajectories in this case are the closed curves shown in Figure 8.14.

**CASE 2**  $|y_0| > 2\omega$  In this case the pendulum makes complete revolutions. The trajectories in this case are the wavy curves on the top and bottom of Figure 8.14.

**CASE 3**  $|y_0| = 2\omega$  In this case the trajectories are the heavy loops that separate the closed and wavy trajectories described in the previous cases. The equations of these trajectories can be found from (19) if we substitute $y_0^2$ by $4\omega^2$, and they are given by

$$y = \pm 2\omega \cos \frac{x}{2}.$$

## EXERCISES

**1.** Linearize system (2) in the biology application in the neighborhood of the critical point $(c/d, a/b)$ and study the nature and stability of the critical point $(0, 0)$ of the linearized system. [*Hint:* Set $x = X + c/d$, $y = Y + a/b$.]

**2. Biology** In the predator–prey interaction that we discussed in the text, we assumed that the prey population has an abundant supply of food. Assume now that the prey has a limited supply of food available. In this case it is reasonable to assume that the predator–prey interaction is described by a system of the form

$$\dot{x} = ax - bxy - \epsilon_1 x^2, \qquad \dot{y} = -cy + dxy - \epsilon_2 y^2,$$

where $a$, $b$, $c$, and $d$ are positive constants and $\epsilon_1$ and $\epsilon_2$ are nonnegative constants. [For $\epsilon_1 = \epsilon_2 = 0$, we obtain system (2) in the biology application.] Find the critical points of the system (negative populations are not permissible). In the special case

$$\dot{x} = 3x - xy - 2x^2, \qquad \dot{y} = -y + 2xy - y^2,$$

where $x$ and $y$ are measured in hundreds of organisms, study the stability of each critical point and determine whether it is a node, saddle, or focus.

**3.** The differential equation

$$\ddot{\theta} + k\dot{\theta} + \omega^2 \sin \theta = 0, \qquad k > 0,$$

describes the motion of a simple pendulum under the influence of a frictional force proportional to the angular velocity $\dot{\theta}$. Transform the differential equation into an equivalent system by setting $x = \theta$ and $y = \dot{\theta}$. Show that $(n\pi, 0)$ for $n = 0, \pm 1, \pm 2, \ldots$ are the only critical points of the system. In each of the following cases, study the stability of the system at $(0, 0)$ and determine whether $(0, 0)$ is a node, saddle, or focus.

(a) $k < 2\omega$            (b) $k = 2\omega$            (c) $k > 2\omega$

**4. Electric Circuits** The differential equation

$$\ddot{x} + \mu(x^2 - 1)\dot{x} + x = 0, \qquad \mu > 0$$

is called the *van der Pol equation* and governs certain electric circuits that contain vacuum tubes. Transform the differential equation into an equivalent system by setting $\dot{x} = y$. Show that $(0, 0)$ is the only critical point of the system. Study the stability of the system at the point $(0, 0)$ when $\mu < 2$ and when $\mu > 2$. In each case, determine whether $(0, 0)$ is a node, saddle, or focus.

## REVIEW EXERCISES

**1.** Show that the IVP

$$\dot{x} = 2x^2 - y + t$$
$$\dot{y} = x + 2y^2 - e^t$$
$$x(0) = 1; \qquad y(0) = 0$$

has a unique solution defined in some interval about the point $t_0 = 0$.

**2.** Show that $(0, 0)$ is the only critical point of the system

$$\dot{x} = 2x^2 - xy, \qquad \dot{y} = x + 2y.$$

**3.** Find the critical points of the system

$$\dot{x} = 2x^2 - xy, \qquad \dot{y} = -2x + y.$$

**4.** Find the critical points and the equations of the trajectories of the solutions of the system

$$\dot{x} = x - 4y, \qquad \dot{y} = 2x - y.$$

**5.** Find the critical points and the equations of the trajectories of the solutions of the system

$$\dot{x} = x, \qquad \dot{y} = -x^2 + 2y.$$

In particular, draw the trajectory corresponding to the solution with initial conditions

$$x(0) = 1, \qquad y(0) = 0$$

and indicate the direction of increasing $t$.

In Exercises 6 through 12, determine whether the critical point $(0, 0)$ of the system is stable, asymptotically stable, or unstable.

**6.** $\dot{x} = -3x + 2y, \dot{y} = 4x - y$      **7.** $\dot{x} = 2x - 8y, \dot{y} = x - 2y$

**8.** $\dot{x} = 3x + 8y, \dot{y} = -x - y$      **9.** $\dot{x} = 4x - 9y, \dot{y} = 5x - 10y$

**10.** $\dot{x} = -3x + 2y + x^2 - y^2, \dot{y} = 4x - y + xy$

**11.** $\dot{x} = 3x + 8y - xy, \dot{y} = -x - y + xy$

**12.** $\dot{x} = 4x - 9y + (x^2 + y^2)^{3/2}, \dot{y} = 5x - 10y - (x^2 + y^2)$

In Exercises 13 through 16, determine whether the critical point $(0, 0)$ is a node, saddle point, focus, or center, and graph a few trajectories.

**13.** $\dot{x} = -3x + 2y, \dot{y} = 4x - y$      **14.** $\dot{x} = 3x + 8y, \dot{y} = -x - y$

**15.** $\dot{x} = 2x - 8y, \dot{y} = x - 2y$      **16.** $\dot{x} = 4x - 9y, \dot{y} = 5x - 10y$

**17.** Determine the nature of the critical point $(0, 0)$ of the system

$$\dot{x} = 4x - 9y + (x^2 + y^2)^2, \qquad \dot{y} = 5x - 10y - (x^2 + y^2)^{3/2}.$$

**18.** Determine[2] the stability of the linear time-invariant network shown in Figure 8.15. Assume that the initial voltage on the capacitor is $v_0$. (The state variables are $x_1 = I$ and $x_2 = Q$.)

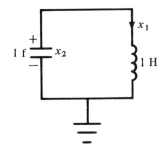

**Figure 8.15**

**19.** Consider[3] the linear time-invariant $RLC$ network shown in Figure 8.16. Let the initial current through the inductor be $i_0$. Determine the stability of this network. (The state variables are $x_1 = -Q$ and $x_2 = I$.)

**Figure 8.16**

[2]This is Example 11.2-2 in Behrouz Peikari, *Fundamentals of Network Analysis and Synthesis* (Englewood Cliffs, N.J.: Prentice-Hall, 1974), © 1974. Reprinted by permission of Prentice-Hall, Inc.
[3]This is Example 11.2-3, ibid., p. 455. © 1974. Reprinted by permission of Prentice-Hall, Inc.

# Determinants and Linear Systems of Equations

Systems of linear algebraic equations occur in many places in this text. For example, to find the values of the constants of integration in an initial value problem requires us to solve a system of linear equations. Further illustrations are the method of undetermined coefficients and variation of parameters for nonhomogeneous differential equations.

In this appendix we provide tools that are sufficient to solve any linear system. However, we make no attempt at a complete discussion of the problem. The main tool presented here is Cramer's rule. For some systems other tools may be more efficient, for example, elimination or matrix methods.

**DEFINITION 1**

*A square matrix, A, of order n is a function whose domain is the set $\{(i, j):$ $i = 1, 2, \ldots, n; j = 1, 2, \ldots, n\}$ and whose range is a subset of the complex numbers.*

**Notation**  The image under $A$ of the ordered pair $(i, j)$ is denoted by $a_{ij}$. That is, $A(i, j) = a_{ij}$. The range of the square matrix $A$ is customarily written in the form

$$
\begin{bmatrix}
a_{11} & a_{12} & \cdots & a_{1n} \\
a_{21} & a_{22} & \cdots & a_{2n} \\
\cdot & \cdot & & \cdot \\
\cdot & \cdot & & \cdot \\
\cdot & \cdot & & \cdot \\
a_{n1} & a_{n2} & \cdots & a_{nn}
\end{bmatrix}. \tag{1}
$$

It is common practice to refer to the expression (1) as *the matrix A*, and each image $a_{ij}$ as an element of $A$. Thus, we speak of the matrix $A$ as consisting of $n$ rows and $n$ columns. For each element $a_{ij}$ the first subscript $i$ refers to the row in which $a_{ij}$ appears, and the second subscript $j$ refers to the column in which $a_{ij}$ appears. Thus, $a_{12}$ appears in the first row, second column.

To every square matrix $A$ we can associate a number (defined below) called the *determinant* of $A$. The determinant of $A$ is denoted by $|A|$ and is written in the form

$$|A| = \begin{vmatrix} a_{11} & a_{12} & \cdots & a_{1n} \\ a_{21} & a_{22} & \cdots & a_{2n} \\ \vdots & \vdots & & \vdots \\ a_{n1} & a_{n2} & \cdots & a_{nn} \end{vmatrix}. \tag{2}$$

Once again we speak of $|A|$ as consisting of $n$ rows and $n$ columns. The expression in (2) is also referred to as a determinant of order $n$.

## DEFINITION 2

*The minor of $a_{ij}$ is the determinant of order $n - 1$ formed by crossing out the $i$th row and the $j$th column of $|A|$. The minor of $a_{ij}$ is denoted by $M_{ij}$.*

**EXAMPLE 1**   If $|A|$ is of order 4, write the minor of $a_{13}$ and the minor of $a_{32}$.

**Solution**   We have

$$|A| = \begin{vmatrix} a_{11} & a_{12} & a_{13} & a_{14} \\ a_{21} & a_{22} & a_{23} & a_{24} \\ a_{31} & a_{32} & a_{33} & a_{34} \\ a_{41} & a_{42} & a_{43} & a_{44} \end{vmatrix}.$$

Thus,

$$M_{13} = \begin{vmatrix} a_{21} & a_{22} & a_{24} \\ a_{31} & a_{32} & a_{34} \\ a_{41} & a_{42} & a_{44} \end{vmatrix}, \qquad M_{32} = \begin{vmatrix} a_{11} & a_{13} & a_{14} \\ a_{21} & a_{23} & a_{24} \\ a_{41} & a_{43} & a_{44} \end{vmatrix}.$$

## DEFINITION 3

*The value of a determinant of order one is $a_{11}$. That is, $|A| = |a_{11}| = a_{11}$. (Be careful not to misinterpret $|\ \ |$ as being absolute value signs; they are not.)*

## DEFINITION 4

*If $|A|$ is a determinant of order $n$, then*

$$|A| = \sum_{i=1}^{n} a_{ij}(-1)^{i+j} M_{ij} \qquad \text{for any fixed } j \tag{3}$$

*or*

$$|A| = \sum_{j=1}^{n} a_{ij}(-1)^{i+j} M_{ij} \qquad \text{for any fixed } i. \tag{4}$$

When we evaluate $|A|$ by formula (3), we are expanding $|A|$ by the minors of the $j$th column. Similarly, formula (4) is spoken of as the expansion of $|A|$ by the minors of the $i$th row.

**EXAMPLE 2**   Find the value of the general determinant of order 2.

**Solution**   We have

$$|A| = \begin{vmatrix} a_{11} & a_{12} \\ a_{21} & a_{22} \end{vmatrix}.$$

We evaluate $|A|$ by taking the expansion of $|A|$ by the minors of the first row. Hence,

$$|A| = a_{11}(-1)^{1+1}|a_{22}| + a_{12}(-1)^{1+2}|a_{21}|$$

$$= a_{11}a_{22} - a_{12}a_{21}. \tag{5}$$

Equation (5) can be viewed as a formula for the value of any determinant of order 2. In fact, one can represent formula (5) by the following schematic:

$$\begin{vmatrix} a_{11} & a_{12} \\ a_{21} & a_{22} \end{vmatrix} \tag{6}$$

In this schematic the arrows indicate multiplication, and the plus and minus signs refer to the algebraic signs needed for formula (5).

**EXAMPLE 3**   Find the value of each of the following determinants of order 2.

$$\text{(a)} \begin{vmatrix} 3 & 4 \\ 2 & 5 \end{vmatrix} \quad \text{(b)} \begin{vmatrix} 3 & 4 \\ -2 & -5 \end{vmatrix}$$

**Solution**

$$\text{(a)} \begin{vmatrix} 3 & 4 \\ 2 & 5 \end{vmatrix} = +(3)(5) - (2)(4) = 15 - 8 = 7$$

$$\text{(b)} \begin{vmatrix} 3 & 4 \\ -2 & -5 \end{vmatrix} = +(3)(-5) - (-2)(4) = -15 + 8 = -7$$

**EXAMPLE 4**   Find the value of the general determinant of order 3.

**Solution**

$$|A| = \begin{vmatrix} a_{11} & a_{12} & a_{13} \\ a_{21} & a_{22} & a_{23} \\ a_{31} & a_{32} & a_{33} \end{vmatrix}.$$

We expand by the minors of the second column.

$$|A| = a_{12}(-1)^{1+2}\begin{vmatrix} a_{21} & a_{23} \\ a_{31} & a_{33} \end{vmatrix} + a_{22}(-1)^{2+2}\begin{vmatrix} a_{11} & a_{13} \\ a_{31} & a_{33} \end{vmatrix}$$

$$+ a_{32}(-1)^{3+2}\begin{vmatrix} a_{11} & a_{13} \\ a_{21} & a_{23} \end{vmatrix}$$

$$= -a_{12}\begin{vmatrix} a_{21} & a_{23} \\ a_{31} & a_{33} \end{vmatrix} + a_{22}\begin{vmatrix} a_{11} & a_{13} \\ a_{31} & a_{33} \end{vmatrix} - a_{32}\begin{vmatrix} a_{11} & a_{13} \\ a_{21} & a_{23} \end{vmatrix}.$$

Each of these determinants of order 2 can be evaluated by use of the schematic (6). Thus,

$$|A| = -a_{12}(a_{21}a_{33} - a_{31}a_{23}) + a_{22}(a_{11}a_{33} - a_{31}a_{13}) - a_{32}(a_{11}a_{23} - a_{21}a_{13}).$$

Finally, after some rearrangement, we can write

$$|A| = a_{11}a_{22}a_{33} + a_{12}a_{23}a_{31} + a_{13}a_{21}a_{32} - a_{31}a_{22}a_{13} - a_{32}a_{23}a_{11} - a_{33}a_{21}a_{12}. \quad (7)$$

Equation (7) can be viewed as a formula for the value of any determinant of order 3. Formula (7) can be represented by the following schematic:

$$(8)$$

In the schematic (8), the first two columns of $|A|$ are written to the right of the determinant. The arrows indicate multiplication and the signs refer to the algebraic signs necessary to obtain formula (7).

**EXAMPLE 5**   Evaluate

$$\begin{vmatrix} 2 & 8 & 7 \\ 1 & 4 & -1 \\ -3 & 6 & -2 \end{vmatrix}.$$

**Solution**

$$\begin{vmatrix} 2 & 8 & 7 \\ 1 & 4 & -1 \\ -3 & 6 & -2 \end{vmatrix}\begin{matrix} 2 & 8 \\ 1 & 4 \\ -3 & 6 \end{matrix} = +(2)(4)(-2) + (8)(-1)(-3) + (7)(1)(6)$$

$$- (-3)(4)(7) - (6)(-1)(2) - (-2)(1)(8)$$

$$= -16 + 24 + 42 + 84 + 12 + 16$$

$$= 162.$$

For determinants of order 4 or larger, there are *no* convenient schematics such as (6) or (8). Therefore, with such determinants we *must* evaluate them by expansion of the minors of some row or column.

**EXAMPLE 6**  Evaluate

$$|A| = \begin{vmatrix} 5 & 4 & -2 & 6 \\ -3 & 1 & 1 & 0 \\ 2 & -3 & 1 & 0 \\ 1 & 2 & -1 & 0 \end{vmatrix}.$$

**Solution**  We evaluate $|A|$ by expansion of the minors of the second row.

$$|A| = (-3)(-1)^{2+1} \begin{vmatrix} 4 & -2 & 6 \\ -3 & 1 & 0 \\ 2 & -1 & 0 \end{vmatrix} + (1)(-1)^{2+2} \begin{vmatrix} 5 & -2 & 6 \\ 2 & 1 & 0 \\ 1 & -1 & 0 \end{vmatrix}$$

$$+ (1)(-1)^{2+3} \begin{vmatrix} 5 & 4 & 6 \\ 2 & -3 & 0 \\ 1 & 2 & 0 \end{vmatrix} + (0)(-1)^{2+4} \begin{vmatrix} 5 & 4 & -2 \\ 2 & -3 & 1 \\ 1 & 2 & -1 \end{vmatrix}$$

$$= (3) \begin{vmatrix} 4 & -2 & 6 \\ -3 & 1 & 0 \\ 2 & -1 & 0 \end{vmatrix} + (1) \begin{vmatrix} 5 & -2 & 6 \\ 2 & 1 & 0 \\ 1 & -1 & 0 \end{vmatrix} - (1) \begin{vmatrix} 5 & 4 & 6 \\ 2 & -3 & 0 \\ 1 & 2 & 0 \end{vmatrix}$$

$$+ (0) \begin{vmatrix} 5 & 4 & -2 \\ 2 & -3 & 1 \\ 1 & 2 & -1 \end{vmatrix}.$$

Each of these determinants of order 3 can be evaluated according to the schematic (8). Thus,

$$\begin{vmatrix} 4 & -2 & 6 \\ -3 & 1 & 0 \\ 2 & -1 & 0 \end{vmatrix} \begin{matrix} 4 & -2 \\ -3 & 1 \\ 2 & -1 \end{matrix} = +(4)(1)(0) + (-2)(0)(2) + (6)(-3)(-1)$$
$$- (2)(1)(6) - (-1)(0)(4) - (0)(-3)(-2)$$
$$= 0 + 0 + 18 - 12 - 0 - 0 = 6$$

$$\begin{vmatrix} 5 & -2 & 6 \\ 2 & 1 & 0 \\ 1 & -1 & 0 \end{vmatrix} \begin{matrix} 5 & -2 \\ 2 & 1 \\ 1 & -1 \end{matrix} = +(5)(1)(0) + (-2)(0)(1) + (6)(2)(-1))$$
$$- (1)(1)(6) - (-1)(0)(5) - (0)(2)(-2)$$
$$= 0 + 0 - 12 - 6 - 0 - 0 = -18$$

$$\begin{vmatrix} 5 & 4 & 6 \\ 2 & -3 & 0 \\ 1 & 2 & 0 \end{vmatrix} \begin{matrix} 5 & 4 \\ 2 & -3 \\ 1 & 2 \end{matrix} = +(5)(-3)(0) + (4)(0)(1) + (6)(2)(2)$$
$$- (1)(-3)(6) - (2)(0)(5) - (0)(2)(4)$$
$$= 0 + 0 + 24 + 18 - 0 - 0 = 42$$

$$\begin{vmatrix} 5 & 4 & -2 \\ 2 & -3 & 1 \\ 1 & 2 & -1 \end{vmatrix} \begin{matrix} 5 & 4 \\ 2 & -3 \\ 1 & 2 \end{matrix} = +(5)(-3)(-1) + (4)(1)(1) + (-2)(2)(2)$$
$$- (1)(-3)(-2) + (2)(1)(5) - (-1)(2)(4)$$
$$= 15 + 4 - 8 - 6 - 10 + 8 = 3. \tag{9}$$

Thus,

$$|A| = (3)(6) + (1)(-18) - (1)(42) + (0)(3) = 18 - 18 - 42 + 0 = -42.$$

The reader no doubt observes that the calculations in (9) could have been disregarded, since the determinant involved was to be multiplied by zero. Indeed, we could have saved considerable labor if originally we had expanded $|A|$ by the minors of the fourth column. In this case we would only have to calculate the minor of the element 6 for the other products would be zero. Hence,

$$|A| = 6(-1)^{1+4} \begin{vmatrix} -3 & 1 & 1 \\ 2 & -3 & 1 \\ 1 & 2 & -1 \end{vmatrix} = (-6) \begin{vmatrix} -3 & 1 & 1 \\ 2 & -3 & 1 \\ 1 & 2 & -1 \end{vmatrix} \begin{matrix} -3 & 1 \\ 2 & -3 \\ 1 & 2 \end{matrix}$$

$$= (-6)[ +(-3)(-3)(-1) + (1)(1)(1) + (1)(2)(2) - (1)(-3)(1)$$
$$- (2)(1)(-3) - (-1)(2)(1)]$$

$$= (-6)(-9 + 1 + 4 + 3 + 6 + 2) = (-6)(7) = -42.$$

On the basis of Example 6 we may formulate a general rule for efficiency of computation: to expand by the minors of the row or column that has the most zeros. Unfortunately, some determinants do not have any zeros. In such a situation we make use of the following theorem about determinants.[1]

### THEOREM 1

*The value of a determinant is unchanged if all the elements of some row (or column) are multiplied by the same nonzero constant and the results added to another row (or column).*

**EXAMPLE 7**   For the determinant

$$|A| = \begin{vmatrix} 1 & 5 & 3 & -2 \\ 7 & -2 & 6 & 1 \\ 8 & 4 & -3 & 5 \\ -3 & 1 & 4 & 2 \end{vmatrix},$$

multiply the second row by $-3$ and add the results to the third row.

**Solution**   To indicate that row 2 is to be multiplied by $-3$ and the results added to row 3, we write $(-3)R_2 + R_3$. Thus,

$$\begin{vmatrix} 1 & 5 & 3 & -2 \\ 7 & -2 & 6 & 1 \\ 8 & 4 & -3 & 5 \\ -3 & 1 & 4 & 2 \end{vmatrix} \overset{(-3)R_2 + R_3}{=} \begin{vmatrix} 1 & 5 & 3 & -2 \\ 7 & -2 & 6 & 1 \\ -13 & 10 & -21 & 2 \\ -3 & 1 & 4 & 2 \end{vmatrix}.$$

Note that row 3 is the *only* row that changes.

[1]For a proof of this theorem, see M. Richardson, *College Algebra,* Alternate Edition (Englewood Cliffs, N.J.: Prentice-Hall, 1958).

**EXAMPLE 8**   For the determinant of Example 7, multiply the first column by 2 and add the results to the fourth column.

**Solution.**

$$
\begin{vmatrix} 1 & 5 & 3 & -2 \\ 7 & -2 & 6 & 1 \\ 8 & 4 & -3 & 5 \\ -3 & 1 & 4 & 2 \end{vmatrix} \overset{2C_1 + C_4}{=} \begin{vmatrix} 1 & 5 & 3 & 0 \\ 7 & -2 & 6 & 15 \\ 8 & 4 & -3 & 21 \\ -3 & 1 & 4 & -4 \end{vmatrix}.
$$

**EXAMPLE 9**   Evaluate the determinant of Example 7.

**Solution**

$$
|A| = \begin{vmatrix} 1 & 5 & 3 & -2 \\ 7 & -2 & 6 & 1 \\ 8 & 4 & -3 & 5 \\ -3 & 1 & 4 & 2 \end{vmatrix} \overset{2C_1 + C_4}{=} \begin{vmatrix} 1 & 5 & 3 & 0 \\ 7 & -2 & 6 & 15 \\ 8 & 4 & -3 & 21 \\ -3 & 1 & 4 & -4 \end{vmatrix}
$$

$$
\overset{-5C_1 + C_2}{=} \begin{vmatrix} 1 & 0 & 3 & 0 \\ 7 & -37 & 6 & 15 \\ 8 & -36 & -3 & 21 \\ -3 & 16 & 4 & -4 \end{vmatrix}
$$

$$
\overset{-3C_1 + C_3}{=} \begin{vmatrix} 1 & 0 & 0 & 0 \\ 7 & -37 & -15 & 15 \\ 8 & -36 & -27 & 21 \\ -3 & 16 & 13 & -4 \end{vmatrix}.
$$

Actually all three of the steps above could have been carried out simultaneously. Thus, we could have saved some time and some writing. We now expand $|A|$ by the minors of the first row, to obtain

$$
|A| = (1)(-1)^{1+1} \begin{vmatrix} -37 & -15 & 15 \\ -36 & -27 & 21 \\ 16 & 13 & -4 \end{vmatrix} = \begin{vmatrix} -37 & -15 & 15 & -37 & -15 \\ -36 & -27 & 21 & -36 & -27 \\ 16 & 13 & -4 & 16 & 13 \end{vmatrix}
$$

$$
= +(-37)(-27)(-4) + (-15)(21)(16) + (15)(-36)(13)
$$
$$
- (16)(-27)(15) - (13)(21)(-37) - (-4)(-36)(-15)
$$
$$
= -3996 - 5040 - 7020 + 6480 + 10{,}101 + 2160
$$
$$
= -16{,}056 + 18{,}741
$$
$$
= 2685.
$$

**EXAMPLE 10**   Evaluate

$$
|A| = \begin{vmatrix} 2 & 1 & 1 & 5 & 1 \\ -3 & 2 & -1 & -4 & 2 \\ 5 & 3 & 1 & -3 & 1 \\ 1 & -2 & 2 & 6 & -1 \\ -1 & 1 & -1 & -5 & 2 \end{vmatrix}.
$$

**Solution**

$$
\begin{matrix}
R_2 + R_1 \\
R_2 + R_3 \\
2R_2 + R_4 \\
-R_2 + R_5
\end{matrix}
$$

$$
|A| = \begin{vmatrix}
-1 & 3 & 0 & 1 & 3 \\
-3 & 2 & -1 & -4 & 2 \\
2 & 5 & 0 & -7 & 3 \\
-5 & 2 & 0 & -2 & 3 \\
2 & -1 & 0 & -1 & 0
\end{vmatrix}.
$$

Expanding by the minors of the third column, we have

$$
|A| = (-1)(-1)^{2+3} \begin{vmatrix}
-1 & 3 & 1 & 3 \\
2 & 5 & -7 & 3 \\
-5 & 2 & -2 & 3 \\
2 & -1 & -1 & 0
\end{vmatrix}
\begin{matrix} -R_1 + R_2 \\ -R_1 + R_3 \\ = \end{matrix}
\begin{vmatrix}
-1 & 3 & 1 & 3 \\
3 & 2 & -8 & 0 \\
-4 & -1 & -3 & 0 \\
2 & -1 & -1 & 0
\end{vmatrix}.
$$

Expanding by the minors of the fourth column, we obtain

$$
|A| = (3)(-1)^{1+4} \begin{vmatrix}
3 & 2 & -8 \\
-4 & -1 & -3 \\
2 & -1 & -1
\end{vmatrix}
\begin{matrix} 2R_2 + R_1 \\ -R_2 + R_3 \\ = (-3) \end{matrix}
\begin{vmatrix}
-5 & 0 & -14 \\
-4 & -1 & -3 \\
6 & 0 & 2
\end{vmatrix}.
$$

Expanding by the minors of the second column, we have

$$
|A| = (-3)(-1)(-1)^{2+2} \begin{vmatrix}
-5 & -14 \\
6 & 2
\end{vmatrix} = (3)(-10 + 84) = (3)(74)
$$

$$
= 222.
$$

Determinants can be used to solve systems of linear algebraic equations. We consider the following system of $n$ equations in $n$ unknowns.

$$
\begin{aligned}
a_{11}x_1 + a_{12}x_2 + \cdots + a_{1n}x_n &= b_1 \\
a_{21}x_1 + a_{22}x_2 + \cdots + a_{2n}x_n &= b_2 \\
&\vdots \\
a_{n1}x_1 + a_{n2}x_2 + \cdots + a_{nn}x_n &= b_n.
\end{aligned}
\tag{10}
$$

In system (10) we understand that each $a_{ij}$ and each $b_i$ is known and that $x_1$, $x_2, \ldots, x_n$ are unknown. The task is to find the value(s) of $x_1, x_2, \ldots, x_n$ that satisfy system (10).

**EXAMPLE 11** Show that $x_1 = 1$, $x_2 = 2$ satisfy the system

$$x_1 + x_2 = 3$$

$$3x_1 + 7x_2 = 17.$$

**Solution** By direct substitution, we have

$$1 + 2 = 3 \Rightarrow 3 = 3$$

$$3(1) + 7(2) = 17 \Rightarrow 17 = 17.$$

**Notation** The determinant of coefficients, denoted by $\Delta$, for the system (10) is

$$\Delta = \begin{vmatrix} a_{11} & a_{12} & \cdots & a_{1n} \\ a_{21} & a_{22} & \cdots & a_{2n} \\ \vdots & \vdots & & \vdots \\ a_{n1} & a_{n2} & \cdots & a_{nn} \end{vmatrix}$$

By $\Delta_k$ we mean the determinant that resembles $\Delta$ in every respect *except* that the $k$th *column* of $\Delta_k$ consists of the numbers $b_1, b_2, \ldots, b_n$. That is, $a_{ik}$ is replaced by $b_i$ for each value of $i$.

We now quote a well known result for the algebraic system (10).

**Cramer's Rule** If $\Delta \neq 0$, the system (10) has a unique solution given by
$$x_k = \Delta_k / \Delta, \ k = 1, 2, \ldots, n.$$

**EXAMPLE 12** Using Cramer's rule, solve the system

$$3x_1 - x_2 + 2x_3 = 1$$

$$5x_1 + 2x_2 - x_3 = -2$$

$$2x_1 - x_2 + x_3 = 3.$$

**Solution**

$$\Delta = \begin{vmatrix} 3 & -1 & 2 \\ 5 & 2 & -1 \\ 2 & -1 & 1 \end{vmatrix} = \begin{vmatrix} 3 & -1 & 2 \\ 5 & 2 & -1 \\ 2 & -1 & 1 \end{vmatrix} \begin{matrix} 3 & -1 \\ 5 & 2 \\ 2 & -1 \end{matrix}$$

$$= + (3)(2)(1) + (-1)(-1)(2) + (2)(5)(-1) - (2)(2)(2)$$

$$- (-1)(-1)(3) - (1)(5)(-1)$$

$$= 6 + 2 - 10 - 8 - 3 + 5 = -8 \neq 0$$

$$x_1 = \frac{\Delta_1}{\Delta} = \frac{\begin{vmatrix} 1 & -1 & 2 \\ -2 & 2 & -1 \\ 3 & -1 & 1 \end{vmatrix}}{-8} = \frac{\begin{vmatrix} 1 & -1 & 2 \\ -2 & 2 & -1 \\ 3 & -1 & 1 \end{vmatrix} \begin{matrix} 1 & -1 \\ -2 & 2 \\ 3 & -1 \end{matrix}}{-8}$$

$$= \frac{2 + 3 + 4 - 12 - 1 - 2}{-8} = \frac{-6}{-8} = \frac{3}{4}$$

$$x_2 = \frac{\Delta_2}{\Delta} = \frac{\begin{vmatrix} 3 & 1 & 2 \\ 5 & -2 & -1 \\ 2 & 3 & 1 \end{vmatrix}}{-8} = \frac{-6 - 2 + 30 + 8 + 9 - 5}{-8}$$

$$= \frac{34}{-8} = -\frac{17}{4}$$

$$x_3 = \frac{\Delta_3}{\Delta} = \frac{\begin{vmatrix} 3 & -1 & 1 \\ 5 & 2 & -2 \\ 2 & -1 & 3 \end{vmatrix}}{-8} = \frac{18 + 4 - 5 - 4 - 6 + 15}{-8}$$

$$= \frac{22}{-8} = -\frac{11}{4}.$$

Thus, the solution is $x_1 = \frac{3}{4}$, $x_2 = -\frac{17}{4}$, and $x_3 = -\frac{11}{4}$.

In case $b_i = 0$ for every $i = 1, 2, \ldots, n$, system (10) is called a *homogeneous* system. Otherwise, the system is called *nonhomogeneous*. The homogeneous system has an obvious solution, $x_1 = 0$, $x_2 = 0, \ldots, x_n = 0$. This solution is referred to as the *trivial solution*. The following theorem is important.

### THEOREM 2

*A homogeneous system has nontrivial solutions (in other words, solutions other than the trivial solution) if and only if $\Delta = 0$.*

**EXAMPLE 13**   Solve the system

$$x_1 - x_2 + x_3 = 0$$
$$x_1 + x_2 + x_3 = 0$$
$$2x_1 \qquad + 2x_3 = 0.$$

**Solution**

$$\Delta = \begin{vmatrix} 1 & -1 & 1 \\ 1 & 1 & 1 \\ 2 & 0 & 2 \end{vmatrix} \overset{R_2 + R_1}{=} \begin{vmatrix} 2 & 0 & 2 \\ 1 & 1 & 1 \\ 2 & 0 & 2 \end{vmatrix} = (1)(-1)^{2+2} \begin{vmatrix} 2 & 2 \\ 2 & 2 \end{vmatrix} = 4 - 4 = 0.$$

Hence, we anticipate nontrivial solutions. From the third equation we obtain

$x_3 = -x_1$. Substituting this result into either the first equation or the second yields $x_2 = 0$. Thus, we have the solution

$$x_1 = a, \qquad x_2 = 0, \qquad x_3 = -a, \text{ where } a \text{ is any number.}$$

Note that there are an infinity of solutions in this case.

# Partial-Fraction Decomposition

The method of partial-fraction decomposition is useful in differential equations when we want to compute the inverse Laplace transform or the integral of functions of such forms as

$$\frac{2x + 3}{x^2(x - 1)}, \frac{3x}{x^2 - 1}, \frac{-4x^3 + 33x^2 - 58x + 31}{(x - 2)^2(x + 1)(x - 1)}, \frac{2x^4 + 1}{x^3(x^2 + 1)^2(x - 2)}.$$

Such functions, being the quotient of two polynomials, are called *rational fractions*. A proper rational fraction is one for which the degree of the numerator is less than the degree of the denominator. Otherwise, the rational fraction is improper.

Using long division, improper fractions can be written as a polynomial plus a proper rational fraction. For this reason we restrict our attention to proper rational fractions.

The reader no doubt is familiar with the algebraic problem of adding the fractions

$$\frac{2}{x - 2} + \frac{5}{(x - 2)^2} + \frac{-7}{x + 1} + \frac{1}{x - 1}. \tag{1}$$

Of course, the method is to find the *least common denominator (L.C.D.)* of the fractions $[(x + 1)(x - 1)(x - 2)^2$ in this illustration], express each fraction as an equivalent fraction having the L.C.D. as denominator, add the resulting numerators, and express the result over the L.C.D. For our illustration the answer would be

$$\frac{-4x^3 + 33x^2 - 58x + 31}{(x - 2)^2(x + 1)(x - 1)}. \tag{2}$$

What about the reverse problem? That is, beginning with the fraction (2), can we write this fraction in the form (1)? The process associated with this reverse problem is known as *decomposing a proper rational fraction into partial fractions*. To decide what *simple* kinds of fractions might occur in the decomposition we look at the denominator in (2). The factor $(x - 2)^2$ suggests two fractions, one with denominator $(x - 2)$ and the other with denominator $(x - 2)^2$ [in the general case of a "repeated factor" $(x - a)^n$, there would be $n$ fractions having, respectively, the denominators $(x - a)$, $(x - a)^2$, $(x - a)^3$, . . . , $(x - a)^n$]; the factor $(x + 1)$ suggests a fraction with this denominator, and the factor $(x - 1)$ suggests a fraction with this denominator. Since the fraction (2) is proper, each of the partial fractions will also be proper.

Thus, the numerators for the fractions having the denominator $(x - 2)$, $(x + 1)$, or $(x - 1)$ will be constants. Since we have no way of knowing (as yet) what values these constants should have, we consider them to be unknowns and denote them by $A$, $B$, and $C$. Since the denominator $(x - 2)^2$ is quadratic, one might suspect that its numerator should be of degree 1, that is, of the form $Dx + E$, where $D$ and $E$ are unknown. Although this suspicion is correct, it is unnecessary. That a linear numerator is unnecessary can be seen as follows. The factor $(x - 2)^2$ would produce the two partial fractions

$$\frac{A}{x - 2} + \frac{Dx + E}{(x - 2)^2} = \frac{A}{x - 2} + \frac{D(x - 2) + (2D + E)}{(x - 2)^2} = \frac{A}{x - 2} + \frac{D(x - 2)}{(x - 2)^2}$$

$$+ \frac{2D + E}{(x - 2)^2}$$

$$= \frac{A + D}{x - 2} + \frac{2D + E}{(x - 2)^2}.$$

Since $\alpha = A + D$ and $\beta = 2D + E$ are unknown constants, it is the same as if we began with constants in both numerators.

Using similar algebraic manipulations, one can establish the following rule.

**Rule 1** *For each partial fraction associated with a repeated factor, the numerator can be taken to be of one less degree than the factor that is repeated.*

For example,

$$\frac{x + 2}{x^2(x - 1)} = \frac{A}{x} + \frac{B}{x^2} + \frac{C}{x - 1}$$

$$\frac{3}{x^2 - 1} = \frac{A}{x - 1} + \frac{B}{x + 1}$$

$$\frac{x^2 - 3}{x(x - 1)^2(x^2 + 1)} = \frac{A}{x} + \frac{B}{x - 1} + \frac{C}{(x - 1)^2} + \frac{Dx + E}{x^2 + 1}$$

$$\frac{x^2 + 1}{x^2 - 1} = 1 + \frac{2}{x^2 - 1} = 1 + \frac{A}{x - 1} + \frac{B}{x + 1}$$

$$\frac{x^3 - 3x^2 + 7}{x^3(x^2 + 1)^2(x - 3)} = \frac{A}{x} + \frac{B}{x^2} + \frac{C}{x^3} + \frac{Dx + E}{x^2 + 1} + \frac{Fx + G}{(x^2 + 1)^2} + \frac{H}{x - 3}$$

For the fraction (2) we would write

$$\frac{-4x^3 + 33x^2 - 58x + 31}{(x - 2)^2(x + 1)(x - 1)} = \frac{A}{x - 2} + \frac{D}{(x - 2)^2} + \frac{B}{x + 1} + \frac{C}{x - 1}. \qquad (3)$$

The problem is to find values for $A$, $B$, $C$, and $D$ so that Eq. (3) holds for all values of $x$. If we clear fractions in Eq. (3), we obtain

$$-4x^3 + 33x^2 - 58x + 31 = A(x - 2)(x + 1)(x - 1) + D(x + 1)(x - 1)$$
$$+ B(x - 2)^2(x - 1) + C(x - 2)^2(x + 1) \qquad (4)$$

$$= A(x^3 - 2x^2 - x + 2) + D(x^2 - 1)$$

$$+ B(x^3 - 5x^2 + 8x - 4) + C(x^3 - 3x^2 + 4)$$

$$= (A + B + C)x^3 + (-2A + D - 5B - 3C)x^2$$

$$+ (-A + 8B)x + (2A - D - 4B + 4C). \qquad (5)$$

Equation (5) is to hold for all values of $x$. Therefore, the coefficient of each power of $x$ should be the same on both sides of the equation; that is,

$$A + B + C = -4$$

$$-2A - 5B - 3C + D = 33 \qquad (6)$$

$$-A + 8B = -58$$

$$2A - 4B + 4C - D = 31.$$

Equations (6) constitute a linear algebraic system of four equations in four unknowns, and so we can solve them by the methods of Appendix A.

The method outlined in Eqs. (3) through (6) is typical of partial-fraction decomposition. The method can be used successfully for the decomposition of any proper rational fraction when the factors of the denominator are known.

In some cases, however, it is possible to introduce some shortcuts. For example, if one were to substitute $x = 2$ in Eq. (4), the terms involving $A$, $B$, and $C$ would disappear and we would have a single equation involving the single unknown $D$:

$$-4(2)^3 + 33(2)^2 - 58(2) + 31 = D(2 + 1)(2 - 1)$$

or

$$D = \frac{-4(2)^3 + 33(2)^2 - 58(2) + 31}{(2 + 1)(2 - 1)} = 5. \qquad (7)$$

It is convenient to write Eq. (7) in the form

$$D = \frac{-4x^3 + 33x^2 - 58x + 31}{(x + 1)(x - 1)}\bigg|_{x=2}. \qquad (8)$$

Formula (8) we note is nothing more than the left-hand side of Eq. (3) with the denominator under $D$ deleted and then evaluated at $x = 2$ (the value that makes the denominator under $D$ equal zero).

Similarly, we can substitute $x = -1$ in Eq. (4) and thus obtain an equation involving the single unknown, $B$. The solution for $B$ can also be written in a form similar to Eq. (8). Thus,

$$B = \frac{-4x^3 + 33x^2 - 58x + 31}{(x - 2)^2(x - 1)}\bigg|_{x=-1} = -7. \qquad (9)$$

Once again formula (9) is the left-hand side of Eq. (3) with the denominator under $B$ deleted and evaluated at $x = -1$ (the value that makes the denominator under $B$ equal zero).

In exactly the same way, we obtain

$$C = \left. \frac{-4x^3 + 33x^2 - 58x + 31}{(x - 2)^2(x + 1)} \right|_{x=1} = 1.$$

Unfortunately, we cannot obtain $A$ by any "convenient" shortcut. [In general, for repeated factors only the numerator of $(x - a)^n$ with the highest exponent can be obtained by this shortcut method; the other numerators can be evaluated by a differentiation process but this process is often more complicated than the method already introduced.] Therefore, we must follow through with the procedure embodied in Eqs. (3) through (6). We notice, however, that since there is only one unknown left to determine, we need only one equation. We could proceed by multiplying Eq. (3) by the L.C.D., thus obtaining Eq. (4). In Eq. (4) it is easy to determine the coefficient of $x^3$ without doing any algebra. Thus, we could immediately write the first of Eqs. (6) and solve this equation for $A$. Hence,

$$A = -4 - B - C = -4 - (-7) - 1 = 2.$$

Combining our results, we have

$$\frac{-4x^3 + 33x^2 - 58x + 31}{(x - 2)^2(x + 1)(x - 1)} = \frac{2}{x - 2} + \frac{5}{(x - 2)^2} + \frac{-7}{x + 1} + \frac{1}{x - 1}.$$

As a second illustration let us decompose the fraction

$$\frac{3x^2 - x + 3}{(x + 1)(x - 1)(x - 3)}$$

into partial fractions. We write

$$\frac{3x^2 - x + 3}{(x + 1)(x - 1)(x - 3)} = \frac{A}{x + 1} + \frac{B}{x - 1} + \frac{C}{x - 3}.$$

Each of $A$, $B$, and $C$ can be evaluated by the shortcut introduced above. Thus,

$$A = \left. \frac{3x^2 - x + 3}{(x - 1)(x - 3)} \right|_{x=-1} = \frac{7}{8},$$

$$B = \left. \frac{3x^2 - x + 3}{(x + 1)(x - 3)} \right|_{x=1} = -\frac{5}{4},$$

and

$$C = \left. \frac{3x^2 - x + 3}{(x - 1)(x + 1)} \right|_{x=3} = \frac{27}{8}.$$

Consequently,

$$\frac{3x^2 - x + 3}{(x + 1)(x - 1)(x - 3)} = \frac{7/8}{x + 1} + \frac{-5/4}{x - 1} + \frac{27/8}{x - 3}.$$

As a final illustration, let us decompose the fraction

$$\frac{18x^2 + 20x + 18}{(x + 3)(x - 3)(x^2 + 1)}$$

into partial fractions. In this case, we write

$$\frac{18x^2 + 20x + 18}{(x + 3)(x - 3)(x^2 + 1)} = \frac{A}{x + 3} + \frac{B}{x - 3} + \frac{Cx + D}{x^2 + 1}. \tag{10}$$

$A$ and $B$ can be obtained by the shortcut method, but $C$ and $D$ cannot. We first obtain $A$ and $B$.

$$A = \frac{18x^2 + 20x + 18}{(x - 3)(x^2 + 1)}\bigg|_{x = -3} = -2$$

and

$$B = \frac{18x^2 + 20x + 18}{(x + 3)(x^2 + 1)}\bigg|_{x = 3} = 4.$$

Clearing fractions in Eq. (10), we have

$$18x^2 + 20x + 18 = A(x - 3)(x^2 + 1) + B(x + 3)(x^2 + 1)$$
$$+ (Cx + D)(x - 3)(x + 3).$$

The coefficient of $x^3$ can be easily read off:

$$A + B + C = 0.$$

Therefore, $C = -A - B = -(-2) - 4 = -2$. Also, the constant term can be easily obtained without doing all the algebra. In this case, we have

$$-3A + 3B - 9D = 18 \Rightarrow D = \tfrac{1}{9}(-3A + 3B - 18)$$
$$= \tfrac{1}{9}[-3(-2) + 3(4) - 18] = 0.$$

Finally, we can write

$$\frac{18x^2 + 20x + 18}{(x + 3)(x - 3)(x^2 + 1)} = \frac{-2}{x + 3} + \frac{4}{x - 3} + \frac{-2x}{x^2 + 1}.$$

These three illustrations incorporate all the features of partial fractions that one needs for the exercises in this text. We repeat that the method outlined in Eqs. (3) through (6) *always* works. Sometimes the amount of labor required can be reduced by using the shortcuts discussed in the illustrations.

# Solutions of Polynomial Equations

In Sections 2.4 and 2.5 we encountered the problem of finding the roots of the characteristic equation

$$a_n\lambda^n + a_{n-1}\lambda^{n-1} + \cdots + a_1\lambda + a_0 = 0. \qquad (1)$$

In this appendix we describe a few techniques that are useful in solving some of the exercises that appear in the text.

Consider first the problem of dividing the polynomial $4\lambda^4 + 4\lambda^3 - 25\lambda^2 - \lambda + 6$ by $\lambda - 3$. Using the usual method of long division we have

$$
\begin{array}{r}
4\lambda^3 - 16\lambda^2 + 23\lambda + 68 \\
\lambda - 3 \enclose{longdiv}{4\lambda^4 + 4\lambda^3 - 25\lambda^2 - \lambda + 6} \\
-4\lambda^4 \pm 12\lambda^3 \\
\hline
16\lambda^3 - 25\lambda^2 \\
-16\lambda^3 \pm 48\lambda^2 \\
\hline
23\lambda^2 - \lambda \\
-23\lambda^2 \pm 69\lambda \\
\hline
68\lambda + 6 \\
-68\lambda \pm 204 \\
\hline
210
\end{array}
$$

Thus we note that the quotient is $4\lambda^3 + 16\lambda^2 + 23\lambda + 68$ and the remainder is 210; in other words,

$$4\lambda^4 + 4\lambda^3 - 25\lambda^2 - \lambda + 6 = (\lambda - 3)(4\lambda^3 + 16\lambda^2 + 23\lambda + 68) + 210.$$

In the special case that the divisor is of the form $\lambda - a$ ($a = 3$ in the above illustration), the long division problem can be carried out very quickly by a process known as *synthetic division*. Suppose we want to divide $a_n\lambda^n + a_{n-1}\lambda^{n-1} + \cdots + a_1\lambda + a_0$ by $\lambda - a$. Symbolically the synthetic division is

$$
\begin{array}{c|cccccc}
a & a_n & a_{n-1} & a_{n-2} & \cdots & a_1 & a_0 \\
& & c_{n-1} & c_{n-2} & \cdots & c_1 & c_0 \\
\hline
& b_n & b_{n-1} & b_{n-2} & \cdots & b_1 & b_0
\end{array}
$$

where

$$b_n = a_n$$

$$c_i = ab_{i+1}, \qquad i = n-1, n-2, \ldots, 1, 0$$

$$b_i = a_i + c_i, \qquad i = n-1, n-2, \ldots, 1, 0.$$

The quotient is $b_n\lambda^{n-1} + b_{n-1}\lambda^{n-2} + \cdots + b_2\lambda + b_1$ and the remainder is $b_0$.
For the above illustration we have

$$
\begin{array}{r|rrrrr}
3 & 4 & 4 & -25 & -1 & 6 \\
  &   & 12 & 48 & 69 & 204. \\
\hline
  & 4 & 16 & 23 & 68 & 210
\end{array}
$$

Thus the quotient is $4\lambda^3 + 16\lambda^2 + 23\lambda + 68$ and the remainder is 210.

As a second illustration let us divide $3\lambda^4 - 2\lambda^3 - 2\lambda^2 + 4$ by $\lambda + 2$, using synthetic division. Note that $\lambda + 2 = \lambda - (-2) \Rightarrow a = -2$. Also observe the presence of 0 in the first line of the synthetic division since the coefficient of $\lambda$ is zero.

$$
\begin{array}{r|rrrrr}
-2 & 3 & -2 & -2 & 0 & 4 \\
   &   & -6 & 16 & -28 & 56 \\
\hline
   & 3 & -8 & 14 & -28 & 60
\end{array}
$$

Thus the quotient is $3\lambda^3 - 8\lambda^2 + 14\lambda - 28$ and the remainder is 60.

Now let us divide $4\lambda^4 + 4\lambda^3 - 25\lambda^2 - \lambda + 6$ by $\lambda - 2$:

$$
\begin{array}{r|rrrrr}
2 & 4 & 4 & -25 & -1 & 6 \\
  &   & 8 & 24 & -2 & -6. \\
\hline
  & 4 & 12 & -1 & -3 & 0
\end{array}
$$

Thus the quotient is $4\lambda^3 + 12\lambda^2 - \lambda - 3$ and the remainder is 0. In this case we can write

$$4\lambda^4 + 4\lambda^3 - 25\lambda^2 - \lambda - 6 = (\lambda - 2)(4\lambda^3 + 12\lambda^2 - \lambda - 3). \tag{2}$$

From Eq. (2) we see that the problem of finding the roots of

$$4\lambda^4 + 4\lambda^3 - 25\lambda^2 - \lambda - 6 = 0 \tag{3}$$

is equivalent to finding the roots of

$$(\lambda - 2)(4\lambda^3 + 12\lambda^2 - \lambda - 3) = 0.$$

Thus one root is $\lambda = 2$ and subsequent roots are found by solving the equation

$$4\lambda^3 + 12\lambda^2 - \lambda - 3 = 0. \tag{4}$$

We conclude that synthetic division provides us with a test to determine whether or not $\lambda = a$ is a root of the polynomial equation (1). Simply perform synthetic division with $a$ and if the remainder is zero, then $\lambda = a$ is a root; if the remainder is not zero, then $\lambda = a$ is not a root.

Thus the synthetic division

$$-3 \,\rfloor \quad 4 \quad\quad 12 \quad -\ 1 \quad -3$$
$$\phantom{-3 \rfloor \quad 4 \quad\quad} -12 \quad\ 0 \quad\quad 3$$
$$\overline{\phantom{-3 \rfloor \quad} 4 \quad\quad 0 \quad -\ 1 \quad\quad 0}$$

shows that $\lambda = -3$ is a root of Eq. (4) (hence a root of Eq. (3)). Subsequent roots are found by solving the equation

$$4\lambda^2 - 1 = 0,$$

in other words, $\lambda = \pm\dfrac{1}{2}$. Thus the roots of Eq. (3) are $\lambda = \dfrac{1}{2}, -\dfrac{1}{2}, 2, -3$.

In the previous illustration it may appear that we made lucky guesses at choices for $a$. This is not the case; the guesses are dictated by the following principle, which we state without proof.

**Possibilities for Rational Roots**  If all the coefficients of Eq. (1) are integers, then $\lambda = p/q$, where $p$ and $q$ are integers, is a root of Eq. (1) only if $p$ is a factor of $a_0$ and $q$ is a factor of $a_n$.

For Eq. (3) $a_n \equiv a_4 = 4$ and $a_0 = -6$. Thus the possibilities for $p$ are $\pm 1$, $\pm 2$, $\pm 3$, $\pm 6$, and the possibilities for $q$ are $\pm 1$, $\pm 2$, $\pm 4$. Therefore, the only rational numbers that can possibly be a root of Eq. (3) are $\pm\dfrac{1}{4}, \pm\dfrac{1}{2}, \pm\dfrac{3}{4}, \pm 1, \pm\dfrac{3}{2}, \pm 2, \pm 3, \pm 6$. We test these numbers and no others. Each time we find a root we decrease the degree of the polynomial to be solved by one. Since any root of the new polynomial must be a root of the original polynomial, there are no additional rational roots possible. In fact, the original list of possibilities will, in general, be diminished. For example, after finding $\lambda = 2$ to be a root of Eq. (3), we search for roots of Eq. (4). The only possible rational roots of Eq. (4) are $\pm\dfrac{1}{4}, \pm\dfrac{1}{2}, \pm\dfrac{3}{4}, \pm 1, \pm\dfrac{3}{2}, \pm 3$.

# Proof of the Existence and Uniqueness Theorem

Here we will prove Theorem 1 of Section 1.2. For a thorough understanding of the proof, the reader should be familiar with some topics usually studied in advanced calculus. For easy reference we restate the theorem here.

**THEOREM 1**

*Consider the initial value problem*

$$y' = F(x, y) \tag{1}$$

$$y(x_0) = y_0. \tag{2}$$

*Assume that the functions $F$ and $\partial F/\partial y$ are continuous in some rectangle*

$$\mathcal{R} = \left\{ (x, y): \begin{array}{l} |x - x_0| \le A \\ |y - y_0| \le B \end{array} \right\} \quad A > 0, B > 0$$

*about the point $(x_0, y_0)$. Then there is a positive number $h \le A$ such that the IVP(1)–(2) has one and only one solution in the interval $|x - x_0| \le h$.*

**Proof**  The proof is long and will be accomplished in several steps.

1. Since $F$ is continuous in $\mathcal{R}$, it is bounded there. That is, there exists a positive number $M$ such that $|F(x, y)| \le M$ for every point $(x, y)$ in $\mathcal{R}$. Let $h$ be the smaller of the two numbers $A$ and $B/M$. That is,

$$h = \min (A, B/M). \tag{3}$$

We will prove that the IVP(1)–(2) has one but no more than one solution in the interval $|x - x_0| \le h$, where $h$ is given by (3). Although we will not be able to find explicitly the solution of the IVP(1)–(2), we will show that it exists, is unique, and how to approximate it. Roughly speaking, the proof consists in showing that the sequence of functions

$$y_0(x) = y_0$$

$$y_1(x) = y_0 + \int_{x_0}^{x} F(s, y_0(s))ds$$

$$y_2(x) = y_0 + \int_{x_0}^{x} F(s, y_1(s))ds$$

$$\cdots\cdots\cdots\cdots\cdots\cdots\cdots\cdots$$

$$y_n(x) = y_0 + \int_{x_0}^{x} F(s, y_{n-1}(s))ds$$

$$\cdots\cdots\cdots\cdots\cdots\cdots\cdots\cdots \tag{4}$$

converges and the limit function is the unique solution of the IVP(1)–(2).

2. Assume that $(x, y_1)$ and $(x, y_2)$ are any two points in $\mathcal{R}$. Applying the mean value theorem to $F$, considering $F(x, y)$ as a function of $y$, we have

$$F(x, y_1) - F(x, y_2) = \frac{\partial F}{\partial y}(x, \bar{y})(y_1 - y_2), \tag{5}$$

where $\bar{y}$ is between $y_1$ and $y_2$. Since $\partial F/\partial y$ is continuous in $\mathcal{R}$, it is bounded there. That is, there exists a positive number $K$ such that

$$\left| \frac{\partial F}{\partial y}(x, y) \right| \leq K \tag{6}$$

for every point $(x, y)$ in $\mathcal{R}$.

From (5) and (6) it follows that for every pair of points $(x, y_1)$ and $(x, y_2)$ in $\mathcal{R}$, the function $F$ satisfies the condition

$$| F(x, y_1) - F(x, y_2) | \leq K | y_1 - y_2 |. \tag{7}$$

A function $F$ which is defined in a set $\mathcal{R}$ and satisfies the condition (7) for some positive constant $K$ and every pair of points $(x, y_1)$ and $(x, y_2)$ in $\mathcal{R}$ is called *Lipschitz continuous in y over $\mathcal{R}$ with Lipschitz constant K*. Thus the hypothesis that $\partial F/\partial y$ is continuous in $\mathcal{R}$ implies that $F$ is Lipschitz continuous in $y$ over $\mathcal{R}$. In this theorem it is the Lipschitz continuity of $F$ rather than the continuity of $\partial F/\partial y$ that is needed for its proof.

3. Consider the sequence of functions $y_0(x), y_1(x), \ldots, y_n(x), \ldots$ defined by (4). We will show that $y_n(x)$ for $n = 0, 1, 2, \ldots$ exists in $| x - x_0 | \leq h$, is continuous there, and satisfies the inequality

$$| y_n(x) - y_0 | \leq B \qquad \text{for} \qquad | x - x_0 | \leq h, \qquad n = 0, 1, 2, \ldots . \tag{8}$$

This will establish the fact that the sequence (4) is well defined. We will prove the above claim by induction. That is, we will first show that our claim is true for $n = 0$. Then we will assume that our claim is true for $n = k$ and show that it is true for $n = k+1$. For $n = 0$, $y_0(x) = y_0$ is continuous in $| x - x_0 | \leq h$ because a constant function is continuous everywhere. Also (8) holds because $| y_0(x) - y_0 | = 0$. Now assume that $y_k(x)$ is continuous in $| x - x_0 | \leq h$ and that (8) holds for $n = k$. Then, $F(x, y_k(x))$ is continuous in $x$ in the interval $| x - x_0 | \leq h$ because $F$ and $y_k$ are. Hence, $y_{k+1}(x)$ is well defined by (4) with $n = k+1$ and is continuous in the interval $| x - x_0 | \leq h$ because the integral of a continuous function is continuous. Finally,

$$| y_{k+1}(x) - y_0 | = \left| \int_{x_0}^{x} F(s, y_k(s))ds \right|$$

$$\leq \left| \int_{x_0}^{x} | F(s, y_k(s)) | ds \right|$$

$$\leq M \left| \int_{x_0}^{x} ds \right| \leq M | x - x_0 | \leq Mh \leq M \frac{B}{M} = B.$$

This establishes our claim for all $n = 0, 1, 2, \ldots$. The very last inequality explains the choice of $h$.

4. Next we will prove, again by induction, that for $|x-x_0| \le h$ the following inequalities hold for $n = 1, 2, \ldots$ .

$$|y_n(x) - y_{n-1}(x)| \le \frac{MK^{n-1}|x-x_0|^n}{n!} \le \frac{MK^{n-1}h^n}{n!}. \qquad (9)$$

For $n = 1$, we have

$$|y_1(x) - y_0(x)| = \left| \int_{x_0}^{x} F(s, y_0(s))\, ds \right| \le M \left| \int_{x_0}^{x} ds \right| = M\,|x-x_0| \le Mh.$$

Hence (9) is true for $n = 1$. Next we assume that (9) holds for $n = m$, that is, we assume that for $|x-x_0| \le h$

$$|y_m(x) - y_{m-1}(x)| \le \frac{MK^{m-1}|x-x_0|^m}{m!} \le \frac{MK^{m-1}h^m}{m!} \qquad (10)$$

and show that (9) holds also for $n = m+1$. In fact, using (4) and (7), we obtain

$$|y_{m+1}(x) - y_m(x)| = \left| \int_{x_0}^{x} F(s, y_m(s))\, ds - \int_{x_0}^{x} F(s, y_{m-1}(s))\, ds \right|$$

$$\le \left| \int_{x_0}^{x} |F(s, y_m(s)) - F(s, y_{m-1}(s))|\, ds \right|$$

$$\le K \left| \int_{x_0}^{x} |y_m(s) - y_{m-1}(s)|\, ds \right|.$$

Now, using (10), we find

$$|y_{m+1}(x) - y_m(s)| \le \frac{MK^m}{m!} \left| \int_{x_0}^{x} |s-x_0|^m\, ds \right|$$

$$= \frac{MK^m|x-x_0|^{m+1}}{(m+1)!} \le \frac{MK^m h^{m+1}}{(m+1)!}$$

and the proof of (9) is complete.

5. The next step in the proof is to show that the sequence of functions $y_n(x)$ converges uniformly to a limit function $y(x)$ on the interval $|x-x_0| \le h$. To this end we make the following observations: First, the $n^{th}$ partial sum of the series

$$y_0(x) + \sum_{n=1}^{\infty} [y_n(x) - y_{n-1}(x)] \qquad (11)$$

is $y_n(x)$. Thus the series (11) and the sequence $y_n(x)$ have the same convergence properties. Second, because of (9), series (11) is dominated by the series of constant terms

$$|y_0| + \sum_{n=1}^{\infty} \frac{MK^{n-1}h^n}{n!}.$$

The latter series converges. In fact,

$$\sum_{n=1}^{\infty} \frac{MK^{n-1}h^n}{n!} = \frac{M}{K} \sum_{n=1}^{\infty} \frac{(Kh)^n}{n!} = \frac{M}{K}(e^{Kh} - 1).$$

Therefore, by the Weierstrass $M$-test the series (11) converges absolutely and uniformly on the interval $|x-x_0| \le h$. It follows now that the sequence $y_n(x)$ converges uniformly on the interval $|x-x_0| \le h$. Let us denote the limit function by $y(x)$. That is,

$$y(x) = \lim_{n\to\infty} y_n(x). \tag{12}$$

6. Next, we will prove that $y(x)$ is a solution of the IVP(1)–(2). First of all $y(x)$ satisfies the initial condition (2). In fact, from (4),

$$y_n(x_0) = y_0, \quad n = 0, 1, 2, \ldots .$$

and taking limits of both sides as $n \to \infty$ we find $y(x_0) = y_0$. Since $y(x)$ is the uniform limit, in the interval $|x-x_0| \le h$, of the continuous functions $y_n(x)$, $n = 0, 1, 2, \ldots$, it follows that $y(x)$ is itself continuous on the interval $|x-x_0| \le h$. Also from (8), taking limits as $n \to \infty$, we find that for $|x-x_0| \le h$,

$$|y(x)-y_0| \le B.$$

Hence, the function $F(x, y(x))$ is well defined and continuous on the interval $|x-x_0| \le h$ and the integral

$$\int_{x_0}^{x} F(s, y(s))ds$$

exists. From (7), we have for $|x-x_0| \le h$,

$$|F(x, y_n(x)) - F(x, y(x))| \le K|y_n(x) - y(x)|.$$

Since the sequence $y_n(x)$ converges uniformly to $y(x)$ on the interval $|x-x_0| \le h$, it follows that the sequence $F(x, y_n(x))$ also converges uniformly to $F(x, y(x))$ and consequently

$$\lim_{n\to\infty} \int_{x_0}^{x} F(s, y_n(x))ds = \int_{x_0}^{x} F(s, y(s))ds. \tag{13}$$

Taking limits on both sides of (4), as $n \to \infty$, and using (12) and (13), we obtain

$$y(x) = y_0 + \int_{x_0}^{x} F(s, y(s))ds. \tag{14}$$

Differentiating both sides of (14) with respect to $x$ (note that the right-hand side is a differentiable function of the upper limit $x$) we find

$$y'(x) = F(x, y(x)),$$

and this completes the proof that $y(x)$ is a solution of the IVP(1)–(2).

7. Finally we will prove that $y(x)$ is the only solution of the IVP(1)–(2). Assume that $\bar{y}(x)$ is another solution of the IVP(1)–(2). Then

$$y'(x) - \bar{y}'(x) = F(x, y(x)) - F(x, \bar{y}(x)).$$

Integrating from $x_0$ to $x$ and using the fact that $y(x_0) = y_0 = \bar{y}(x_0)$, we find

$$y(x) - \bar{y}(x) = \int_{x_0}^{x} [F(s, y(s)) - F(s, \bar{y}(s))] \, ds. \tag{15}$$

Now assume that $x \geq x_0$. The case $x < x_0$ is treated in a similar way. From (15), using (7), we obtain

$$| y(x) - \tilde{y}(x) | \leq K \int_{x_0}^x | y(s) - \tilde{y}(s) | \, ds. \tag{16}$$

Setting

$$w(x) = \int_{x_0}^x | y(s) - \tilde{y}(s) | \, ds \tag{17}$$

and using (16), we find

$$w'(x) = | y(x) - y(x) | \leq K \int_{x_0}^x | y(s) - y(s) | \, ds = Kw(x)$$

or

$$w'(x) - Kw(x) \leq 0. \tag{18}$$

Multiplying both sides of the inequality (18) by the positive integrating factor $\exp(-Kx)$ we obtain

$$\frac{d}{dx} \left[ w(x)e^{-Kx} \right] \leq 0. \tag{19}$$

Integrating both sides of (19) from $x_0$ to $x$ we find

$$w(x)e^{-Kx} - w(x_0)e^{-Kx_0} \leq 0.$$

But from (17), $w(x_0) = 0$ and $w(x) \geq 0$. It follows then that

$$0 \leq w(x)e^{-Kx} \leq 0 \quad \text{or} \quad w(x) \equiv 0.$$

Hence, from (16), we conclude that $y(x) = \tilde{y}(x)$, and the proof is complete.

# Answers to
# Odd-Numbered Exercises

**1.** True.  **3.** False.  **5.** True.  **7.** False.  **21.** $y' = -\dfrac{x}{y}$; $y^2 + x^2 = k^2$.

**23.** True.  **25.** False.  **27.** False.

**29.** Differentiating both sides of the first family, we obtain

$$2x + 2yy' = 2c = \frac{x^2 + y^2}{x} \Rightarrow y' = \frac{y^2 - x^2}{2xy}.$$

Differentiating both sides of the second family, we find that

$$y' = -\frac{2xy}{y^2 - x^2}.$$

Since the slopes are negative reciprocals, we conclude that every member of the family $x^2 + y^2 = 2cx$ cuts every member of the family $x^2 + y^2 = 2ky$ at right angles, and vice versa.

**1.** Yes.  **3.** Yes.  **5.** No, because $F_y(x, y)$ is not continuous for $y = 0$.

**7.** The solution through the point $(-3, 0)$ exists throughout the interval $\left(-\infty, -\dfrac{3}{2}\right)$; the solution through $(-1, 5)$ exists throughout the interval $\left(-\dfrac{3}{2}, 0\right)$; the solution through the point $(1, -7)$ exists throughout the interval $(0, 2)$; the solution through the point $(3, 0)$ exists throughout $(2, \infty)$.

**9.** Show by direct substitution that $y = 2x - 1$ is a solution of the IVP. Thus, by Example 2, $y = 2x - 1$ is the only solution of the IVP [in the interval $(0, \infty)$].

**11.** True.  **13.** False.

**15.** The isoclines are the hyperbolas $xy = c$ for $c \neq 0$ and the lines $x = 0$ and $y = 0$ for $c = 0$. The heavy polygonal line shown here is a graphical approximation of the solution through $(1, 2)$.

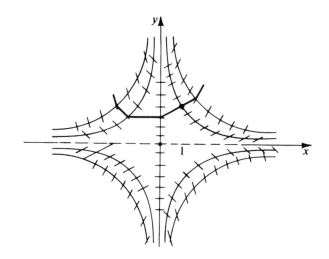

**Section 1.3**

Page 27

**1.** Separable; $y = \dfrac{c}{xe^x}$. **3.** Separable; $y = 0$ and $y = \dfrac{1}{4}(x + c)^2$.

**5.** It is not separable. **7.** It is not separable.

**9.** Separable; $y = 1 - ce^{-(1/2)x^2 - x}$.

**11.** $y = \cos^{-1}(ce^{-x} - 1)$ is the general solution while the curves $x = n\pi + \pi/2$, $n = 0, \pm 1, \pm 2, \ldots$ are singular solutions.

**13.** It is not separable. **15.** $y = e^x$.

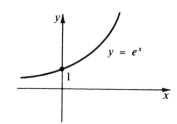

**17.** $y^{a_1}e^{a_2y} = cx^{b_1}e^{b_2x}$

**19.** $x(t) = \left[\dfrac{a}{3B(b - Bt)^{1/2}} + c\right]^6$. **21.** $I(w) = 2.4 + ce^{-0.088w}$. **23.** 12 hours.

**25.** 0.58 second. **27.** $\dfrac{100}{6}\ln 2 = 11.5$ years.

**29.** $y = ke^{-x/C}$, where $k$ is an arbitrary constant. **31.** $y = k(1 - x)^{(1-A)/C}x^{A/C}$.

**33.** $\dfrac{1000}{55}\ln 4 = 25.2$ years. **35.** $10\ln 3 = 10.99\%$. **37.** 86.3 years; 57.6 years.

**1.** Linear; $y = 3 + \dfrac{c}{x}$.　**3.** It is not linear.　**5.** Linear; $y = 2x^2 + cx$.

**7.** Linear; $y = \dfrac{1}{2}(x - 1)^5 + c(x - 1)^3$.　**9.** It is not linear.

**11.** Linear; $y = \dfrac{1}{6}x^5 + \dfrac{c}{x}$.　**13.** It is not linear.　**15.** $y = x + \dfrac{1}{x}e^x$.

**17.** $y(x) = \begin{cases} e^x, & x < 0 \\ 2e^x - x - 1, & x \geq 0. \end{cases}$

**21.** 0.19 amperes. After a long time the current will be 1.6 amperes.

**23.** $I(t) = (10.2)e^{-2t} + (0.4)\sin t - (0.2)\cos t$.

**25.** The amount of salt in the tank after 5 minutes will be $30\left(1 - \dfrac{1}{e}\right) \approx 18.96$ pounds. After a very long time, the amount will be 30 pounds.

**29.** $y = \dfrac{1}{2x^2 + cx}$.　**31.** $y = \dfrac{1}{-\dfrac{2}{3}\ln x + c(\ln x)^{-1/2}}$.　**33.** $y = x$.　**35.** $y = -1$.

**37.** $y(x) = \dfrac{1}{2}(x - 1)^5 - 4(x - 1)^3$.

**41.** (a) $t = \dfrac{3\ln 3}{\ln 3 - \ln 2} = 8.13$ minutes.

　　 (b) $40 + 30\left(\dfrac{2}{3}\right)^{5/3} = 55.26°F$.

**43.** $y = 1 + \dfrac{10}{t(\ln t + c)}$.

**1.** It is exact; $x^2 + y^2 + 4y - 2xy = c$.

**3.** It is exact; $x^2 - 2x + y^2 + 6y - 2xy = c$.

**5.** It is exact; $x^3 + 3xy^2 = c$.　**7.** It is exact; $y + e^{xy} = c$.

**9.** It is exact; $\sin x - \cos y + xy = c$.　**11.** It is not exact.

**13.** $-y^3 + y + 3x^2y = k$.　**17.** $x^2 + y^2 + 4y - 2xy = 4$.

**19.** $x^3 + 3xy^2 = 4$.

**25.** $\mu = x^{-3}$ is an integrating factor; $\ln x + x^{-2}y = c$.

**27.** $\mu = x^{-2}$ is an integrating factor; $2 \ln x + yx^{-1} = c$.

**29.** $\mu = x^{-5}$ is an integrating factor; $\ln x - \dfrac{1}{4}x^{-4}y^4 = c$.

**31.** The solution is Eq. (4) of Section 1.4.   **33.** $y = cx$.   **35.** $y = cx$.

**37.** $y = cx$.   **39.** $2x + y + \dfrac{1}{2}\ln(x^2 + y^2) = c$.

---

**Section 1.6**
Page 54

**1.** It is homogeneous; $y = cx^{1/3} - \dfrac{5}{2}x$.   **3.** It is homogeneous; $x^2 - y^2 = cx$.

**5.** It is homogeneous; $-\dfrac{x^3}{3y^3} + \ln y = c$.   **7.** It is homogeneous; $y^7(y - 10x)^3 = c$.

**9.** It is homogeneous; $\ln y + \dfrac{x^2}{2y^2} = c$.

**11.** $\ln\sqrt{y^2 + xy + x^2} + \sqrt{3}\tan^{-1}\dfrac{2\sqrt{3}}{3}\left(\dfrac{y}{x} + \dfrac{1}{2}\right) = c$.

**13.** $(x + y)(x - y)^5 = c$.

**17.** $\sqrt{2}\tan^{-1}\dfrac{y - 2}{\sqrt{2}(x - 1)} - \ln[2(x - 1)^2 + (y - 2)^2] = c$.

**19.** $\left(y + \dfrac{11}{3}\right)^2 + 2\left(x + \dfrac{1}{3}\right)\left(y + \dfrac{11}{3}\right) - 5\left(x + \dfrac{1}{3}\right)^2 = c$.

**21.** $[(y + 1)^2 - 2x(y + 1) - x^2][(y + 1) + (\sqrt{2} - 1)x]^{\sqrt{2}} = $
$c[(y + 1) - (\sqrt{2} + 1)x]^{\sqrt{2}}$.

**23.** $\dfrac{2}{\sqrt{7}}\tan^{-1}\dfrac{2(y - 1) - x}{x\sqrt{7}} = c + \ln[(y - 1)^2 - x(y - 1) + 2x^2]$.

**25.** $y = x$.

**27.** $\sqrt{2}\tan^{-1}\dfrac{y - 2}{\sqrt{2}(x - 1)} - \ln[2(x - 1)^2 + (y - 2)^2] = -\ln 2$.

**Section 1.7**
Page 57

**1.** $y(x) = c_1 + c_2 e^{-x} + 3x.$   **3.** $y(x) = \dfrac{1 + (-x + 1)^2}{2}.$

**9.** $y(x) = c_1 x^5 + c_2 x^3 + c_3 x^2 + c_4 x + c_5.$   **11.** $y(x) = c_1 \sinh(c_2 \pm x).$

**13.** $2c_1 y + \ln \dfrac{c_1 y - 1}{c_1 y + 1} = 2c_1(c_1^2 x + c_2).$ Also, $y =$ constant is a solution.

**15.** $y = c_1 + \sin^{-1}(c_2 e^x).$

**REVIEW
EXERCISES
CHAPTER 1**
Page 58

**1.** $y(x) = \dfrac{2e}{xe^x}.$   **3.** $y(x) = x(1 + x^2).$

**5.** $y(x) = -\left(\dfrac{5 - 2x^3}{3x}\right)^{1/2}$   **7.** $y(x) = \dfrac{1}{2}x^2 + \dfrac{3}{2} - \ln x.$

**9.** $x - \ln|x - 2| = t + c.$ Also $x = 2$ is a solution.

**11.** $\sin x = \sin t + c.$   **13.** $y = x + \dfrac{1}{1 + ce^x}.$

**15.** $(y - x)x^2 = cy$   **17.** $y = cx + xe^x$

**19.** $x^5 - y^5 + 5xy^4 = c$   **21.** $x^2 + y^2 - 4xy = c$

**23.** $(0, \infty)$

**25.** $y(x) = \begin{cases} \dfrac{x^2 - 1}{2x}, & x \le 1 \\ \dfrac{x - 1}{x}, & x > 1 \end{cases}.$

**27.** The current satisfies Eq. (7), Section 1.4.

**29.** $\left(\dfrac{\ln 2}{\ln 5 - \ln 2}\right)$ hours $\approx 45$ minutes.

**31.** $-\dfrac{5568\,(\ln 0.85)}{\ln 2} \approx 1305.5$ years.

**33.** (a) $y(t) = 2(30 + t) - 50\left(\dfrac{30}{30 + t}\right)^2$; (b) $51.87$ pounds.

**35.** 2183 years.　　**37.** $10\,e^{-1000t}$.

**39.** 6.2 ft.

**41.** Show that the motion satisfies an IVP of the form $\dot{v} + Kv = -g$, $v(0) = v_0$, where $K$ is a constant.

**43.** $(x + y)e^{[(x-y)^2]/2} = c$; $xy(x + y - 1) = c$.　　**45.** $I(t) = 4.5e^{-2t} + 0.5e^{-t}$.

---

**Section 2.1**

Page 71

**1.** It is linear, homogeneous, with variable coefficients, and of order 2.

**3.** It is nonlinear.

**5.** It is linear, nonhomogeneous, with constant coefficients, and of order 5.

**7.** It is linear, nonhomogeneous, with variable coefficients, and of order 1.

**9.** It is linear, nonhomogeneous, with variable coefficients, and of order 2.

**11.** It is linear, nonhomogeneous, with variable coefficients, and of order 2.

**13.** It is linear, nonhomogeneous, with constant coefficients, and of order 3.

**15.** It is linear, homogeneous, with constant coefficients, and of order 2.

**17.** It is linear, nonhomogeneous, with variable coefficients, and of order 3.

**19.** It is linear, homogeneous, with variable coefficients, and of order 2.

**21.** $y(t) = \dfrac{1}{2}\cos 16t - \dfrac{1}{4}\sin 16t$.

**23.** $y(t) = 10 \sin \sqrt{5}\,t \Rightarrow$ amplitude $= 10$ cm and period $= 2\pi/\sqrt{5}$.

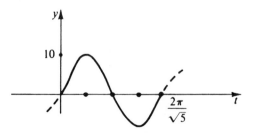

**1.** 2.      **3.** $-3$.      **5.** $e^x\left(\ln x - \dfrac{1}{x}\right)$.      **7.** $-e^{5x}$.      **9.** $x^3$.      **11.** $e^{14x}$.

**13.** $(\lambda_2 - \lambda_1)x^{\lambda_1 + \lambda_2 - 1}$      **15.** 0      **17.** $y = c_1x^2 + c_2x + c_3$.

**19.** $y = -\sin x + c_1x + c_2$.      **21.** $y = c_1x^3 + c_2x^2 + c_3x + c_4$.

**23.** $y = c_1x^4 + c_2x^3 + c_3x^2 + c_4x + c_5$.      **25.** $y = c_1e^{2x} + c_2e^{-2x}$.

**27.** $y = \dfrac{c_1}{x}$      **29.** $y = c_1e^{-2x} + c_2xe^{-2x}$.

**31.** $y = c_1(\cos x + \sin x) + c_2(\cos x - \sin x)$.      **33.** $y = c_1 + c_2 \cos x + c_3 \sin x$.

**35.** $W = ax^3$.      **37.** $W = c$.      **43.** $p(v) = c_1 + c_2 \ln v$.

**1.** Yes.      **3.** Yes.

**5.** No, because the leading coefficient vanishes at $\pi$ and $\pi$ belongs to the interval $3\pi/4 \le x \le 5\pi/4$.

**7.** Yes.      **9.** Yes.      **11.** $y(x) \equiv 0$.

**13.** Any interval not containing the point $x = 0$.

**15.** Any interval not containing any of the points $x = k\pi + \pi/2, k = 0, \pm 1, \pm 2, \ldots$.

**17.** Any interval at all.

**19.** A spring at rest. Its solution is $y = 0$.      **21.** $W(y_1, y_2; t) = -2ai \ne 0$.

**23.** $y = \dfrac{c_1x + c_2}{(x - a)(x - b)}$.

**1.** $4\lambda^3 - 2\lambda^2 + 6\lambda - 7 = 0$.      **3.** $\lambda^3 - \lambda + 2 = 0$.

**5.** $5\lambda^3 - 5\lambda^2 + \lambda - 2 = 0$.      **7.** $3\lambda^3 + 4\lambda^2 = 0$.

**9.** $3\lambda^5 - 2\lambda^4 + \lambda^2 - 2\lambda = 0$.

**11.** No.      **13.** No.      **15.** Yes.      **17.** No.      **19.** Yes.

**21.** $\lambda_1 = 4i, \lambda_2 = -4i; y(x) = c_1 e^{4ix} + c_2 e^{-4ix}$ or $y(x) = \alpha_1 \cos 4x + \alpha_2 \sin 4x$.

**23.** $\lambda_{1,2} = \dfrac{-1 \pm i\sqrt{3}}{2}; y(x) = c_1 e^{[(-1+i\sqrt{3})/2]x} + c_2 e^{[(-1-i\sqrt{3})/2]x}$ or

$$y(x) = \alpha_1 e^{-(x/2)} \cos \frac{x\sqrt{3}}{2} + \alpha_2 e^{-(x/2)} \sin \frac{x\sqrt{3}}{2}$$

**25.** $\lambda_{1,2} = 1 \pm \dfrac{1}{2}i; y(x) = c_1 e^{[1-(i/2)]x} + c_2 e^{[1-(i/2)]x}$ or

$$y(x) = \alpha_1 e^x \cos \frac{x}{2} + \alpha_2 e^x \sin \frac{x}{2}.$$

**27.** $\lambda_{1,2} = \pm 2i; y(x) = c_1 e^{2ix} + c_2 e^{-2ix}$ or $y(x) = \alpha_1 \cos 2x + \alpha_2 \sin 2x$.

**29.** $\lambda_{1,2} = \pm 5i; y(x) = c_1 e^{5ix} + c_2 e^{-5ix}$ or $y(x) = \alpha_1 \cos 5x + \alpha_2 \sin 5x$.

**31.** $y(x) = c_1 e^{2x} + c_2 e^{-2x} = \alpha_1 \cosh 2x + \alpha_2 \sinh 2x$.

**33.** $y(x) = c_1 e^{5x} + c_2 e^{-5x} = \alpha_1 \cosh 5x + \alpha_2 \sinh 5x$.

**35.** $y(x) = c_1 e^{9x} + c_2 e^{-9x} = \alpha_1 \cosh 9x + \alpha_2 \sinh 9x$.

**37.** $y(x) = c_1 e^{3x} + c_2 e^{-3x} = \alpha_1 \cosh 3x + \alpha_2 \sinh 3x$.

**39.** $\lambda_{1,2} = \pm \sqrt{\beta}; D(t) = c_1 e^{t\sqrt{\beta}} + c_2 e^{-t\sqrt{\beta}}$ or $D(t) = \alpha_1 \cosh t\sqrt{\beta} + \alpha_2 \sinh t\sqrt{\beta}$.

**Section 2.5**
Page 99

**1.** $y(x) = c_1 e^{-x/2} + c_2 e^{-x}$.    **3.** $y(x) = c_1 e^{-x/2} \cos \dfrac{x\sqrt{3}}{2} + c_2 e^{-x/2} \sin \dfrac{x\sqrt{3}}{2}$.

**5.** $y(x) = c_1 \cos x\sqrt{6} + c_2 \sin x\sqrt{6}$.    **7.** $y(x) = c_1 e^{3x}$.

**9.** $y(x) = c_1 e^{-x} + c_2 e^x + c_3 x e^x + c_4 e^{-2x}$.

**11.** $y(x) = c_1 e^{3x} + c_2 e^{-2x} + c_3 x e^{-2x} + c_4 e^x \cos 2x + c_5 e^x \sin 2x + c_6 e^{-2x} \cos x$

$\qquad + c_7 e^{-2x} \sin x + c_8 x e^{-2x} \cos x + c_9 x e^{-2x} \sin x$. Order 9.

**13.** $y(x) = c_1 e^{(3/2)x} + c_2 e^{-(4/5)x} + c_3 \cos 2x + c_4 \sin 2x + c_5 x \cos 2x$

$\qquad + c_6 x \sin 2x + c_7 e^{[(1+\sqrt{5})/2]x} + c_8 e^{[(1-\sqrt{5})/2]x}$. Order 8.

**15.** $y(x) = c_1 \cos 2x + c_2 \sin 2x + c_3 \cos x + c_4 \sin x + c_5 x \cos x$

$\qquad + c_6 x \sin x$. Order 6.

**17.** $y(x) = c_1 e^{7x} + c_2 e^{(7/2)x} + e^{-(1/2)x}\left[c_3 \cos\dfrac{\sqrt{3}}{2}x + c_4 \sin\dfrac{\sqrt{3}}{2}x\right]$

$\qquad + e^{-4x}[c_5 + c_6 x + c_7 x^2]$. Order 7.

**19.** $y(x) = c_1 e^{-2x} + e^{5x}[c_2 + c_3 x] + e^{4x}[c_4 + c_5 x + c_6 x^2]$

$\qquad + \cos 2x[c_7 + c_8 x + c_9 x^2] + \sin 2x[c_{10} + c_{11}x + c_{12}x^2]$. Order 12.

**21.** $y(x) = e^{-x/2}\left[\cos\dfrac{x\sqrt{3}}{2} + \left(2 + \dfrac{1}{\sqrt{3}}\right)\sin\dfrac{x\sqrt{3}}{2}\right].$

**23.** $y(x) = 3\cos x\sqrt{3} - 6\sin x\sqrt{3}$.     **25.** $y(x) = -1 + x + 2e^x$.

**27.** $y(x) = (2 + x)e^x - 2e^{-x}$.     **29.** $y(x) = 2e^x(\sin x - \cos x)$.

**31.** $y(x) = 3\cos x + 2\sin x$.

**33.** $y(t) = (15\sqrt{6} - 18)e^{-(2/3)t} + (12 - 15\sqrt{6})e^{-t} \to 0$ as $t \to \infty$.

**35.** Overdamped.     **37.** Underdamped.     **39.** Underdamped.

**41.** Critically damped.     **43.** Underdamped.     **45.** $f(x) = c(\cos x + \sin x)$.

---

**1.** $y(x) = c_1 x^{3/5} + c_2 x$.

**3.** $y(x) = (-x)^{-1/6}\left[c_1 \cos\left(\dfrac{\sqrt{11}}{6}\ln(-x)\right) + c_2 \sin\left(\dfrac{\sqrt{11}}{6}\ln(-x)\right)\right].$

**5.** $y(x) = c_1(-x) + c_2(-x)^{-4}$.     **7.** $y(x) = c_1(-x)^{3/2} + c_2(-x)^{-1}$.

**9.** $y(x) = c_1 x + c_2 x \ln x + c_3 x^{-2}$.

**11.** $y(x) = (-x - 3)^{-1}[c_1 \cos 2\ln(-x - 3) + c_2 \sin 2\ln(-x - 3)]$.

**13.** $y(x) = c_1(-x)^3 + c_2(-x)^{-2}$.     **15.** $y(x) = c_1 x + c_2 x^2 + c_3 x^{-4}$.

**17.** $y(x) = -x + x \ln(-x)$.     **19.** $y(x) = \dfrac{3}{7}(-x)^{1/2} + \dfrac{64}{7}(-x)^{-3}$.

**21.** (b) Apply l'Hospital's rule and use the formula

$$\frac{d}{dx}(a^x) = a^x \ln a.$$

**23.** $p(r) = c_1 + c_2 \ln r.$     **25.** $y(x) = c_1 x + c_2 x^2 + c_3 x^{4 + \sqrt{13}} + c_4 x^{4 - \sqrt{13}}.$

**27.** $y(x) = c_1(-x)^{-1} + c_2(-x)^{(7 + \sqrt{65})/4} + c_3(-x)^{(7 - \sqrt{65})/4}.$

**29.** $y(x) = c_1 x + c_2 x^2 + c_3 x^{(23 + \sqrt{445})/14} + c_4 x^{(23 - \sqrt{445})/14}.$

**31.** Characteristic equation: $\lambda^6 - 15\lambda^5 + 73\lambda^4 - 143\lambda^3 + 142\lambda^2 - 48\lambda = 0;$
characteristic roots: $\lambda_1 = 0,\ \lambda_2 = \lambda_3 = 1,\ \lambda_4 = 2,\ \lambda_5 = 3,\ \lambda_6 = 8.$
$y(x) = c_1 + (c_2 + c_3 \ln x) x + c_4 x^2 + c_5 x^3 + c_6 x^8.$

**33.** $y(x) = c_1 x + c_2 x^5.$     **35.** $y(x) = c_1 + c_2(-x) + c_3(-x)^2 + c_4(-x)^4.$

**37.** $y(x) = c_1(x - 4) + c_2(x - 4)^{-4}.$     **39.** $y(x) = [c_1 + c_2 \ln (1 - x)] (1 - x)^{-2}.$

---

**Section 2.8**
**Page 114**

**1.** $y_2(x) = x^{-1}.$     **3.** $y_2(x) = (x + 1) e^{-3x}.$     **5.** $y_2(x) = x - 2 + \dfrac{2}{x}.$

**7.** $y_2(x) = \ln x.$     **9.** $y_2(x) = x^{-1/2} \cos x.$     **11.** $y_2(x) = x^{-2} e^{-(3/4)x^{2/3}}.$

**13.** $y_2(x) = \ln x.$     **15.** $y_2(x) = e^{2x}.$

**17.** $y_2(x) = (8x^3 - 12x) \displaystyle\int_{x_0}^{x} \frac{e^{t^2}\, dt}{(8t^3 - 12t)^2}.$

**19.** $y_2(x) = \dfrac{5x^3 - 3x}{2} \displaystyle\int_{x_0}^{x} \frac{4\, dt}{(t^2 - 1)(5t^3 - 3t)^2}.$

**21.** $y_2(x) = (2 - 4x + x^2) \displaystyle\int_{x_0}^{x} \frac{e^t\, dt}{t(2 - 4t + t^2)^2}.$

**23.** $y_2(x) = x \displaystyle\int_{x_0}^{x} \frac{dt}{t^2 \sqrt{1 - t^2}}.$

**25.** $y_2(x) = (4x^3 - 3x) \displaystyle\int_{x_0}^{x} \frac{dt}{(4t^3 - 3t)^2 \sqrt{1 - t^2}}.$     **27.** $y_2(x) = e^x \ln x.$

**35.** $y_2(x) = -\dfrac{1}{4}(3x^2 - 1) \ln \dfrac{x - 1}{x + 1} - \dfrac{3}{2} x.$

**1.** $y(x) = 1 - \frac{1}{2}x^2 + \frac{1}{8}x^4 + 0 \cdot x^5 - \cdots .$

**3.** $y(x) = 2 + x - \frac{3}{2}x^2 - \frac{1}{24}x^4 + \frac{1}{24}x^5 + \cdots .$

**5.** $y(x) = 1 - \frac{3}{2}x^2 - \frac{1}{6}x^3 + \frac{3}{8}x^4 + \frac{1}{10}x^5 + \cdots .$

**7.** $y(x) = 2 + x - x^2 + \frac{1}{6}x^3 - \frac{1}{4}x^4 + \frac{1}{24}x^5 + \cdots .$

**9.** $y(x) = 1 + x - \frac{1}{3}x^3 + \frac{1}{3}x^4 - \frac{7}{30}x^5 + \cdots .$

**11.** $y(x) = a_0\left(1 - \frac{x^2}{2} + \frac{x^4}{24} - \cdots\right) + a_1\left(x - \frac{x^3}{6} + \frac{x^5}{120} - \cdots\right).$

**13.** $y(x) = a_0\left(1 - \frac{1}{6}x^3 + \cdots\right) + a_1\left(x - \frac{1}{6}x^4 + \cdots\right).$

**15.** $y(x) = a_0(1 - x^2 + \cdots) + a_1\left(x - \frac{1}{12}x^3 + \frac{1}{480}x^5 + \cdots\right).$

**17.** $y(x) = a_0\left(1 + \frac{1}{6}x^3 - \frac{1}{8}x^4 - \frac{1}{40}x^5 + \cdots\right)$

$$+ a_1\left(x - \frac{3}{2}x^2 + \frac{3}{2}x^3 - \frac{13}{24}x^4 - \frac{13}{20}x^5 + \cdots\right).$$

**19.** $y(x) = a_0\left(1 + \frac{1}{12}x^4 + \cdots\right) + a_1\left(x + \frac{1}{30}x^5 + \cdots\right) + a_2(x^2 + \cdots).$

**21.** For $a \neq -1$, we have $a_2 = -\frac{\lambda a_0}{1 + a}; a_3 = -\frac{\lambda a_1}{6(1 + a)}; a_4 = \frac{\lambda(\lambda - 4a)}{12(1 + a)^2}a_0;$

$a_5 = \frac{\lambda(\lambda - 12a)}{120(1 + a)^2}a_1.$

**1.** $y_h(x) = c_1e^{2x} + c_2e^{3x}.$   **3.** $y_h(x) = c_1e^{-x} + c_2e^{2x}.$

**5.** $y_h(x) = c_1 \cos 3x + c_2 \sin 3x.$   **7.** $y_h(x) = c_1e^{4x} + c_2e^{9x}.$

**9.** $y_h(x) = c_1 e^{5x} + c_2 x e^{5x}$.　　**11.** $y_h(x) = c_1 e^x + c_2 e^{-x} + c_3 \cos x + c_4 \sin x$.

**13.** $y_h(x) = c_1 x^{8/3} + c_2 x^{-1}$.　　**15.** $y_h(x) = c_1 x \sin x + c_2 x \cos x$.

**17.** True.　　**19.** True.　　**21.** False.　　**23.** True.

---

**Section 2.11**

Page 131

**7.** $y_p(x) = Ax^3 + Bx^2 + Cx + D$.

**9.** $y_p(x) = Ax^2 + Bx + C + Dx \cos x + Ex \sin x$.

**11.** $y_p(x) = Ax^6 + Bx^5 + Cx^4 + Dx^3 + Ex^2 + Fx + G \cos x + H \sin x$.

**13.** $y_p(x) = x + e^{-x}$.　　**15.** $y_p(x) = \frac{1}{2}x^2 e^{2x}$.　　**17.** $y_p(x) = -\frac{1}{72}x^4$.

**19.** $y_p(x) = -\frac{1}{6}x^2 e^{-2x}$.　　**21.** $y_p(x) = \frac{1}{2}xe^x$.　　**23.** $y_p(x) = -\frac{1}{2}x \sin x - \frac{1}{2}\cos x$.

**25.** $y_p(x) = \left(\frac{1}{2}x^2 - \frac{5}{2}x + \frac{19}{4}\right)e^x$.　　**27.** $y(x) = c_1 \cos x + c_2 \sin x - \frac{1}{2}x \cos x$.

**29.** $y(x) = c_1 e^{2x} + c_2 e^{-x} - \frac{1}{3}x^2 e^{-x} - \frac{2}{9}xe^{-x} - \frac{1}{2}x^2 + \frac{1}{2}x - \frac{3}{4}$.

**31.** $y(x) = c_1 e^x + c_2 e^{-x} + xe^x - xe^{-x}$.　　**33.** $y(x) = c_1 x + c_2 x^{-4} + \frac{1}{14}x^3$.

**35.** $y(x) = c_1 e^{2x} + c_2 e^{3x} + \frac{1}{10}e^x(\sin x + 3 \cos x)$.

**37.** $y(x) = c_1 e^x + c_2 e^{4x} - \frac{1}{5}e^{2x}(\cos x + 2 \sin x)$.

**39.** $y(x) = -2 \sin x + x + e^{-x}$.　　**41.** $y(x) = \frac{1}{2}x^2 e^{2x}$.　　**43.** False.　　**45.** True.

**47.** False.

**49.** $Q(t) = \frac{1}{104}[12e^{-3t} - 13e^{-2t} + \cos 2t + 5 \sin 2t]$.

$I(t) = \frac{1}{52}[13e^{-2t} - 18e^{-3t} - \sin 2t + 5 \cos 2t]$.

**51.** $u_p = \dfrac{\beta}{b\tau}$.    **53.** $u(t) = \dfrac{3}{2}e^{-2t} - \dfrac{4}{3}e^{-3t} + \dfrac{5}{6}$; $\lim\limits_{t \to \infty} u(t) = \dfrac{5}{6}$.    **55.** $x_p = \dfrac{K}{\beta}$.

**59.** $u(x) = c_1 + c_2 x + c_3 x^2 + c_4 x^3 + \cos x$.

**61.** $u(x) = c_1 + c_2 x + c_3 x^2 + c_4 x^3 + \sinh x$.

**63.** $u(x) = c_1 + c_2 x + c_3 x^2 + c_4 x^3 + \dfrac{1}{360}x^6$.

---

**1.** $y(x) = c_1 \cos x + c_2 \sin x - (\cos x) \ln |\sec x + \tan x|$.

**3.** $y(x) = c_1 e^x + c_2 x e^x + x e^x \ln x$.    **5.** $y(x) = c_1 e^{-3x} + c_2 x e^{-3x} + \dfrac{1}{2x}e^{-3x}$.

**7.** $y(x) = c_1 e^{6x} + c_2 x e^{6x} + \dfrac{1}{4}x^2 e^{6x}(2 \ln x - 3)$.

**9.** $y(x) = c_1 \cos x + c_2 \sin x + \dfrac{1}{2}\sin x \tan x$.

**11.** $y(x) = c_1 e^{-x} + c_2 x e^{-x} + e^{-x}\left[-\dfrac{1}{2}(\ln x)^2 - \ln x - 1\right]$

**13.** $y(x) = c_1 x^{3/5} + c_2 x + 4x^{1/2}$.    **15.** $y(x) = c_1 x^{3/2} + c_2 x^{-1} + \dfrac{1}{18}x^{-3}$.

**17.** $y(x) = c_1 x^{-2} + c_2 x^{-2} \ln x + x^{-3}(\ln x + 2) + \dfrac{1}{27}x^{-5}(3 \ln x + 2)$.

**19.** $y(x) = \cos x + \sin x - (\cos x) \ln |\sec x + \tan x|$.

**21.** $y(x) = e^{-3x}\left(\dfrac{1}{2x} + \dfrac{21}{2}x - 7\right)$.    **23.** $y(x) = e^{2x}\left(\dfrac{4}{3}x + \dfrac{1}{6}x^{-2} - \dfrac{3}{2}\right)$.

**25.** $y(x) = 24x^{3/2} - \dfrac{59}{36}x^{-1} + \dfrac{1}{18}x^{-3}$.

**27.** $y(x) = x[c_1 e^x + c_2 e^{-x} + c_3 \cos x + c_4 \sin x] + c_5 x[e^x \int x^{-5}e^{-x}dx$

$\qquad - e^{-x}\int x^{-5}e^x dx + 2 \cos x \int x^{-5} \sin x \, dx + 2 \sin x \int x^{-5} \cos x \, dx]$.

**29.** $x(t) = c_1 e^{-t} + c_2 t e^{-t} + e^{-t} \ln(t + 1)$.

1.  Nonhomogeneous, variable coefficients, order 3. The general solution is $y = y_h +$ $y_p$. To determine $y_h$ use the method associated with the Euler equation and to determine $y_p$ use the method of undetermined coefficients.

3.  Homogeneous, variable coefficients, order 2. The general solution can be obtained by the reduction-of-order method after recognizing that $y = x$ is a solution. Alternatively the general solution can be approximated by using the power-series method.

5.  $W(x) = W(x_0)e^{-8(x-x_0)}$, $x_0$ an arbitrary value of $x$.

7.  $y = (c_1 + c_2x + c_3x^2)e^{7x} + c_4e^{\sqrt{3}x} + c_5e^{-\sqrt{3}x}$
    $+ e^{-x}(c_6 \cos x + c_7 \sin x)$. Order 7.

9.  $y(x) = 0 + 3(x-1) - \dfrac{1}{2}(x-1)^2 + \dfrac{11}{18}(x-1)^3 + \dfrac{11}{72}(x-1)^4 + \cdots$.

11. $y = c_1e^{-x} + c_2e^{-16x} + \dfrac{1}{34}e^x + \dfrac{4}{15}xe^{-x}$.

13. $y = c_1x^3 + c_2x^{-4/3}$.

15. $y = c_1e^{2x} + c_2e^{8x} + c_3e^{-x}$.

17. $y = c_1 \tan x + c_2e^x$,     $0 < x < 1$.

19. $y = c_1e^{-3x} + e^{-(3/2)x}\left(c_2 \cos \dfrac{3\sqrt{3}}{2}x + c_3 \sin \dfrac{3\sqrt{3}}{2}x\right) + \dfrac{1}{27}x$.

21. $y = c_1 \cos x + c_2 \sin x + (\sin x) \ln |\csc x - \cot x|$.

23. $y = c_1 \cos x + c_2 \sin x - \dfrac{1}{4}x^2 \cos x + \dfrac{1}{4}x \sin x$.

25. $y = x^{-4}[c_1 \cos (\ln x) + c_2 \sin (\ln x)]$.

27. (a) 2.991 sec.     (b) 2392.8 ft.     (c) $-96.31$ ft/sec.

29. Yes, the ball will clear the fence; $y = 28.1$ ft when $x = 320$ ft.

31. 6.969 mi/sec.

33. Velocity $= 2576$ ft/sec, height $= 51,520$ ft.

35. Range $= 42.32$ ft, maximum height $= 14.42$ ft, time of flight $= 1.62$ sec.

37. The ellipse is given by $\left(\dfrac{x}{a}\right)^2 + \left(\dfrac{\dot{x}}{a\omega}\right)^2 = 1$, and the constant energy is $ma^2\omega^2/2$.

**41.** $v_c = \cos 2t + \dfrac{5}{2}\sin 2t - t\cos 2t$.

**43.** Time of flight $= \left[ v_0 \sin \alpha + \sqrt{(v_0 \sin \alpha)^2 + 2gh} \right]/g$.

**47.** $y = \left(\dfrac{a}{4}\right)^{1/3}$ or $y = \left(\dfrac{a}{27c^2}\right)e^{2x} + ce^{-x}$, where $c$ is an arbitrary constant.

---

**1.** (b) $x_1(t) = c_1 e^{3t} + c_2 e^{5t}$, $x_2(t) = c_1 e^{3t} + 3c_2 e^{5t}$.　　**7.** True.

**9.** False (they are not solutions).　　**11.** True.　　**15.** $\alpha = \beta = \gamma = \dfrac{1}{3}$.

**17.** A second linearly independent solution is $\begin{bmatrix} t\ln t + 1 \\ \ln t \end{bmatrix}$ and the general solution is

$x_1(t) = c_1 t + c_2(t \ln t + 1)$, $x_2(t) = c_1 + c_2 \ln t$.

**19.** "The $n$ solutions

$$\begin{bmatrix} x_{11}(t) \\ x_{21}(t) \\ \cdot \\ \cdot \\ \cdot \\ x_{n1}(t) \end{bmatrix}, \begin{bmatrix} x_{12}(t) \\ x_{22}(t) \\ \cdot \\ \cdot \\ \cdot \\ x_{n2}(t) \end{bmatrix}, \cdots, \begin{bmatrix} x_{1n}(t) \\ x_{2n}(t) \\ \cdot \\ \cdot \\ \cdot \\ x_{nn}(t) \end{bmatrix}$$

of the system (21) are linearly independent on an interval $I$ if and only if the determinant

$$\begin{vmatrix} x_{11}(t) & x_{12}(t) & \cdots & x_{1n}(t) \\ x_{21}(t) & x_{22}(t) & \cdots & x_{2n}(t) \\ \cdot & \cdot & & \cdot \\ \cdot & \cdot & & \cdot \\ \cdot & \cdot & & \cdot \\ x_{n1}(t) & x_{n2}(t) & \cdots & x_{nn}(t) \end{vmatrix}$$

is never zero on $I$."

**21.** "There exist $n$ linearly independent solutions of the system (21). Furthermore, if the $n$ column vectors

$$\begin{bmatrix} x_{11}(t) \\ x_{21}(t) \\ \cdot \\ \cdot \\ \cdot \\ x_{n1}(t) \end{bmatrix}, \begin{bmatrix} x_{12}(t) \\ x_{22}(t) \\ \cdot \\ \cdot \\ \cdot \\ x_{n2}(t) \end{bmatrix}, \cdots, \begin{bmatrix} x_{1n}(t) \\ x_{2n}(t) \\ \cdot \\ \cdot \\ \cdot \\ x_{nn}(t) \end{bmatrix}$$

are linearly independent solutions of the system (21), then the general solution of the system (21) is given by

$$\begin{bmatrix} x_1(t) \\ x_2(t) \\ \vdots \\ x_n(t) \end{bmatrix} = c_1 \begin{bmatrix} x_{11}(t) \\ x_{21}(t) \\ \vdots \\ x_{n1}(t) \end{bmatrix} + c_2 \begin{bmatrix} x_{12}(t) \\ x_{22}(t) \\ \vdots \\ x_{n2}(t) \end{bmatrix} + \cdots + c_n \begin{bmatrix} x_{1n}(t) \\ x_{2n}(t) \\ \vdots \\ x_{nn}(t) \end{bmatrix},$$

where $c_1, c_2, \ldots, c_n$ are arbitrary constants."

**23.** True.  **25.** True.

**27.** False. (The given column vector is a solution, but it fails to satisfy all the initial conditions.)

**29.** True.

**Section 3.2**
Page 168

**1.** $x_1(t) = c_1 e^{2t} + c_2 e^{3t}$, $x_2(t) = 2c_1 e^{2t} + c_2 e^{3t}$.

**3.** $x_1(t) = c_1 e^{2t} + c_2 e^{3t} - 3te^{2t} - \dfrac{1}{3}t - \dfrac{5}{18}$,

$x_2(t) = 2c_1 e^{2t} + c_2 e^{3t} - 6te^{2t} - \dfrac{4}{3}t + 6e^{2t} - \dfrac{7}{9}$.

**5.** $x(t) = c_1 e^{-t} + c_2 te^{-t} + 1$, $y(t) = c_1 e^{-t} + c_2\left(te^{-t} - \dfrac{1}{6}e^{-t}\right) + 1$.

**7.** $x(t) = c_1 e^{2t} \cos 3t + c_2 e^{2t} \sin 3t - \dfrac{2}{13}t^2 + \dfrac{36}{169}t + \dfrac{534}{2197}$,

$y(t) = \dfrac{1}{2}c_1 e^{2t}(\cos 3t + 3\sin 3t) + \dfrac{1}{2}c_2 e^{2t}(\sin 3t - 3\cos 3t)$

$\qquad + \dfrac{10}{13}t^2 + \dfrac{80}{169}t + \dfrac{567}{2197}$.

**9.** $I_1(t) = \dfrac{1}{10}e^{-3t} - \dfrac{1}{10}\cos t - \dfrac{3}{10}\sin t$,

$I_2(t) = -\dfrac{1}{20}e^{-3t} + \dfrac{1}{20}\cos t - \dfrac{3}{20}\sin t$.

**11.** $x_1(t) = 2e^{2t} - e^{3t}$, $x_2(t) = 4e^{2t} - e^{3t}$.

**13.** $x_1(t) = -3te^{2t} - \dfrac{1}{3}t - \dfrac{5}{18}$, $x_2(t) = -6te^{2t} - \dfrac{4}{3}t + 6e^{2t} - \dfrac{7}{9}$.

**15.** $x(t) = -e^{-t} + 1, y(t) = -e^{-t} + 1.$

**17.** $x(t) = -\dfrac{2}{13}t^2 + \dfrac{36}{169}t + \dfrac{534}{2197}, y(t) = \dfrac{10}{13}t^2 + \dfrac{80}{169}t + \dfrac{567}{2197}.$

**19.** $x(t) = c_1 e^{-t}, y(t) = c_2 e^{-t}.$    **21.** $x(t) = c_1 e^{-t} + c_2 e^{2t}, y(t) = 2c_1 e^{-t} + \dfrac{1}{2}c_2 e^{2t}.$

**23.** $x(t) = c_1 e^{-t} \cos t + c_2 e^{-t} \sin t,$
$\quad y(t) = -c_1 e^{-t} \sin t + c_2 e^{-t} \cos t.$

**25.** $x(t) = c_1 e^t + c_2 t e^t, y(t) = c_1 e^t + c_2(t + 1)e^t.$

**27.** $x(t) = c_1 e^{3t} + c_2 e^{5t}, y(t) = c_1 e^{3t} + 3c_2 e^{5t}.$

**29.** $c_1(t) = A + Be^{-[(K/V_1) + (K/V_2)]t}$
$\quad c_2(t) = A - (V_1/V_2) Be^{-[(K/V_1) + (K/V_2)]t},$ $A$ and $B$ are arbitrary constants.

**31.** $y(t) = \dfrac{abd}{(a + d)(a + cf)} + \dfrac{bd}{a + d}\left[1 - \dfrac{a}{(a + d)(a + cf)}\right]e^{-(a-cf)t},$

$\quad \lim\limits_{t\to\infty} y(t) = \dfrac{abd}{(a + d)(a + cf)} > 0.$

---

**1.** $\lambda = 1, 2; x_1(t) = c_1 e^t + c_2 e^{2t}, x_2(t) = -\dfrac{3}{2}c_1 e^t - c_2 e^{2t}.$

**3.** $\lambda = 1, 1; x_1(t) = c_1 e^t + c_2\left(t + \dfrac{1}{2}\right)e^t, x_2(t) = c_1 e^t + c_2 t e^t.$

**5.** $\lambda = -2 \pm 3i; x_1(t) = c_1 e^{-2t} \cos 3t + c_2 e^{-2t} \sin 3t,$

$\quad x_2(t) = c_1\left(\dfrac{5}{2}e^{-2t}\cos 3t + \dfrac{3}{2}e^{-2t}\sin 3t\right) + c_2\left(\dfrac{5}{2}e^{-2t}\sin 3t - \dfrac{3}{2}e^{-2t}\cos 3t\right).$

**7.** $x_1(t) = -e^{2t} + 2e^{3t}, x_2(t) = -e^{2t} + e^{3t}.$

**9.** $x_1(t) = -2e^{-2t}\sin 3t, x_2(t) = -e^{-2t}(5\sin 3t - 3\cos 3t).$

**11.** $\lambda = 2, -3; x_1(t) = c_1 e^{2t}, x_2(t) = c_2 e^{-3t}.$

**13.** $\lambda = 1, 3; x_1(t) = c_1 e^t + c_2 e^{3t}, x_2(t) = -c_1 e^t - 3c_2 e^{3t}.$

**15.** $\lambda = 1, 3; x_1(t) = c_1 e^t + 3c_2 e^{3t}, x_2(t) = -c_1 e^t - c_2 e^{3t}.$

**17.** $\lambda = 1, 2, 3;\ x_1(t) = c_1 e^t + 2c_2 e^{2t}$

$$x_2(t) = -c_2 e^{2t} + c_3 e^{3t}$$

$$x_3(t) = -c_1 e^t - c_3 e^{3t}.$$

**19.** $\lambda = -1, 2, 2;\ x_1(t) = c_1 e^{-t} + c_2 e^{2t} + c_3(t + 1)e^{2t}$

$$x_2(t) = -3c_1 e^{-t} + c_3 e^{2t}$$

$$x_3(t) = 4c_1 e^{-t} + c_2 e^{2t} + c_3(t + 3)e^{2t}.$$

**21.** $\lambda = 0, 1, -2, -3;\ x_1(t) = -2c_1 + c_2 e^t + c_3 e^{-2t} + 3c_4 e^{-3t}$

$$x_2(t) = -2c_3 e^{-2t} - 4c_4 e^{-3t}$$

$$x_3(t) = c_1 + c_3 e^{-2t}$$

$$x_4(t) = -c_1 + 4c_4 e^{-3t}.$$

**23.** $\lambda = 1, 2, 3;\ x_1(t) = 2c_1 e^t + 3c_2 e^{2t} + c_3 e^{3t}$

$$x_2(t) = c_1 e^t + c_2 e^{2t} - c_3 e^{3t}$$

$$x_3(t) = 3c_1 e^t + 5c_2 e^{2t} + 2c_3 e^{3t}.$$

**25.** $x_1(t) = e^t,\ x_2(t) = 0,\ x_3(t) = -e^t.$

**27.** $x_1(t) = (t + 1)e^{2t},\ x_2(t) = e^{2t},\ x_3(t) = (t + 3)e^{2t}.$

**29.** $x_1(t) = 2e^t + 3e^{2t} + e^{3t}$

$x_2(t) = e^t + e^{2t} - e^{3t}$

$x_3(t) = 3e^t + 5e^{2t} + 2e^{3t}.$

**31.** A particular solution is $\begin{bmatrix} x_{1p} \\ x_{2p} \end{bmatrix} = \begin{bmatrix} \dfrac{3}{2} e^{5t} \\ \dfrac{7}{2} e^{5t} \end{bmatrix}.$

The general solution is

$$x_1(t) = c_1 e^{3t} + \left( c_2 + \frac{3}{2} \right) e^{5t}$$

$$x_2(t) = c_1 e^{3t} + \left( 3c_2 + \frac{7}{2} \right) e^{5t}.$$

**33.** A particular solution is $\begin{bmatrix} x_{1p} \\ x_{2p} \end{bmatrix} = \begin{bmatrix} 1 \\ t \end{bmatrix}.$

The general solution is

$$x_1(t) = c_1 e^t + c_2 e^{3t} + 1$$

$$x_2(t) = -c_1 e^t - 3c_2 e^{3t} + t.$$

**35.** A particular solution is $\begin{bmatrix} x_p \\ y_p \end{bmatrix} = \begin{bmatrix} 1 \\ 4 \end{bmatrix}$.

The general solution is

$x(t) = c_1 e^{-3t} + c_2 t e^{-3t} + 1$

$y(t) = c_1 e^{-3t} + c_2 (t+1) e^{-3t} + 4$.

**37.** $\lambda = 0, \ -\left(\dfrac{K}{V_1} + \dfrac{K}{V_2}\right); c_1(t) = A + Be^{-[(K/V_1) + (K/V_2)]t}; A, B$ constants;

$$c_2(t) = A - B\frac{V_1}{V_2} e^{-[(K/V_1)+ (K/V_2)]t}.$$

REVIEW
EXERCISES
CHAPTER 3
Page 187

**1.** $x(t) = c_1 e^t + c_2 e^{2t}, \ y(t) = c_1 e^t + 2c_2 e^{2t}$.

**3.** $x(t) = c_1 e^t + c_2 e^{2t}, \ y(t) = c_1 e^t + 2c_2 e^{2t} - e^{3t}$.

**5.** $x_1(t) = c_1 e^t + c_2 e^{-t} + c_3 \cos t + c_4 \sin t$

$x_2(t) = c_1 e^t - c_2 e^{-t} - c_3 \sin t + c_4 \cos t - t^2$

$x_3(t) = c_1 e^t + c_2 e^{-t} - c_3 \cos t - c_4 \sin t - 2t$

$x_4(t) = c_1 e^t - c_2 e^{-t} + c_3 \sin t - c_4 \cos t - 2$.

**7.** $x(t) = 3c_1 e^{-t} + 2c_3 e^{4t}$

$y(t) = 6c_1 e^{-t} - c_3 e^{4t}$

$z(t) = -c_1 e^{-t} + c_2 e^{2t} + c_3 e^{4t}$.

**9.** $x_1(t) = c_1 e^t + \dfrac{1}{3} c_2 e^{-2t} + \dfrac{1}{10} c_3 e^{3t}$

$x_2(t) = c_2 e^{-2t} - \dfrac{1}{5} c_3 e^{3t}$

$x_3(t) = c_3 e^{3t}$.

**11.** $x(t) = -e^t + 2e^{2t}, \ y(t) = -e^t + 4e^{2t}$.

**13.** $x_1(t) = e^t, \ x_2(t) = e^t - t^2, \ x_3(t) = e^t - 2t$

$x_4(t) = e^t - 2$.

**15.** $x_1(t) = \dfrac{35}{6} e^t - \dfrac{14}{15} e^{-2t} + \dfrac{1}{10} e^{3t}$

$x_2(t) = -\dfrac{14}{5} e^{-2t} - \dfrac{1}{5} e^{3t}$

$x_3(t) = e^{3t}$.

**17.** Explosion.

**19.** $x(t) = e^{2t}(\sin t - 2\cos t) + 2$
$y(t) = e^{2t}(3\sin t - \cos t) + 1.$

---

**Section 4.2**
Page 197

**13.** $\cos 4t.$     **15.** $\sinh 5t.$     **17.** $\sin 2t.$

**19.** $\dfrac{5}{s} - \dfrac{48}{s^4}.$     **21.** $\dfrac{8s}{64s^2 + 9}.$     **23.** $\dfrac{s-3}{(s-3)^2 + 4} - \dfrac{5}{(s-1)^2 - 25}.$

**25.** $\dfrac{s-1}{s^2 + 1}.$

**27.** $\dfrac{7!}{s^8} - \dfrac{4!}{s^5} + \dfrac{5 \cdot 2!}{s^3}.$     **29.** $\dfrac{2s}{(s^2 - 1)^2}.$     **31.** $\dfrac{s}{(s-5)^2}.$

**33.** $\dfrac{s^3}{s^2 + 1} + \dfrac{s^2}{(s-1)^2} - s - 1.$     **35.** $\dfrac{s^2}{s^4 - 1}.$     **37.** $\dfrac{s-1}{s[(s-1)^2 + 4]}.$

**39.** $\dfrac{2}{k}\sin kt.$     **41.** $t^n e^{kt}.$     **43.** $\cosh kt.$     **45.** $\sin t - t\cos t.$

**47.** $2e^{-9t} - 2e^{-4t} + 1.$     **49.** $e^{-2t} + 2te^{-2t} - e^{2t} + 3te^{2t}.$     **51.** $e^{-2t} + \dfrac{2}{7}\sin 7t.$

**53.** $3e^{3t} - 2.$     **55.** $3 - 2t + \dfrac{1}{2}t^2 - 3e^{-t}.$     **57.** $e^{-t}\left(2\cos 4t - \dfrac{5}{4}\sin 4t\right).$

---

**Section 4.3**
Page 205

**1.** $y(t) = e^{-2t}.$     **3.** $y(t) = e^{-4t} - e^{t}.$     **5.** $y(t) = \dfrac{1}{3}t^3 e^{-t} + 3e^{-t}.$

**7.** $y(t) = e^{-t} + e^{t} + 3te^{t}.$     **9.** $y(t) = t^2 e^{-5t} - te^{-5t}.$

**11.** $y(t) = 3 - 5e^{3t} + 2e^{6t}.$     **13.** $y(t) = -\dfrac{1}{10}\cos 3t - \dfrac{1}{30}\sin 3t + \dfrac{1}{10}e^{t}.$

**15.** $y(t) = e^{t} + e^{-5t}\sin t.$     **17.** $y(t) = 1 - 2e^{-t} + 2e^{(1/2)t}\cos\dfrac{\sqrt{3}}{2}t.$

**19.** $y(t) = -2 + 3t + 3e^{-3t} + 6te^{-3t}.$     **21.** $y(t) = -e^{-t} - te^{-t} - \dfrac{5}{2}t^2 e^{-t} + e^{4t}.$

**23.** $y(t) = 1 - te^{-t} + e^{-5t} - e^{3t}.$          **25.** $y(t) = -e^{-2t} + te^{-2t} + e^{-5t} + 2te^{-5t}.$

**27.** $y(t) = 3e^{-t} + e^{t} + 2te^{t} + 2e^{-(1/2)t} \cos \dfrac{\sqrt{3}}{2} t.$

**29.** $y(t) = -2e^{t} \cos t + 3e^{t} \sin t + e^{-2t} \cos t.$

**31.** $y(t) = -1 + e^{2t} - 2te^{2t} + \dfrac{1}{2}t^{3}e^{2t}.$

**33.** $y(t) = -11e^{-2t} - te^{-2t} + 3e^{t} \cos t\sqrt{3}.$

**35.** $y(t) = -19e^{-t} + 8e^{-6t} + 13e^{t} \cos t + 9e^{t} \sin t.$

**37.** $x_{1}(t) = 2e^{2t} - e^{3t}, x_{2}(t) = 4e^{2t} - e^{3t}.$

**39.** $x_{1}(t) = -e^{2t} + 2e^{3t}, x_{2}(t) = e^{3t} - e^{2t}.$

**41.** $x_{1}(t) = -2e^{-2t} \sin 3t, x_{2}(t) = -e^{-2t} (5 \sin 3t - 3 \cos 3t).$

**43.** $x_{1}(t) = -3te^{2t} - \dfrac{1}{3}t - \dfrac{5}{18}, x_{2}(t) = 6(1 - t)e^{2t} - \dfrac{4}{3}t - \dfrac{7}{9}.$

**45.** $x_{1}(t) = (t + 1)e^{2t}, x_{2}(t) = e^{2t}, x_{3}(t) = (t + 3)e^{2t}.$

---

**1.** $y(t) = 1 - e^{-t} - 4te^{-t}.$

**3.** $y(t) = h_{1}(t - 2)[4e^{(t-2)} - 5 + e^{-4(t-2)}].$

**5.** $y(t) = 1 + 3t + 5e^{3t} - 3e^{6t}.$

**7.** $y(t) = e^{-t} + e^{t}(2t - 1) + h_{1}(t - 3)[e^{(t-3)} + e^{-(t-3)} - 2].$

**9.** $y(t) = h_{1}(t - 1)[2 - 2 \cos 2(t - 1)] + h_{1}(t - 2)[\cos 2(t - 2) - 1] - \cos 2t.$

**11.** $y(t) = e^{-5t} \cos t + h_{1}(t - 1)[2 - 2e^{-5(t-1)} \cos (t - 1) - 10e^{-5(t-1)} \sin (t - 1)]$
$\qquad\qquad + h_{1}(t - 3)[1 - e^{-5(t-3)} \cos (t - 3) - 5e^{-5(t-3)} \sin (t - 3)].$

**13.** $y(t) = -5 + e^{4t} + 4e^{-t} + h_{1}(t - 1)[-10 + 2e^{4(t-1)} + 8e^{-(t-1)}]$
$\qquad\qquad + h_{1}(t - 2)[-5 + e^{4(t-2)} + 4e^{-(t-2)}].$

**Section 4.5**

Page 213

**1.** $y(t) = e^{4t} - e^{-t}$.

**3.** $y(t) = te^{-3t} + h_1(t - 2)[(t - 2)e^{-3(t-2)}]$.

**5.** $y(t) = 6 - e^t + e^{2t}$.

**7.** $y(t) = e^t - e^{-t} + h_1(t - 3)[e^{-(t-3)} - e^{-2(t-3)}]$.

**9.** $y(t) = -te^{-5t} + h_1(t - 1)[(t - 1)e^{-5(t-1)}] - 3h_1(t - 2)[(t - 2)e^{-5(t-2)}]$.

**11.** $y(t) = 9e^{3t} - e^{-3t} + h_1(t - 2)[e^{3(t-2)} - e^{-3(t-2)}]$.

**13.** $y(t) = e^{4t} - e^{-2t} - 9 \cos t - 2 \sin t + h_1(t - 5)[e^{4(t-5)} - e^{-2(t-5)}]$.

**Section 4.6**

Page 219

**5.** $r = 4n(0) + 6 + [4n(0) + 24]t + 12t^2$.

**13.** (a)

$$i(t) = \frac{(\omega^2 - \omega_0^2)v_0\omega}{L[(\omega_0^2 - \omega^2) + 2a\omega^2]}e^{-at} + \left\{\frac{v_0\omega a(\omega_0^2 - \omega^2)}{L[(\omega_0^2 - \omega^2)^2 + 2a\omega^2]}\right.$$

$$\left. - \frac{v_0\omega\omega_0^2}{L[(\omega_0^2 - \omega^2)^2 + 2a\omega^2]}\right\} te^{-at} + \frac{v_0\omega(\omega_0^2 - \omega^2)}{L[(\omega_0^2 - \omega^2)^2 + 2a\omega^2]} \cos \omega t$$

$$+ \frac{v_0\omega^2}{L[(\omega_0^2 - \omega^2)^2 + 2a\omega^2]} \sin \omega t.$$

(b)

$$i(t) = \frac{v_0\omega(\omega^2 - \omega_0^2)}{L[(\omega_0^2 - \omega^2)^2 + 2a\omega^2]}e^{-at} \cos \sqrt{\omega_0^2 - a^2}\,t$$

$$+ \frac{1}{\sqrt{\omega_0^2 - a^2}}\left\{\frac{v_0\omega a(\omega_0^2 - \omega^2)}{L[(\omega_0^2 - \omega^2)^2 + 2a\omega^2]} - \frac{v_0\omega\omega_0^2}{L[(\omega_0^2 - \omega^2)^2 + 2a\omega^2]}\right\}$$

$$\cdot e^{-at} \sin \sqrt{\omega_0^2 - a^2}\,t$$

$$+ \frac{v_0\omega(\omega_0^2 - \omega^2)}{L[(\omega_0^2 - \omega^2)^2 + 2a\omega^2]} \cos \omega t + \frac{v_0\omega^2}{L[(\omega_0^2 - \omega^2)^2 + 2a\omega^2]} \sin \omega t.$$

(c)

$$i(t) = \frac{-v_0\omega(\omega_0^2 - \omega^2)}{L[(\omega_0^2 - \omega^2)^2 + 2a\omega^2]} e^{-at} \cosh \sqrt{a^2 - \omega^2} t$$

$$+ \frac{1}{\sqrt{a^2 - \omega_0^2}} \left\{ \frac{v_0\omega a(\omega_0^2 - \omega^2)}{L[(\omega_0^2 - \omega^2)^2 + 2a\omega^2]} - \frac{v_0\omega\omega_0^2}{L[(\omega_0^2 - \omega^2)^2 + 2a\omega^2]} \right\}$$

$$\cdot e^{-at} \sinh \sqrt{a^2 - \omega_0^2} t$$

$$+ \frac{v_0\omega(\omega_0^2 - \omega^2)}{L[(\omega_0^2 - \omega^2)^2 + 2a\omega^2]} \cos \omega t + \frac{v_0\omega^2}{L[(\omega_0^2 - \omega^2)^2 + 2a\omega^2]} \sin \omega t.$$

**REVIEW
EXERCISES
CHAPTER 4**
Page 221

**1.** $\dfrac{480}{s^6} + \dfrac{6s}{(s^2 + 1)^2} \cdot$　　**3.** $1 + \dfrac{e^{-2s}}{s} \cdot$　　**5.** $\dfrac{1}{8} e^{9t} - \dfrac{1}{8} e^{t}.$

**7.** $h_1(t - 3)e^t.$　　**9.** $y(t) = 91 - 48t + e^{8t} - 43e^{-t}.$

**11.** $y(t) = 17 - 6t - 17 \cos 2t - 7 \sin 2t + 5e^{-2t/5} - 5e^{-3t}.$

**13.** $y(t) = 3e^t - e^{-t} - 2e^{(1/2)t} \cos \dfrac{\sqrt{3}}{2} t - 2\sqrt{3} e^{(1/2)t} \sin \dfrac{\sqrt{3}}{2} t.$

**15.** $y(t) = 5h_1(t - 2) - 8h_1(t - 2)e^{3(t-2)} + 3h_1(t - 2)e^{-2(t-2)} + 8e^{3t} - 3e^{8t}.$

**17.** See Exercise 16, Section 4.6, and the convolution theorem.

**19.** $x(t) = Ae^{-bt} + Bte^{-bt} + Ce^{\lambda_1 t} + De^{\lambda_2 t},$

where $\lambda_1 = [-R + \sqrt{R^2 - 4Km}]/2m, \lambda_2 = [-R - \sqrt{R^2 - 4Km}]/2m,$

$A = Q(2mb - R)/m^2(b + \lambda_1)^2(b + \lambda_2)^2,$

$B = Q/[mb^2 - Rb + K],$

$C = Q/[(b + \lambda_2)^2 \sqrt{R^2 - 4Km}],$

and

$D = -Q/[(b + \lambda_1)^2 \sqrt{R^2 - 4Km}].$

**21.** $x_1(t) = e^{-t}, x_2(t) = 5e^{-t}.$

**23.** $x_1(t) = e^{-t} - 2e^{2t}, x_2(t) = 2e^{-t} - e^{2t}.$

**25.** $x_1(t) = 7e^{-t} \cos t + 3e^{-t} \sin t$

$x_2(t) = -7e^{-t} \sin t + 3e^{-t} \cos t.$

**27.** $x_1(t) = e^t - 6te^t, \ x_2(t) = e^t - 6(t + 1)e^t.$

**29.** $x_1(t) = e^{3t} - e^{5t}, \ x_2(t) = e^{3t} - 3e^{5t}.$

---

**Section 5.2**

Page 228

**1.** $R = \dfrac{1}{3}.$ **3.** $R = +\infty.$ **5.** $R = 1$

**7.** First derivative: $\displaystyle\sum_{n=1}^{\infty} 3^n n(x - 1)^{n-1}; R = \dfrac{1}{3}.$

Second derivative: $\displaystyle\sum_{n=2}^{\infty} 3^n n(n - 1)(x - 1)^{n-2}; R = \dfrac{1}{3}.$

**9.** First derivative: $\displaystyle\sum_{n=1}^{\infty} \dfrac{(-1)^n}{(2n - 1)!} x^{2n-1}; R = +\infty.$

Second derivative: $\displaystyle\sum_{n=1}^{\infty} \dfrac{(-1)^n}{(2n - 2)!} x^{2n-2}; R = +\infty.$

**11.** First derivative: $\displaystyle\sum_{n=1}^{\infty} \dfrac{n}{n + 1}(x + 1)^{n-1}; R = 1.$

Second derivative: $\displaystyle\sum_{n=2}^{\infty} \dfrac{n(n - 1)}{n + 1}(x + 1)^{n-2}; R = 1.$

**15.** Solution: $y(x) = a_0 \displaystyle\sum_{n=0}^{\infty} \dfrac{3^n(x - 1)^n}{n!} = a_0 e^{3(x - 1)}.$

**17.** $\cos x = \displaystyle\sum_{n=0}^{\infty} \dfrac{(-1)^n \ x^{2n}}{(2n)!}; R = +\infty;$ converges for all $x.$

**19.** $\sin x = \displaystyle\sum_{n=0}^{\infty} \dfrac{(-1)^n(x - \pi)^{2n-1}}{(2n + 1)!}; R = +\infty;$ converges for all $x.$

**21.** $\dfrac{1}{x + 3} = \displaystyle\sum_{n=0}^{\infty} \dfrac{(-1)^n(x - 1)^n}{4^{n-1}}; R = 4;$ converges for $-3 < x < 5.$

**23.** $\dfrac{1}{x} = \displaystyle\sum_{n=0}^{\infty} -\dfrac{(x + 3)^n}{3^{n-1}}; R = 3;$ converges for $-6 < x < 0.$

**Section 5.3**
Page 232

1. $x = 0$ is a regular singular point. Every other finite value of $x$ is an ordinary point.

3. $x = 1$ is a regular singular point. Every other finite value of $x$ is an ordinary point.

5. $x = 1$ is a regular singular point. Every other finite value of $x$ is an ordinary point.

7. $x = 0$ is an irregular singular point. $x = 1$ and $x = -1$ are regular singular points. Every other finite value of $x$ is an ordinary point.

9. There are no singular points.     11. $x = 1$ and $x = -1$ are regular singular points.

13. There are no singular points.     15. $x = 1$ and $x = -1$ are regular singular points.

17. False.     19. True.     21. True.

**Section 5.4**
Page 244

1. $y(x) = 1 - 2x^2$.     3. $y(x) = -7 - 8x - 2x^2$.     5. $y(x) = x$.

7. $y(x) = x$.     9. $y(x) = -x$.     11. $a_0 = 0$, $a_1 = 1$, $a_2 = 0$, $a_3 = 0$.

13. $a_0 = -20$, $a_1 = -2$, $a_2 = -1$, $a_3 = 1$.     15. $a_0 = 0$, $a_1 = 1$, $a_2 = 0$, $a_3 = 0$.

17. $a_0 = 0$, $a_1 = 1$, $a_2 = 0$, $a_3 = -\dfrac{2}{3}$.     19. $a_0 = 1$, $a_1 = 0$, $a_2 = -1$, $a_3 = 0$.

21. Apply Theorem 1.     23. Apply Theorem 1.     25. $P_1(x) = x$.

27. $P_3(x) = \dfrac{5}{2}x^3 - \dfrac{3}{2}x$.     31. $T_1(x) = x$.     33. $T_3(x) = 4x^3 - 3x$.

35. $H_2(x) = 4x^2 - 2$.     37. $2a_2 + 3a_1 + 2a_0 = 0$;

$$a_{n+2} = -\frac{3(n + 1)a_{n+1} + (n + 2)a_n}{(n + 2)(n + 1)},$$

$$n = 1, 2, 3, \ldots.$$

**Section 5.5**
Page 264

1. $y_1(x) = x^{1/3} \displaystyle\sum_{n=0}^{\infty} a_n x^n; y_2(x) = x^{-1/3} \displaystyle\sum_{n=0}^{\infty} b_n x^n.$

3. $y_1(x) = \displaystyle\sum_{n=0}^{\infty} a_n x^n; y_2(x) = y_1(x) \ln|x| + \displaystyle\sum_{n=0}^{\infty} b_n x^n.$

**5.** $y_1(x) = \displaystyle\sum_{n=0}^{\infty} a_n x^n; y_2(x) = y_1(x) \ln |x| + \displaystyle\sum_{n=0}^{\infty} b_n x^n.$

**7.** $y_1(x) = |x|^{1/4} \displaystyle\sum_{n=0}^{\infty} a_n x^n; y_2(x) = |x|^{-1/4} \displaystyle\sum_{n=0}^{\infty} b_n x^n.$

**9.** $y_1(x) = \displaystyle\sum_{n=0}^{\infty} a_n x^n; y_2(x) = y_1(x) \ln |x| + \displaystyle\sum_{n=0}^{\infty} b_n x^n.$

**11.** $y_1(x) = \displaystyle\sum_{n=0}^{\infty} a_n x^n; y_2(x) = C y_1(x) \ln |x| + x^{-2} \displaystyle\sum_{n=0}^{\infty} b_n x^n.$

**13.** $y_1(x) = x^2 \left[ 1 + \displaystyle\sum_{n=1}^{\infty} \dfrac{(-1)^n}{2^{2n} n! \, 3 \cdot 4 \cdots (2+n)} x^{2n} \right].$

**15.** $y_1(x) = 1 + 2x + \dfrac{8}{3} x^2 + \cdots.$      **17.** $y_1(x) = 1 - 2x + \dfrac{1}{2} x^2.$

**19.** $y_1(x) = 1 - x; y_2(x) = (1 - x) \ln |x| + (1 + 2x + \cdots).$

**21.** True.     **23.** False.     **25.** True.     **27.** False.

**29.** $y_1(x) = 1 - \dfrac{1}{2} x^4 + \dfrac{1}{24} x^8 - \cdots,$      **31.** $a_n = \dfrac{1}{n(n+3)} a_{n-1}, n = 1, 2, 3, \ldots.$

$y_2(x) = x^2 - \dfrac{1}{6} x^6 + \dfrac{1}{120} x^{10} - \cdots.$

**33.** $a_n = \dfrac{1}{2n+3} a_{n-1}, n = 1, 2, 3, \ldots.$      **37.** $y_1(x) = \displaystyle\sum_{n=0}^{\infty} \dfrac{(-1)^n}{(n!)^2} \left( \dfrac{x}{2} \right)^{2n}.$

**39.** $L_1(x) = 1 - x.$      **41.** $L_3(x) = 6 - 18x + 9x^2 - x^3.$

**43.** $F\left( 1, 2, \dfrac{3}{4}; x \right) = 1 + \displaystyle\sum_{n=1}^{\infty} \dfrac{4^n (n+1)!}{3 \cdot 7 \cdot 11 \cdots (4n-1)} x^n.$

**45.** $t(1 - t)\ddot{y} + \left( \dfrac{1}{2} - t \right)\dot{y} - 2y = 0.$

**REVIEW**
**EXERCISES**
**CHAPTER 5**
Page 266

**1.** $y(x) = -1 - \displaystyle\sum_{n=1}^{\infty} \dfrac{1 \cdot 4 \cdots (3n-2)}{(3n)!} x^{3n}.$

**3.** $y(x) = -3x + 4x^3$

**5.** $a_0 = 1, a_1 = 0, a_2 = -\dfrac{1}{2}, a_3 = \dfrac{1}{6}.$

**7.** $a_0 = 4, a_1 = 1, a_2 = \dfrac{1}{7}, a_3 = \dfrac{6}{245}.$

**9.** It has two linearly independent solutions, each of the form

$$y(x) = \sum_{n=0}^{\infty} a_n x^n; \, R = \infty.$$

**11.** It has two linearly independent solutions, each of the form

$$y(x) = \sum_{n=0}^{\infty} a_n (x - 1)^n; \, R \geq 1.$$

**13.** $y_1(x) = |x + 1|^{1/2} \sum_{n=0}^{\infty} a_n (x + 1)^n$

$$y_2(x) = \sum_{n=0}^{\infty} b_n (x + 1)^n; \, R \geq 2.$$

**15.** $y(x) = a_0 x^3 \left[ 1 + \sum_{n=1}^{\infty} \frac{(-1)^n}{2^{2n} \cdot n! \cdot 4 \cdot 5 \cdot 6 \cdots (3 + n)} x^{2n} \right].$

**17.** $y(x) = a_0 (6 - 18x + 9x^2 - x^3).$

**19.** $\cos x = 1 - \dfrac{x^2}{2!} + \dfrac{x^4}{4!} - \dfrac{x^6}{6!} + \cdots.$

**1.** True.　　**3.** False (compare Exercise 3 with Exercises 1 and 2).

**5.** False $[y(\pi/2) = +1, \text{ not } -1]$.　　**7.** $y(x) = \cos 3x + c_2 \sin 3x.$

**9.** $y(x) = \cos 3x + \sin 3x.$　　**11.** $y(x) = 0.$　　**13.** $y(x) = xe^x.$

**15.** $y_1(x) = x$ and $y_2(x) = x^3$ are linearly independent solutions of the homogeneous differential equation. Also,

$$\begin{vmatrix} y_1(1) & y_2(1) \\ y_1(2) & y_2(2) \end{vmatrix} = \begin{vmatrix} 1 & 1 \\ 2 & 8 \end{vmatrix} = 8 - 2 = 6 \neq 0.$$

Therefore, by Theorem 1, part (a), there exists one and only one solution.

**17.** $y(x) = c_1 + \dfrac{c_2 - c_1}{l} x.$ Note that $y(0) = c_1$ and $y(l) = c_2.$

**1.** $\lambda_n = 4\pi^2 n^2, \, y_n = \sin 2\pi nx; \, n = 1, 2, 3, \ldots.$

**3.** $\lambda_n = n^2 \pi^2, \, n = 1, 2, 3, \ldots, y_n(x) = \begin{cases} \sin n\pi x & \text{if } n \text{ is even} \\ \cos n\pi x & \text{if } n \text{ is odd} \end{cases} \quad n = 1, 2, 3, \ldots.$

**5.** $\lambda_n = -2 + \dfrac{n^2 \pi^2}{4}, n = 1, 2, 3, \ldots, y_n(x) = \sin \dfrac{n\pi x}{2}, n = 1, 2, 3, \ldots.$

**7.** $\lambda_n = -2 + n^2 \pi^2, \, n = 1, 2, 3, \ldots, y_n = \sin n\pi x, \, n = 1, 2, 3, \ldots.$

**11.** Step 1. Solve the EVP $X'' - \lambda X = 0$

$$X(0) = X(l) = 0$$

to obtain $\lambda_n = -\dfrac{n^2\pi^2}{l^2}, n = 1, 2, 3, \ldots$

$$X_n(x) = \sin\frac{n\pi x}{l}, \; n = 1, 2, 3, \ldots$$

Step 2. Solve $T'' + \dfrac{c^2 n^2 \pi^2}{l^2} T = 0$ to obtain

$$T_n(t) = c_1 \cos\frac{cn\pi}{l}t + c_2 \sin\frac{cn\pi}{l}t.$$

Step 3. $u_n(x, t) = X_n(x)T_n(t)$ and $\displaystyle\sum_{n=1}^{\infty} \alpha_n u_n(x, t)$ are solutions of the DE.

Step 4. Choose $\beta_n$ and $\gamma_n$, where $\beta_n = \alpha_n c_1$, $\gamma_n = \alpha_n c_2$ such that

$$\sum_{n=1}^{\infty} \beta_n \sin\frac{n\pi x}{l} = f(x)$$

and

$$\sum_{n=1}^{\infty} \gamma_n \sin\frac{n\pi x}{l} = g(x).$$

**REVIEW EXERCISES CHAPTER 6**
Page 285

**1.** It has no solution.

**3.** It has exactly one solution: $y(x) = 2x - \dfrac{\pi}{4}\sin 4x$.

**5.** No solution if $A \ne B - 2\pi$; infinitely many solutions if $A = B - 2\pi$. It never has exactly one solution.

**7.** $\lambda_n = -n^2$, $\quad n = 1, 2, \ldots$ ; $y_n(x) = \sin nx$, $\quad n = 1, 2, \ldots$

**9.** $\lambda_n = n^2$, $n = 0, 1, 2, \ldots$ ; $y_n(x) = \cos nx$, $n = 0, 1, 2, \ldots$

**Section 7.2**
Page 293

**1.** $\{(0, 3), (0.2, 1.8), (0.4, 1.12), (0.6, 0.752), (0.8, 0.5712), (1, 0.5027)\}$.

**3.** $\{(0, 0), (0.2, 0), (0.4, 0.024), (0.6, 0.1104), (0.8, 0.2822), (1, 0.5533)\}$.

**5.** $\{(0, 0), (0.2, 0.2), (0.4, 0.392), (0.6, 0.5587), (0.8, 0.6772), (1, 0.7129)\}$.

**7.** $\{(0, 3), (0.1, 2.4), (0.2, 1.93), (0.3, 1.564), (0.4, 1.2812), (0.5, 1.0650), (0.6, 0.9020),$ $(0.7, 0.7816), (0.8, 0.6953), (0.9, 0.6362), (1, 0.5990)\}$.

**9.** {(0, 0), (0.1, 0.1), (0.2, 0.199), (0.3, 0.2950), (0.4, 0.3857), (0.5, 0.4685), (0.6, 0.5405), (0.7, 0.5977), (0.8, 0.6354), (0.9, 0.6478), (1, 0.6278)}.

**11.** $y(0.85) \approx 0.2344$; actual value $= 0.2544$.

**13.** {(0, 0), (0.1, 0), (0.2, 0.01), (0.3, 0.031), (0.4, 0.0641), (0.5, 0.1105), (0.6, 0.1716), (0.7, 0.2487), (0.8, 0.3436), (0.9, 0.4579), (1, 0.5937)}; actual values $= \{(0, 0), (0.1, 0.0052), (0.2, 0.0214), (0.3, 0.0499), (0.4, 0.0918), (0.5, 0.1487), (0.6, 0.2221), (0.7, 0.3137), (0.8, 0.4255), (0.9, 0.5596), (1, 0.7183)\}$.

---

**1.** {(0, 0), (0.2, 0), (0.4 0.0168), (0.6, 0.0717), (0.8, 0.1907), (1, 0.4054)}.

**3.** {(0, 1), (0.2, 0.84), (0.4, 0.7448), (0.6, 0.7027), (0.8, 0.7042), (1, 0.7415)}.

**5.** {(0, 1), (0.2, 1.02), (0.4, 1.0820), (0.6, 1.1937), (0.8, 1.3694), (1, 1.6334)}.

**7.** {(0, 2), (0.2, 2), (0.4, 2.0321), (0.6, 2.1306), (0.8, 2.3407), (1, 2.7344)}.

**9.** {(0, 0), (0.2, 0.02), (0.4, 0.0884), (0.6, 0.2158), (0.8, 0.4153), (1, 0.7027)}; actual values—see the answer to Exercise 13, Section 7.2.

**11.** {(0, 0), (0.1, 0.005), (0.2, 0.0210), (0.3, 0.0492), (0.4, 0.0909), (0.5, 0.1474), (0.6, 0.2204), (0.7, 0.3116), (0.8, 0.4228), (0.9, 0.5562), (1, 0.7141)}; actual values—see the answer to Exercise 13, Section 7.2.

**13.** {(0, 0), (0.1, 0), (0.2, 0.0021), (0.3, 0.0085), (0.4, 0.0218), (0.5, 0.0449), (0.6, 0.0809), (0.7, 0.1331), (0.8, 0.2056), (0.9, 0.3024), (1, 0.4282)}.

**15.** {(0, $-1$), (0.1, $-1.205$), (0.2, $-1.4221$), (0.3, $-1.6540$), (0.4, $-1.9038$), (0.5, $-2.1757$), (0.6, $-2.4743$), (0.7, $-2.8057$), (0.8, $-3.1769$), (0.9, $-3.5969$), (1, $-4.0762$)}.

---

**1.** {(0, 1), (0.2, 1.5587), (0.4, 2.5387), (0.6, 4.1467), (0.8, 6.6909), (1, 10.6307)}.

**3.** {(0, 0), (0.2, 0.0147), (0.4, 0.0493), (0.6, 0.0926), (0.8, 0.1397), (1, 0.1885)}.

**5.** {(0, 0), (0.2, $-0.0398$), (0.4, $-0.1576$), (0.6, $-0.3446$), (0.8, $-0.5810$), (1, $-0.8391$)}.

**7.** {(0, 0), (0.2, $-0.0179$), (0.4, $-0.0638$), (0.6, $-0.1170$), (0.8, $-0.1401$), (1, $-0.0649$)}.

**9.** {(0, 1), (0.2, 1.5605), (0.4, 2.5442), (0.6, 4.1591), (0.8, 6.7157), (1, 10.6769)}.

11. $\{(0, 0), (0.2, -0.0399), (0.4, -0.1580), (0.6, -0.3452), (0.8, -0.5819), (1, -0.8398)\}$.

13. $\{(0, 0), (0.2, 0.0213), (0.4, 0.0917), (0.6, 0.2218), (0.8, 0.4250), (1, 0.7175)\}$; actual values—see the answer to Exercise 13, Section 7.2.

15. $\{(0, 0), (0.1, 0.0052), (0.2, 0.0214), (0.3, 0.0498), (0.4, 0.0918), (0.5, 0.1487), (0.6, 0.2221), (0.7, 0.3137), (0.8, 0.4255), (0.9, 0.5595), (1, 0.7182)\}$; actual values—see the answer to Exercise 13, Section 7.2.

17. $\{(0, 1), (0.1, 1.2373), (0.2, 1.5604), (0.3, 1.9882), (0.4, 2.5438), (0.5, 3.2556), (0.6, 4.1582), (0.7, 5.2938), (0.8, 6.7139), (0.9, 8.4815), (1, 10.6735)\}$.

**Section 7.5**
Page 307

1. $\{(0, 0, 0), (0.2, 0.2, 0), (0.4, 0.44, 0.008), (0.6, 0.7216, 0.0483), (0.8, 1.0513, 0.1621), (1, 1.4437, 0.4156)\}$.

3. $\{(0, 0, 0), (0.2, 0.2201, 0.0034), (0.4, 0.4828, 0.0316), (0.6, 0.7969, 0.1281), (0.8, 1.1829, 0.3696), (1, 1.6820, 0.8961)\}$.

5. $\{(0, 1, 0), (0.2, 1.2, 0.2), (0.4, 1.44, 0.52), (0.6, 1.704, 1.016), (0.8, 1.9616, 1.7632), (1, 2.1613, 2.8608)\}$.

7. $\{(0, 1, 1), (0.2, 1.4, 1), (0.4, 1.88, 1.04), (0.6, 2.4723, 1.128), (0.8, 3.2213, 1.2769), (1, 4.1916, 1.5057)\}$.

9. $\{(0, 0, 1), (0.2, -0.12, 1), (0.4, -0.3984, 1.04), (0.6, -0.6019, 1.1232), (0.8, -0.8110, 1.2580), (1, -1.0251, 1.4599)\}$.

11. From Example 1, Section 3.2 $x(1) \approx 7.8666$, $y(1) \approx 13.2448$; actual values: $x(1) = 10.1073$, $y(1) = 17.4964$.

13. $\{(0, 2, 3), (0.2, 2.712, 4.2027), (0.4, 3.7137, 5.9358), (0.6, 5.1342, 8.4466), (0.8, 7.1627, 12.1004), (1, 10.0779, 17.4384)\}$; actual values—see the answer to Exercise 11.

**Section 7.6**
Page 309

1. $\{(0, 22.3), (0.2, 23.1675), (0.4, 24.1069), (0.6, 25.1273), (0.8, 26.2395), (1, 27.4563)\}$.

3. $\{(0, 22.3), (0.2, 23.2042), (0.4, 24.1883), (0.6, 25.2633), (0.8, 26.4426), (1, 27.7418)\}$.

5. $\{(0, 9.45), (0.2, 9.4071), (0.4, 9.3646), (0.6, 9.3224), (0.8, 9.2807), (1, 9.2393)\}$.

**7.** $\{(0, 9.45), (0.2, 9.4073), (0.4, 9.3650), (0.6, 9.3230), (0.8, 9.2814), (1, 9.2402)\}$.

**9.** $\{(0, 30), (0.2, 29.52), (0.4, 29.0704), (0.6, 28.6479), (0.8, 28.2497), (1, 27.8736)\}$.

**11.** $\{(0, 30), (0.2, 29.5200), (0.4, 29.0704), (0.6, 28.6479), (0.8, 28.2497), (1, 27.8736)\}$.

**13.** (a) $\dfrac{1}{4T_1^3} \ln \left| \dfrac{T - T_1}{T + T_1} \right| - \dfrac{1}{2T_1^3} \tan^{-1}\left(\dfrac{T}{T_1}\right) = -\dfrac{A}{M} t + C$, where $C$ is a constant of in-
tegration.

**15.** $n = \dfrac{e^{2\sqrt{Akt}} - 1}{e^{2\sqrt{Akt}} + 1} \sqrt{\dfrac{A}{k}}$. From Exercise 14, we have $\{(0, 0), (0.2, 20{,}000), (0.4, 39{,}600),$
$(0.6, 58{,}032), (0.8, 74{,}664), (1, 89{,}089)\}$; actual values (using the formula above)
$= \{(0, 0), (0.2, 19{,}867.7248), (0.4, 38{,}966.3962), (0.6, 56{,}641.5992), (0.8, 72{,}433.6982),$
$(1, 86{,}105.172)\}$.

**17.** $\{(1, 1), (1.2, 0.9822), (1.4, 0.9227), (1.6, 0.8124), (1.8, 0.6250), (2, 0.1793)\}$.

**19.** $c_1 = a/(b + ace^{at})$, where $c$ is a constant of integration.

**REVIEW
EXERCISES
CHAPTER 7**
Page 312

**1.** $\{(1, 0), (1.2, 0.2), (1.4, 0.488), (1.6, 0.9132), (1.8, 1.5827), (2, 2.7602)\}$.

**3.** $\{(1, 0), (1.2, 0.2452), (1.4, 0.5564), (1.6, 1.056), (1.8, 1.916), (2, 3.8013)\}$.

**5.** $\{(1, 0), (1.2, 0.2466), (1.4, 0.6294), (1.6, 1.2936), (1.8, 2.7854), (2, 9.9533)\}$.

**7.** $\{(1, 1, -1), (1.2, 1.3745, -0.5829), (1.4, 1.7924, -0.0720), (1.6, 1.6219, 0.5609),$
$(1.8, 1.1283, 1.3056), (2, 0.6285, 2.0004)\}$.

**9.** $\{(1, 1, 1), (1.2, 1.3748, -0.5803), (1.4, 1.5792, -0.1035), (1.6, 1.4895, 0.3972), (1.8,$
$1.1607, 0.8374), (2, 0.7659, 1.2441)\}$.

**11.** The equation $\dot{y} = x^2 + y^2$ indicates that $y$ is an increasing function of $t$. Thus the
relationship of Exercise 10 cannot be valid for all $t \geq 1$ since in each of Exercises
6 through 9 (see the answers for Exercises 7 and 9) we note that $y(t) > 0$ for all
$t > t_0$, where $1.4 < t_0 < 1.6$ and positive values for $y$ cannot be obtained from the
relation given in Exercise 10.

**13.** $\{(1, 4.5), (1.2, 4.3812), (1.4, 4.3211), (1.6, 4.2795), (1.8, 4.2436), (2, 4.2076)\}$.
Actual values (see Example 2, Section 1.5): $\{(1, 4.5), (1.2, 4.4112), (1.4, 4.3553),$
$(1.6, 4.3124), (1.8, 4.2734), (2, 4.2333)\}$.

**15.** $\{(1, 4.5), (1.2, 4.4113), (1.4, 4.3562), (1.6, 4.3132), (1.8, 4.2741), (2, 4.2339)\}$.
Actual values—see the answer to Exercise 13.

**Section 8.2**
Page 317

1.  In the region $-1 \le t \le 1, 0 \le x_1 \le 1, 0 \le x_2 \le 1$, choose $x_1 = 1/n^2, x_2 = y_2$, $y_1 = 1/(n + 1)^2$. In this case, the assumption that the Lipschitz condition is satisfied leads to the impossible requirement that the constant $L_1$ satisfy the inequality

    $$\frac{1}{n(n + 1)} \le L_1 \frac{2n + 1}{n^2(n + 1)^2}.$$

    In the region $-1 \le t \le 1, 3 \le x_1 \le 4, 0 \le x_2 \le 1$, we can write

    $$| f(t, x_1, x_2) - f(t, y_1, y_2) | \le | x_1^{1/2} - y_1^{1/2} | + | x_2 - y_2 |$$

    $$= \frac{| x_1 - y_1 |}{\sqrt{x_1} + \sqrt{y_1}} + | x_2 - y_2 | \le \frac{| x_1 - y_1 |}{3 + 0} + | x_2 - y_2 |.$$

    Thus, we can choose $L_1 = \frac{1}{3}$ and $L_2 = 1$.

3.  The general solution of the IVP is $y(t) = A \cos t + B \sin t$. If $\overline{y}(t) = \overline{A} \cos t + \overline{B} \sin t$ is the true value of the solution, then

    $$| \overline{y}(t) - y(t) | \le | \overline{A} - A | + | \overline{B} - B | \le 2 \times 10^{-3}.$$

**Section 8.3**
Page 320

1.  $(0, 0)$ is the only critical point. For $xy \ne 0$, the trajectories are the half-lines $y = cx, c \ne 0$. The direction for increasing $t$ is away from the origin.

3.  $(0, 0)$ is the only critical point. For $xy \ne 0$, the trajectories are the hyperbolas $x^2 - y^2 = c, c \ne 0$. For $c = 0$, there are four half-lines $y = x, x > 0; y = x, x < 0; y = -x, x > 0; y = -x, x < 0$.

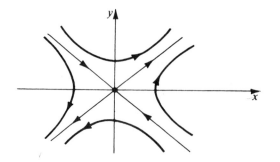

**5.** $x = e^t$, $y = -e^{2t}$. The trajectory is $y = -x^2$, $x > 0$.

**7.** The trajectories are given by $y = x \ln |x|$, $x \geq 1$.

**9.** $x(t) = e^{-t}(-\cos t + \sin t)$, $y(t) = e^{-t}(\sin t + \cos t)$. This curve is a logarithmic spiral approaching the origin as $t \to \infty$.

**11.** The only critical point is $(0, 0)$. The trajectories are given by $y(y^2 - 3x^2) = C$.

**13.** The critical points are at $(n\pi, 0)$, $n = 0, \pm 1, \pm 2, \ldots$. The trajectories are given by $\frac{1}{2}y^2 - 4\cos x = C$.

**15.** Set $\dot{x} = y$; then $\dot{y} = -x$. The trajectories are the circles $x^2 + y^2 = C$.

**17.** Set $\dot{x} = y$; then $\dot{y} = x$. The trajectories are the hyperbolas $y^2 - x^2 = C$.

**19.** (a) $ax + by = 0$, $cx + dy = 0$. From Appendix A, the solution of this homogeneous system is $(0, 0)$ if and only if the determinant of coefficients is not zero (if $ad - bc \neq 0$).
(b) If $ad - bc = 0$, the two equations coincide and every point on the line $ax + by = 0$ is a critical point.

---

**1.** Stable.     **3.** Asymptotically stable.     **5.** Unstable.     **7.** Unstable.

**9.** Asymptotically stable.     **11.** $(0, 0)$ is asymptotically stable.

**13.** Asymptotically stable.     **15.** Unstable.     **17.** Asymptotically stable.

**19.** Asymptotically stable.     **21.** Asymptotically stable.

**23.** Asymptotically stable.     **25.** Stable.

---

**1.** $\lambda = 1, 3$. Unstable node.     **3.** $\lambda = -1, 1$. Saddle point.

**5.** $\lambda = 2, 2$. Unstable node.     **7.** $\lambda = -1, -1$. Stable node.

**9.** $\lambda = -2 \pm i$. Stable focus.     **11.** $\lambda = -1 \pm i$. Stable focus.

**13.** $\lambda = -2, -3$. Stable node.     **15.** $\lambda = \pm \sqrt{7}$. Saddle point.

**17.** $\lambda = \pm 1$. Saddle point.     **19.** $\lambda = \pm 1$. Saddle point.     **21.** Stable node.

**23.** Stable node.     **25.** Stable focus.

**27.** $s = -2$, $p = -4$, $s^2 - 4p > 0$. Saddle point.

**29.** $s = 0$, $p = -13$, $s^2 - 4p > 0$. Saddle point.

**31.** $s = 2$, $p = 1$, $s^2 - 4p = 0$. Unstable node.

**33.** $s = -2$, $p = 1$, $s^2 - 4p = 0$. Asymptotically stable node.

---

**Section 8.6**

Page 345

**1.** The linearized system for $X$, $Y$ is $\dot{X} = -(bc/d)\, Y$, $\dot{Y} = (ad/b)\, X$. The point $(X, Y)$ $= (0, 0)$ is a center.

**3.** The linearized system is $\dot{x} = y$, $\dot{y} = -w^2 x - ky$.

    (a) Stable focus.
    (b) Stable node.
    (c) Stable node.

**REVIEW
EXERCISES
CHAPTER 8**

Page 346

**1.** The hypotheses of Theorem 1, Section 8.1 are satisfied.

**3.** The line $y = 2x$.

**5.** $(0, 0)$ is the only critical point; the equation of trajectories is $y(x) = x^2(c - \ln x)$; the trajectory through $(1, 0)$ is $y(x) = -x^2 \ln x$.

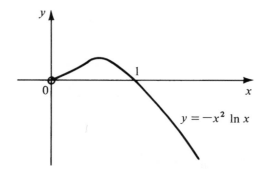

**7.** Stable.

**9.** Asymptotically stable.

**11.** Unstable.

**13.** Saddle point.

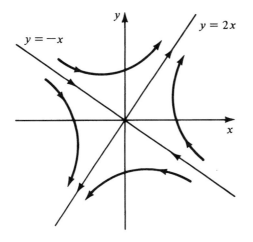

**15.** Center.

**17.** Node.

**19.** Asymptotically stable focus.

---

**1.** Linear, homogeneous, variable coefficients, order two.

**3.** Nonlinear.

**5.** Linear, nonhomogeneous, constant coefficients, order two.

**7.** Linear, nonhomogeneous, variable coefficients, order one.

**9.** Linear, nonhomogeneous, variable coefficients, order one.

**11.** Linear, nonhomogeneous, variable coefficients, order two.

# Index